PHYSIOLOGICAL PROCESSES
LIMITING PLANT PRODUCTIVITY

Proceedings of Previous Easter Schools in Agricultural Science, published by Butterworths, London

*SOIL ZOOLOGY   Edited by D. K. McE. Kevan (1955)

*THE GROWTH OF LEAVES   Edited by F. L. Milthorpe (1956)

*CONTROL OF THE PLANT ENVIRONMENT   Edited by J. P. Hudson (1957)

*NUTRITION OF THE LEGUMES   Edited by E. G. Hallsworth (1958)

*THE MEASUREMENT OF GRASSLAND PRODUCTIVITY   Edited by J. D. Ivins (1959)

*DIGESTIVE PHYSIOLOGY AND NUTRITION OF THE RUMINANT   Edited by D. Lewis (1960)

*NUTRITION OF PIGS AND POULTRY   Edited by J. T. Morgan and D. Lewis (1961)

*ANTIBIOTICS IN AGRICULTURE   Edited by M. Woodbine (1962)

*THE GROWTH OF THE POTATO   Edited by J. D. Ivins and F. L. Milthorpe (1963)

*EXPERIMENTAL PEDOLOGY   Edited by E. G. Hallsworth and D. V. Crawford (1964)

*THE GROWTH OF CEREALS AND GRASSES   Edited by F. L. Milthorpe and J. D. Ivins (1965)

*REPRODUCTION IN THE FEMALE MAMMAL   Edited by G. E. Lamming and E. C. Amoroso (1967)

*GROWTH AND DEVELOPMENT OF MAMMALS   Edited by G. A. Lodge and G. E. Lamming (1968)

*ROOT GROWTH   Edited by W. J. Whittington (1968)

*PROTEINS AS HUMAN FOOD   Edited by R. A. Lawrie (1970)

*LACTATION   Edited by I. R. Falconer (1971)

*PIG PRODUCTION   Edited by D. J. A. Cole (1972)

*SEED ECOLOGY   Edited by W. Heydecker (1973)

HEAT LOSS FROM ANIMALS AND MAN: ASSESSMENT AND CONTROL   Edited by J. L. Monteith and L. E. Mount (1974)

*MEAT   Edited by D. J. A. Cole and R. A. Lawrie (1975)

*PRINCIPLES OF CATTLE PRODUCTION   Edited by Henry Swan and W. H. Broster (1976)

*LIGHT AND PLANT DEVELOPMENT   Edited by H. Smith (1976)

PLANT PROTEINS   Edited by G. Norton (1977)

ANTIBIOTICS AND ANTIBIOSIS IN AGRICULTURE   Edited by M. Woodbine (1977)

CONTROL OF OVULATION   Edited by D. B. Crighton, N. B. Haynes, G. R. Foxcroft and G. E. Lamming (1978)

POLYSACCHARIDES IN FOOD   Edited by J. M. V. Blanshard and J. R. Mitchell (1979)

SEED PRODUCTION   Edited by P. D. Hebblethwaite (1980)

PROTEIN DEPOSITION IN ANIMALS   Edited by P. J. Buttery and D. B. Lindsay (1981)

These titles are now out of print but are available on microfiche

This book is to be returned on
or before the date stamped below

UNIVERSITY OF PLYMOUTH

ACADEMIC SERVICES
PLYMOUTH LIBRARY
Tel: (0752) 232323
This book is subject to recall if required by another reader
Books may be renewed by phone
CHARGES WILL BE MADE FOR OVERDUE BOOKS

A book may be renewed by telephone or by personal visit
if it is not required by another reader.
CHARGES WILL BE MADE FOR OVERDUE BOOKS

Physiological Processes Limiting Plant Productivity

C. B. JOHNSON, PhD
Department of Botany, University of Reading

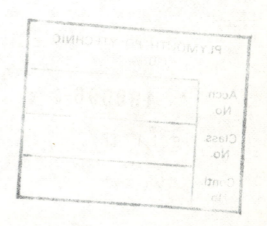

BUTTERWORTHS
London Boston Sydney Wellington Durban Toronto

First published 1981

ISBN 0 408 10649 2

© The several contributors named in the list of contents 1981

British Library Cataloguing in Publication Data

Physiological processes limiting plant productivity
  1. Plants, Cultivated—Congresses
  2. Plant physiology—Congresses
  I. *Johnson, C.*
  631.5′4   SB16   80–41941

  ISBN 0–408–10649–2   ✓

Typeset by CCC in Great Britain by
William Clowes (Beccles) Limited, Beccles and London
Printed and bound by The Camelot Press Southampton

PREFACE

The contents of this volume represent the proceedings of the Thirtieth University of Nottingham Easter School in Agricultural Science, which was held at Sutton Bonington on 2–5 April 1979. The subject 'Physiological Processes Limiting Plant Productivity', in line with the overall concept of the Easter Schools, was selected so as to set the fundamental research in the field of plant growth and development in the context of agricultural productivity. The meeting was planned with the aim of bringing together workers in fundamental research in relevant aspects of plant biochemistry, physiology and genetics in order that an assessment might be made of the extent and conditions under which the various processes that contribute to 'plant productivity' are in fact constraints to productivity. Thus, the proceedings are concerned primarily with the fundamental mechanisms which underlie crop production, and their control, rather than with crop production as such.

It is appropriate that a substantial proportion of the proceedings is devoted to the utilization of light by crop plants, thus recognizing the fact that light is the ultimate source of all biological energy as well as providing plants with essential information about their environment. Further sections are concerned with other aspects of plant nutrition, with water relations and with the effects of an adaptation to unfavourable conditions including those imposed by atmospheric pollution.

The success of the conference, which was attended by 130 participants from all over the world was, as so often, substantially due to the efficient, helpful and conscientious efforts of Miss Edna Lord, the conference secretary. Her task was made even more onerous than usual by my departure to Reading University in January 1979, prior to the conference. Without her persistence many essentials would have been neglected.

Professor Harry Smith and Professor W. J. Whittington were of considerable assistance in planning the conference. I am grateful, too, for the assistance of Mr John Blunt and Mrs Valerie Blunt who organized the visual aids, and to Dr Valerie Black and Mr Jonathan Hughes for assistance with the poster sessions. Finally, my thanks are due to the chairmen of the various sessions: Professors P. G. Jarvis, T. A. Mansfield, J. L. Monteith, M. G. Pitman and H. Woolhouse.

C. B. JOHNSON

CONTENTS

1

CROP PHYSIOLOGY IN RELATION TO AGRICULTURAL PRODUCTION: THE GENETIC LINK

H. W. WOOLHOUSE*
Department of Plant Sciences, University of Leeds, UK

Introduction

Most agricultural experimental stations around the world include on their staff physiologists and biochemists concerned with research into the improvement of yield and quality in crops. In many universities there are academic staff who are in receipt of research grants for the same purposes. It is my aim here to ask some questions about the past and future roles of plant physiologists in this endeavour. The essence of my argument is that if, in areas with well-developed agricultural systems, the supporting efforts of plant breeders, entomologists, phytopathologists and specialists in crop protection were to be suddenly withdrawn, catastrophic losses in crop production would soon follow. If, on the other hand, the physiologists and biochemists were suddenly deleted from the scene, the effect on our success in sustaining and improving yields would scarcely be noticed.

On occasional exposure of this assessment of the contribution of plant physiology and biochemistry it has been variously branded as perverse, foolish or missing the point. It is evident that knowledge of photoperiodism has revolutionized floriculture; hormone research has yielded enormous benefits in controlling production in fruit crops, and design of herbicides in plant propagation; and research into plant nutrition was a prerequisite for the success of fertilizers in raising productivity. These benefits are, of course, quite undeniable but most of them arise primarily as spin-off from basic research that was not initiated with agricultural objectives in view. Moreover, these are achievements which are now consolidated, whereas our greatest concern must be with the contemporary scene in crop physiology; past success is not of itself a sufficient justification for future investment in a subject. We should confront the problem of assessing the extent to which work in the physiology and biochemistry of crop plants is really contributing to increasing their yields; my conclusions are not optimistic.

If due allowance is made for the enhancement of yields which have accrued from improved agronomic practices, as for example irrigation, fertilizer applications and the control of pests and diseases, then the rest of the increases have resulted primarily from the work of plant breeders. Plant breeders have in the past used basic knowledge from the field of cytogenetics to enable them to manipulate levels of ploidy and to construct particular

---

* Present address: John Innes Institute, Norwich.

genotypes as a basis for their programmes of hybridization and selection, but the extent of the contributions of physiology to such work has been relatively slight. I therefore wish to enquire what crop physiologists ought to be doing, what they are doing and to see if there are any indications as to how they might make a more positive contribution in the future to improvements of crop yields. It is the basis of my argument that if work in crop physiology is going to make an impact on crop production it must be in a style, or of a kind, which provides the breeder with information that he can handle. I propose, therefore, to review briefly the range of problems which plant physiologists ought to be tackling; to follow this by an assessment of what is happening in crop physiology and to conclude with some examples of approaches which hold promise of having something useful to offer to the plant breeder.

The Problems Confronting Crop Physiology

The problems which confront contemporary crop physiologists may be conveniently summarized under six headings as follows.

THE NEED TO UNDERSTAND THE PROCESSES INVOLVED IN THE DEVELOPMENT OF CROP PLANTS

Some of the problems of development may be considered as basic issues with a wide general significance, as for example the origins of phyllotaxis and the initiation of leaves and roots. Others are unique to particular crops, as for example the problems of bulb formation in onions, bolting in sugar beets and tuberization in potatoes.

Perhaps the most basic problem of development is to provide an accurate detailed description of the timing, cytological and morphological sequence of events in development. It may appear surprising to many developmental biologists that for none of our major crop plants does there exist a sufficiently detailed account of this kind. The picture is fairly good for wheat and maize, less so for the other cereals and quite inadequate for many of the legumes and other dicotyledonous crops.

Alongside the description of developmental processes goes the need to study their control. This embraces a range of problems, which considered sequentially through the life cycle include:

(1) the need to understand the physiology of seed survival and germination;
(2) the development and senescence of leaves in relation to the duration of the crop canopy and its efficiency in photosynthesis;
(3) the development and subsequent pattern of root systems in relation to their efficiency in the uptake of water and nutrients;
(4) the factors which control the number of flowers initiated, the number which subsequently develop and the number which mature;
(5) the factors which control the size of developing fruits (vegetative storage organs may also be included here) and their capacity for the storage of carbohydrate, lipid and protein reserves;
(6) the control of ripening processes;

(7) interwoven with many of the preceding catalogue of problems is the
    basic need for a better knowledge of the mechanisms of hormone action
    and of how these relate to development.

A further aspect of developmental work in crop physiology is the use of
protoplast fusion techniques along with cell and tissue cultures. There can be
no doubting the enormous current enthusiasm for this area of study. It ranges
from the use of deoxyribonucleic acid (DNA) viruses and plasmids of
Agrobacterium as potential vehicles for the transfer of specific segments of
DNA as part of the genetic engineering programme, to the use of tissue
cultures for bulk propagation of specific genotypes; the bulk screening of
specific genotypes for particular attributes; the preservation of germ plasm
over long periods for the benefit of plant breeders and the study of
organogenesis from undifferentiated masses of tissue.

THE NEED TO UNDERSTAND THE EFFECTS OF ENVIRONMENTAL FACTORS
WHICH LIMIT CROP PRODUCTION

The first part of the task in this area is to identify when temperature, light,
CO_2, nutrients and water supply are limiting for growth and how these
deficits may best be ameliorated, given that over vast areas of the earth there
will be relatively little that can be done. For example, CO_2 enrichment on a
field scale, cooling procedures in hot regions, heating regimes in cold areas
and fertilizer treatments in high rainfall areas are all practices which are
never likely to be feasible on a large scale. In these circumstances we must
look instead to optimizing local management practices to reduce as far as
possible the extremes of stress and, in particular, to intensify the search for
genotypes which will show the highest resilience in the face of a given stress.
This is a vast and complex subject, as may be seen by reference to the two
most ubiquitous sources of environmental stress: water supply and
temperature.

Water stress

Ever since plants came to occupy terrestrial environments the process of
adaptation to conserve moisture has been in progress. The study of plant
adaptation to varying degrees of water stress remains an outstanding problem
in plant physiology; in agricultural practice there are two main solutions:
irrigation and generalized selection of genotypes that have greater drought
tolerance. The physiologist finds himself grappling with near-intractable
problems in the significance of stomatal control (Cowan, 1977), regulation of
osmotic potentials, studies of root penetration and extremely difficult
problems which concern the effects of water stress on metabolic processes. It
is right and proper that many physiologists will choose to work where they
see the greater challenges, but whether this particular area of study is likely
to offer much to the plant breeder or to the cause of increased crop production
seems doubtful for the foreseeable future.

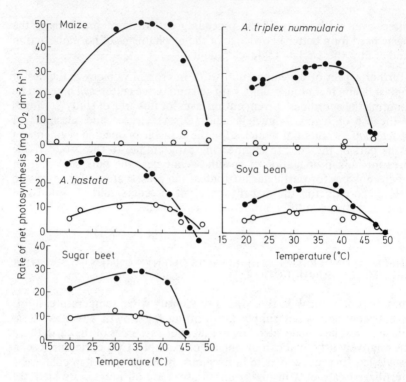

Figure 1.1   *The net photosynthesis (•) and enhancement of photosynthesis in oxygen-free air (○) of the leaves of five species at different temperatures. (After Hofstra and Hesketh, 1969)*

Temperature stress

For the majority of C_3 plants, and particularly those of the temperate zone, the optimum temperature for net photosynthesis is in the range of 20–25°C under conditions of atmospheric CO_2 (*Figure 1.1*); when the CO_2 concentration is artificially increased the optimum temperature may be raised to 30°C or even higher. This result is now known to arise from the process of photorespiration, the light-driven efflux of CO_2, which is extremely sensitive to temperature (*Figure 1.2*) and to relative concentrations of oxygen and CO_2 (*Figure 1.3*; Ogren and Bowes, 1971). As the biochemistry of photorespiration has been elucidated there has been much speculation and some claims of intraspecific genetic variation in photorespiration (Zelitch, 1975). However, this work awaits confirmation and there are indications that there may be insurmountable problems in this area (Woolhouse, 1978), particularly if, as now seems likely, the photorespiratory process has an important role in preventing the formation of superoxide radicals under conditions of low CO_2 supply (Tolbert, 1971); this is the case, for example, when closure of stomata is induced by water stress.

    In agricultural systems in the temperate zone it is low temperatures which afford the greatest single limitation to plant growth (Wareing, 1971). A great deal of effort has been put into the evaluation and selection of genotypes of

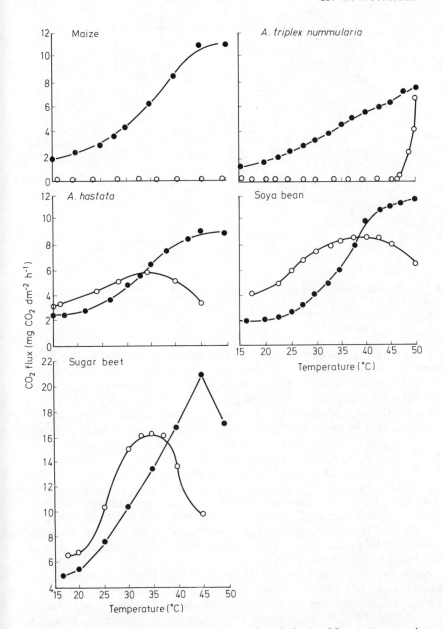

Figure 1.2 *The evolution of CO_2 into CO_2-free air from the leaves of five species exposed to an illuminance of 10 000 fc* (o) *and in darkness* (•) *at different temperatures. (After Hofstra and Hesketh, 1969)*

many crop plants in order to obtain forms which are able to survive, grow and develop at low temperatures. With many species there has been significant success in the selection of frost-resistant genotypes but the ability to grow at low temperatures seems to be a more deep-seated characteristic less prone to intraspecific variation. At the present time much interest

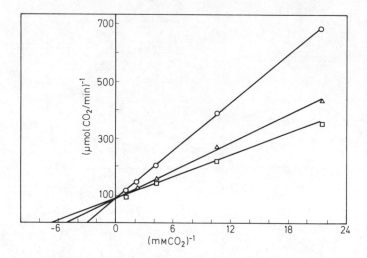

Figure 1.3   *A double reciprocal plot of rate of incorporation of CO_2 (μmol of CO_2 per minute) by ribulose bisphosphate carboxylase as a function of CO_2 concentration in 0 per cent oxygen ($\square$), 21 per cent oxygen ($\triangle$) and 100 per cent oxygen ($\circ$). (After Ogren and Bowes, 1971)*

attaches to the development of maize as a forage crop in the UK. Significant progress has been made with the breeding of varieties with desirable foliage characteristics and the ability to make rapid growth during the summer months. The really substantial limitation to the advancement of the crop is the slow rate of growth under the cool conditions of an average English spring. It is of interest to note that the problem appears to reside primarily with the capacity for cell division and expansion at low temperatures, rather than with the slower rate of photosynthesis (Monteith and Elston, 1971). Attempts to find varieties with better low temperature characteristics have been made by searching for stocks grown at high altitude in the Andes; close investigation of these genotypes in their natural surroundings, however, reveals that they take much longer to grow—often 12–15 months from sowing to harvest; that is, they survive but do not grow significantly in the cooler periods and so have little to offer with respect to the contracted growing seasons of the temperate zones.

Some degree of intraspecific variation in the ability to grow at low temperatures has been found in the forage grasses, particularly *Lolium perenne*, *Festuca arundinacea* and *Dactylis glomerata* (Cooper, 1964; Maccoll and Cooper, 1967; Robson, 1967). It is important to emphasize, however, that these observations derive primarily from the work of plant breeders with physiological interests and from studies based largely on selection procedures. We know almost nothing of the nature of the rate-limiting steps to growth at low temperatures. The work of Raison (1973) and others on phase changes in the membrane lipids of species that show different temperature optima for growth, offers some hope of deeper insights but at the present time it cannot be said that there is any significant activity among physiologists which might afford positive leads to the breeder. The physiologist is, as in so many instances, offering little more than *post hoc* explanations of differences unearthed by the breeders—and in this case he would seem to be a very long way behind.

Clearly, there is a sense in which crop physiologists should be encouraged to devote more effort to understanding the basic mechanisms involved in damage from environmental stresses and in the evidence of and resistance to such stresses. The possibility of obtaining some measure of chemical control of stress susceptibility may prove a sufficient impetus to continue effort in this field, but the prospects for significant progress at a level which might provide mechanistic guidelines for the breeder seem remote at the present time.

THE NEED TO UNDERSTAND THE PHYSIOLOGICAL AND BIOCHEMICAL BASIS OF RESISTANCE TO PESTS AND DISEASES

This is an aspect of the study of crops in which the physiologist has become involved somewhat later, and one which it is hoped may have a significant impact in the future. The example of fungal diseases in cereals will suffice to illustrate the subject. Plant breeders have for years been playing a lively game of leap-frog with a number of fungal diseases which diminish substantially the yields of the crops that they infect. In some instances a very thorough understanding of the complexities of the host and fungus have been obtained (Day, 1974) and the plant breeder enjoys a reasonable success in manipulating the genetic structure of his crop to counter the spectrum of virulent genotypes thrown up by the pathogen. Plant pathologists have afforded solid support in the description of the pathogens and in the assaying of virulence, but the duel which has resulted is a very expensive operation which would be better avoided if that were possible. It would be desirable if the plant breeder could be released from this treadmill for it would permit him to concentrate on other aspects of yield. For some pests and diseases highly effective sprays are now available, although the costs involved in their manufacture and application are often high, environmental effects have to be carefully monitored and the emergence of toxin-resistant pathogens presents a continuing challenge.

Plant physiologists have contributed relatively little in the past to the protection of crops from pests and diseases but there are indications that they may have more to offer in the future. As the genetics of disease resistance are becoming clearer, test systems involving 'gene-for-gene' resistance in otherwise isogenic lines are providing the physiologist with assay systems with which to study elicitor-phyto-alexin interactions, the nature of the cellular recognition systems involved in host–pathogen interactions and the critical events in hypersensitive responses and other defence mechanisms (Albersheim and Anderson-Prouty, 1975). It would seem a reasonable hope that studies of this kind might ultimately provide blueprints for better informed approaches to both the breeding of resistant forms and the development of pesticides to replace the relative clumsy empiricism of mass-screening procedures which is at the heart of present practice. It must be noted, however, that this is no more than speculative optimism concerning the potential impact of the physiologist in this vital aspect of crop production.

THE REQUIREMENT FOR AN UNDERSTANDING OF THE CARBON
BALANCE OF CROPS

If resolute defence were to be made against the charge of ineffectiveness on
the part of crop physiologists the chosen ground might well be photosynthesis
and the carbon balance problem. Certainly, the magnitude of the effort is
impressive.

The basic problems which concern carbon balance fall essentially into
three categories.

(a) There is the need to analyse the photochemical and carbon assimilatory
aspects of photosynthesis in order to define the rate-limiting steps under
particular conditions and to be able to provide screening procedures, which
enable plant breeders to search for variation between genotypes for those
factors that limit the rate of photosynthesis. The complexity of photosynthesis
means that this objective alone provides a mandate for enormous efforts in
increasing interception and utilization of light, improving stomatal characteristics and CO_2 uptake, and analysing the enzymic aspects of CO_2
assimilation. On the debit side of the carbon balance is the need to measure
accurately the magnitude of dark and photorespiration and to identify factors
which control them.

(b) Closely allied to the study of input and output aspects of carbon
balance is the need for a much clearer understanding of the factors which
control the partitioning of carbon compounds between their sites of
production or sources and sites where they are stored or utilized, the sinks.
The relationships of sources and sinks are only understood in a general way.
It is agreed by most physiologists that translocation always takes place along
a gradient from a region of high concentration where sugars are loaded into
the system (the source) to a region of lower concentration (the sink) where
unloading takes place. Much of the background knowledge of the patterns of
partitioning comes from growth analysis (Warren–Wilson, 1972), but the
more detailed aspects of how partitioning occurs are not known. We do not
know, for example, how far growth is dependent on amounts of photosynthate
available for distribution and how far the rate of photosynthesis is determined
by the capacity of different parts of the plant to accept photosynthate, i.e. the
sink strength (Whittingham, 1972).

In so far as the solution of this problem rests with understanding the
mechanism of translocation *per se* the prospects for progress are not
encouraging, but techniques are becoming available which should make it
possible to understand in much greater detail what controls the strengths of
sources and sinks in terms of their size and activity. Although the mechanism
of translocation is not understood, there is a widely held view that the 'mass
flow' hypothesis of Münch is able to explain quite a number of the facts,
although certain awkward problems remain. Acceptance of a mechanism of
this kind focusses attention on the processes of loading and unloading, for, on
a 'neo-Münchian' view of the mechanism of translocation, it is the capacity
of these systems that will largely determine the fluxes of photosynthates out
of the sources and into the sinks. Recent progress in the isolation of
membranes from plant cells and the study of solute transport across cell
membranes (Poole, 1978) takes us to a point at which the results may be
applied in the investigation of loading and unloading of the phloem. A

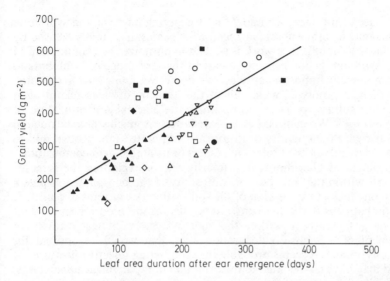

Figure 1.4   *The relationship between grain yield and leaf area duration after ear emergence for wheat crops in a range of environments. (△) UK (Watson et al., 1963); (◇) UK (Thorne, 1966); (▽) UK (Welbank et al., 1966); (□) UK (Welbank et al., 1968); (○) UK (Thorne et al., 1969); (◆) Australia (Davidson, 1965); (●) Australia (Turner, 1966); (▲) Australia (Fischer and Kohn, 1966); (■) Mexico (Fischer, unpublished data). (From Evans, Wardlaw and Fischer, 1975)*

coherent picture begins to emerge, which links studies of neutral ion pumps (Hodges, 1976), electrogenic pumps (Higginbotham and Anderson, 1974), uptake kinetics (Epstein, 1972), ionophores as carriers in fluid lipid membranes (Reed and Lardy, 1972) and transport systems for sugars, amino acids and organic acids (e.g. Komor and Tanner, 1974) involving co-transport of protons (Raven and Smith, 1974; Bentrup, 1978). Application of the concepts derived from these studies to the partitioning problem should lead us to a better understanding of the nature of loading and unloading processes and of how they might be manipulated as components of the source and sink strengths which contribute to yield.

(c)   There is the need to understand the relationships between development and the processes of respiration and photosynthesis. It is well known that the photosynthetic capacity of leaves changes with the stage of development (Woolhouse, 1967), and it becomes clear that for many crops there are relationships between the turnover of individual leaves, the duration of the leaf canopy of the crop and the yield (*Figure 1.4*).

THE REQUIREMENT FOR IMPROVED INPUT OF NITROGEN AND PHOSPHORUS TO CROPS

Over vast areas in which agriculture is practised, supplies of nitrogen and phosphorus limit agricultural production. The solution to this problem may lie in part with the achievement of cheaper supplies of nitrogen and phosphorus fertilizers, but the fundamental problems of sources of raw materials and thermodynamic aspects of the reactions involved in manufacture are likely to provide major barriers to progress in this direction.

In the case of nitrogen, we know that free-living bacteria and symbiotic organisms may fix atmospheric nitrogen which can subsequently be used by higher plants. Much further work is needed to optimize the combination of host and symbiont genotypes and to extend the host range of symbionts to other economically important species. We do not yet know how to combine features of host genotype, which affect the capacity of roots for nitrogen uptake, with cultural practices to optimize the efficiency of utilization of nitrogen by crops. When uptake is achieved, there is much still to be done in order to increase the quantity of nitrogen incorporated into protein and to improve the quality of the protein in respect of its amino-acid composition.

The problem of phosphorus is primarily one of exploiting the reserves which exist within the soil. The low diffusivity of the phosphate ion puts a premium on efficient exploration of the soil by the root systems of plants. It frequently happens that other constraints, such as the need for an efficient system of gas transport within the root system, militates against the development of a finely divided root system that would be optimal for phosphorus uptake. It seems probable that conflicting selective pressures of this kind may have contributed to the evolution of mycorrhizal associations, in which the fungus provides an extension of the root system in order to achieve a wider exploitation of the soil (Tinker, 1975). There is an increasing awareness that the endogenous mycorrhizas may contribute significantly to the phosphorus uptake by crop plants. Clearly, it is important that much more work be done to study the development of root systems and mycorrhizal associations which have maximum efficiency with respect to uptake of phosphorus from the soil.

PROBLEMS RELATING TO THE SUPPLY OF WATER AND MINERAL NUTRIENTS TO CROPS

The water and nutrient requirements of most crops are known; the demand for water and nutrients may vary according to the genotype of the crop, the stage of development and the environmental conditions.

Intraspecific variation in the response of crops to nutrient supply has often been recorded but the subject is not one which has claimed a great deal of attention from crop physiologists or plant breeders, probably because most plant breeding work has been carried out under circumstances where high levels of fertilizer treatment may be assured. This is unfortunate because it probably means that many of the successful agricultural varieties are dependent on much higher levels of nutrients than would have been the case had the procedures by which they were selected included a screening for capacity to crop well at low levels of fertilizer application.

As fossil fuels become scarcer and hence more expensive the use of fossil fuels in agriculture will have to be thoroughly investigated so that possible economies can be assessed. The management of soils for arable crops is one of the aspects of agriculture where the fuel bill is high. Under conventional agricultural systems the principal costs for soil management are for the extraction of raw materials and manufacturing processes for fertilizers, and the oil and petrol costs involved in transport to the farms and distribution on the land; additional fuel is needed for ploughing and subsequent cultivations.

In the attempt to reduce these costs more physiological work is needed. The move to minimal cultivation practices reduces fuel costs but we know little of whether there is genetic variation in the ability of crop plants to develop root systems which can most effectively exploit the more compacted soils that develop under these conditions. There is a substantial body of academic work on the development of roots and the mechanisms by which they take up nutrients, but there is at present little evidence of effort to relate the findings to crop performance and to explore the implications for plant breeding.

The last topic in this brief resumé of the problems which challenge the crop physiologist is that of water relations. For most crops in most parts of the world there is some period in the life cycle when water deficits limit plant growth. If water deficits occur at crucial stages of development, the consequences for yield may be catastrophic. It is scarcely surprising that with water limiting productivity of so many crops there should be such an enormous amount of research on the subject; I would nevertheless question whether this is entirely justified. Almost all the progress with plant water relations up to the present time has come from improving the techniques of irrigation. It is true that some crop varieties have been identified which are better able to tolerate dry conditions by virtue of such mechanisms as deep root systems, tighter stomatal control or a capacity to synthesize high levels of osmoregulatory solutes. Little of this work, however, seems likely to bring knowledge which might in the foreseeable future be directly exploited by the plant breeder, other than assisting marginally in providing improved criteria for selection. At the centre of the difficulty is the problem that so many processes contribute to water use characteristics of a plant that there seems little prospect for subjecting them to genetic analysis at the present time. If one takes a single aspect of the problem, that of stomatal control, it is clear that there is little prospect at the present time for understanding the complex relationship between stomatal structure, the external factors which affect stomatal movement, the components of the leaf water potential and the other internal factors which influence stomatal movement (Cowan, 1977). For this reason it would seem realistic to sustain the effort on water relations as long-term strategic work at an academic level rather than to develop this as front-line crop physiology with the expectation of immediate pay-offs.

Some Practical Considerations Relating to Crop Physiology

In the foregoing discussion I have endeavoured to summarize the main areas where it seems that a deeper understanding of the physiological and metabolic processes involved might enable farmers to increase the yields and quality of their crops. There is another reason for including the physiologist in work with crops, be it in plant breeding, cultivation trials, pathological work or most other branches of the study of crops. This arises from the continuing need to make measurements in order that all aspects of the progress of a crop may be recorded. Crop physiologists must always be involved with techniques, providing means of measuring parameters which for one reason or another are considered important, as for example measurements of stomatal movements, growth of roots, shoots and leaves, gas exchange and water potential. The carrying out of repeated measurements involves much routine and repetition, which may often appear to offer limited scope for originality, but a brief consideration of the inadequacy of present methods for measuring such parameters as root growth, respiration, photorespiration and translocation under field conditions shows that there is a great deal yet to be done.

Few crop physiologists would be likely to accept a permanent role as super technicians whose primary mandate was to make endless exacting measurements on behalf of those involved in other approaches to crop improvement such as plant breeders or pathologists. I shall, therefore, conclude with a brief consideration of two aspects of work in crop physiology where progress offers closer links between the physiology and breeding of crops.

CASE STUDIES OF PHYSIOLOGY IN RELATION TO CROP IMPROVEMENT

Photosynthesis and leaf duration

A large part of the photosynthesis in most crops is carried out in the leaves and it is scarcely surprising, therefore, that a great deal of work has been done on the growth of leaves and particularly on the effects of environmental factors such as light and temperature.

Herbaceous plants of the forest floor produce a single canopy of leaves which they then retain for as long as possible, ranging from a few weeks in *Hyacinthoides non-scriptus*, during which period there is most light penetrating the canopy above, to the whole growing season in such species as *Mercurialis perennis* and periods of several years in ground cover species of evergreen forests. By contrast, species of open habitats, from which almost all of our crops are derived, are much more profligate in their behaviour as they continuously generate new layers of leaves and discard the older ones so that the leaf area index measured at any one time represents a mixture of developing, mature and senescing leaves (Brougham, 1958).

In the management of forage grasses the aim is to exploit the developing and mature leaves and to minimize the proportion which escape to the senescent phase, but in arable crops this is not possible and represents a substantial loss of carbon and other nutrients in the course of a season. In winter wheat, over 90 per cent of the dry matter recovered in the grain is assimilated after anthesis (Austin *et al.*, 1977) and the primary challenge is to manipulate the division of growth between the vegetative and the reproductive periods. Indeed, there is strong evidence that much of the improvement in yields of cereals, which have resulted from the work of plant breeders over the past 50 years, comes primarily from an increase in the harvest index, that is, in the proportion of the total dry matter assimilated which is distributed to the grain, rather than from any significant increase in the total assimilation in the course of the season (Austin, 1978). By comparison, indeterminate crops such as potatoes and sugar beet are able to achieve their higher yields by renewal of the canopy throughout a longer period of the potential growing season. It is well known that nitrogen stress hastens leaf senescence in maturing cereals and it is at this grain filling period that the delaying of senescence is crucial.

The finding of close relationships between the duration and integrity of the photosynthetic canopy and yield has placed a new emphasis on the factors which control leaf senescence (Woolhouse, 1974). At the same time it must be recognized that if ceiling yields are to be raised then something must be done to increase the photosynthetic potential of the canopy once it is developed and as long as it may be caused to persist. There are many ramifications to this problem and the emphasis has shifted several times in recent years. Much attention was formerly given to the geometrical features of the foliage in relation to the trapping of light (Austin *et al.*, 1976) but, while the importance of establishing a complete cover of leaves as rapidly as possible is generally conceded, it now seems that detailed characteristics of the geometry of the canopy may be of less importance.

Over the past 80 years it has been shown repeatedly for many species that the rate of photosynthesis increases linearly with photon flux density at low flux densities, but that the rate of increase then falls off to approach asymptotically to a limiting value at high flux densities. If the ambient CO_2 concentration is increased, the rate of photosynthesis at light saturation is also increased. It became widely accepted from these observations that at saturating photon flux densities physical resistances to CO_2 diffusion were the primary determinants of photosynthetic rate (Gaastra, 1959).

In the course of experiments to determine the maximum photosynthetic rate as a function of leaf age in *Perilla frutescens* we found that the rate decreased gradually from the time that leaf expansion was completed (Hardwick, Wood and Woolhouse, 1968). Concomitant measurements of water vapour efflux showed that resistance to diffusion in the gaseous phase of the leaf was not changing significantly during this period, but that the internal or mesophyll resistance was rising. These findings were not easily reconciled with the view that physical resistances to diffusion of CO_2 could account for the declining photosynthetic rate as a leaf grows older, and the suggestion was made that a biochemical or carboxylation resistance was limiting photosynthesis (Woolhouse, 1967). Further support for this hypothesis comes from the finding that when *Perilla* plants were decapitated to leave just the basal pair of leaves in a yellowed state, these leaves regreened and increased their fraction 1 protein content and photosynthetic capacity to higher values than had been obtained previously (Woolhouse, 1967). A similar effect of partial defoliation on photosynthetic rate was demonstrated for *Zea mays*, *Phaseolus vulgaris* and *Salix spp.* (Wareing, Khalifa and Treharne, 1968) and shown to be associated with an increased ribulose 1,5-bisphosphate carboxylase activity in the leaves that were left on the plants. Subsequent developments of the work on photosynthesis in the course of senescence and after partial defoliation, have revealed a great many changes taking place in the photosynthetic apparatus which contribute to the changing values of maximum photosynthetic rate.

The work of Correns (1909) and others subsequently showed that various mutations which affect plastid structure showed non-Mendelian maternal inheritance, which was subsequently shown to be associated with a specific genome in the plastids. Plastids and mitochondria have subsequently been shown to contain DNA in a circular duplex form (Manning *et al.*, 1971) together with the appropriate enzymic apparatus for DNA replication and transcription and a 70S ribosome system which carries out synthesis of proteins within the organelles (Boulter, Ellis and Yarwood, 1972).

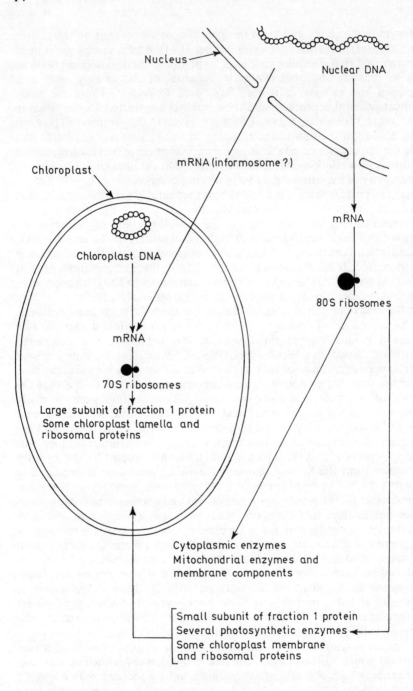

Figure 1.5 *Suggested interdependence of nucleus, cytoplasmic ribosomes, and chloroplast in the synthesis of chloroplast photosynthetic enzymes, and ribosomal and membrane proteins*

Cloning and hybridization of DNA are now being applied to produce a genetic map of the plastid genome (Gillham, 1978). The plastid genome is found to code for some components of the thylakoid membrane system and of the stromal compartment of the plastid, although the majority of the proteins of both elements of the plastid are coded for in the nucleus of the cell. Of the plastid proteins coded for in the nuclear genome the majority of those studied to date are translated on the cytoplasmic ribosome system (Ellis, 1975) and the proteins are then processed and transported into the organelle. There is increasing evidence, however, that some plastid proteins encoded in the nuclear genome may be synthesized on ribosomes of the plastid (Hoober and Stegeman, 1975), which implies that there must exist a system for transporting messenger ribonucleic acid (mRNA) from the nucleus to the plastid (*Figure 1.5*).

In *Phaseolus vulgaris* and *Perilla spp.* the capacity for protein synthesis in cell-free plastids decreases sharply after the completion of leaf expansion (Callow, Callow and Woolhouse, 1972). At the same time the polysome content of the plastids decreases, although the cytoplasmic polysome content does not change significantly. Experiments in which leaves of different ages are pulsed with $^{32}PO_4$ show that the decreased polysome content of the plastids is associated with a loss of the capacity to synthesize plastid ribosomal ribonucleic acid (rRNA) (Callow, Callow and Woolhouse, 1972). It has subsequently been shown that a plastid polymerase enzyme, responsible for RNA synthesis, becomes inactive at the time that leaf expansion is completed (Ness and Woolhouse, 1980a, 1980b), and it appears that this may be the essential mechanism by which the renewal of plastid components is switched off. Indirect support for this hypothesis is provided by the finding that the decreasing photosynthetic rate in mature leaves is correlated with a loss in activity of those enzymes that are synthesized wholly or in part on the plastid ribosomes (Batt and Woolhouse, 1975). The capacity for photosynthetic electron transport also declines in mature leaves and recent work suggests that this may be associated with a rate-limiting step in the PQ-cytochrome f (*Figure 1.6*) component of the electron transport chain (Jenkins and Woolhouse, 1981). This is of particular interest in view of the recent finding that cytochrome f is also coded for in the plastid and synthesized on the plastid ribosome system (Gray, 1978). Under decapitation or partial defoliation treatments a complex series of changes occurs, which involves extra DNA synthesis in the cells of the remaining leaves (Ness and Woolhouse, 1980b), a reactivation of the plastid RNA polymerase system and renewed synthesis of all the major plastid constituents. The adaptive significance of this system is probably that it allows a plant which becomes partially defoliated by grazing, pest attack or fungal infection to compensate for the loss of photosynthetic tissue by an increased capacity for photosynthesis in the leaves that remain.

As one delves into mechanisms of this complexity it all comes to look rather far removed from agricultural production and indeed at the present time it is. But it is the author's conviction that the analysis of these systems to the point of identification of the rate-limiting steps and of the elements which regulate the synthesis and renewal of these constituents should shortly lead us to design experiments in which the regulatory mechanisms that govern these systems might be manipulated. If, for example, it were to prove

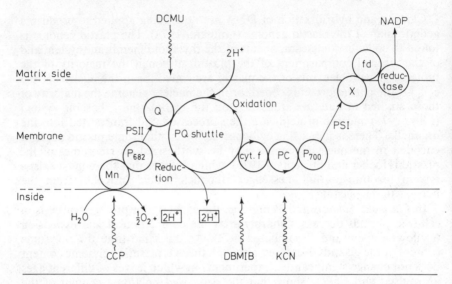

Figure 1.6   *Photosynthetic electron flow from water to triphosphopyridine nucleotide (NADP) in a zigzag across the membrane. Two proton-releasing sites inside and the points of inhibition of some inhibitors of electron flow are indicated*

possible to identify and delete or modify an operator locus for the plastid RNA polymerase then this might afford a key to the genetic manipulation of leaf duration. These are, of course, speculative propositions but they suggest directions towards a greater degree of biochemical analysis and genetics linked to breeding work which must form a major component of plant breeding programmes of the future.

Nitrogen metabolism and storage proteins

For my final example of the potential linking of analytic physiology to plant breeding I will turn very briefly to nitrogen metabolism and the legumes.

Many people have worked on leguminous crops but one may say in general terms that they are of more recent domestication than the cereals. They tend to be less uniform, they often present awkward problems in their reproductive biology because of elaborate incompatibility systems, indeterminate inflorescences and flower abortion, to mention but a few of the technical difficulties. Notwithstanding this, their high protein content and capacity for symbiotic nitrogen fixation demands that they be grown to the highest possible levels of productivity.

Now, while the breeders have been getting to know the legumes, seeking mutants with determinate inflorescences, wrestling with in-breeding self-incompatible species and the relationships of host strain to *Rhizobium* strain, a good deal of other work has been in progress on the physiology.

Firstly, there is the description of development in quantitative terms—from a vast literature, the work of Pate (1975) and his colleagues will serve our purposes. These investigators have been concerned to describe the carbon

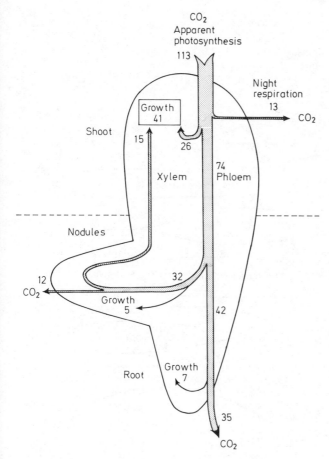

Figure 1.7 *Flow sheet for carbon in effectively nodulated plants of* Pisum sativum *(cv. 'Meteor'), relying solely on the atmosphere for their nitrogen. The study period covered is 21–30 days after sowing, a time when the roots and nodules are still growing actively. Data are expressed on the basis of a net gain by the shoot of 100 units of carbon from the atmosphere. (From Pate, 1975)*

and nitrogen economy of *Pisum sativum. Figure 1.7* gives an example of the carbon economy in the growing plant between days 21 and 30 from the time of sowing. In a more detailed extension of this work, Pate's group have described and analysed the growth curve for the seeds; *Figure 1.8* shows the timing of formation of the various constituents of the seed against the backcloth of the progressive increase of the seed fresh weight.

This is a good and solid physiology but it does not, as yet, tell us anything which the breeder can use in any fundamental way. I would suggest, however, that the importance of possessing this knowledge will become apparent in the next decade for the following reason. Several groups in Australia, the USA and the UK (Rothamsted and Durham) have been purifying the storage proteins, sequencing and making antibodies to them with which to 'titrate' in a definitive way, small amounts of those proteins. Using these techniques, Boulter's group has recently succeeded in obtaining *in vitro* synthesis of the

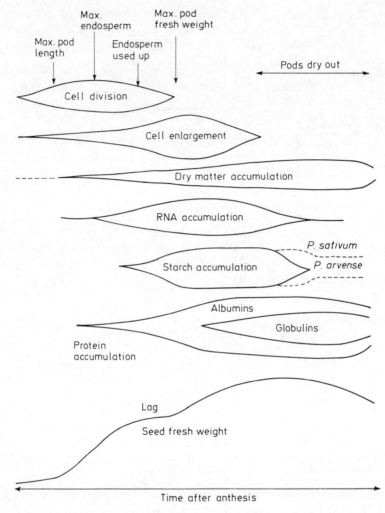

Figure 1.8   *The cardinal events in seed maturation of* Pisum spp. *The common time scale is the growth curve for seed fresh weight. All events depicted refer to rates within the developing cotyledons, divergent lines denoting periods of increasing rate, convergent lines times of decreasing rate. Differences between* P. sativum *and* P. arvense *are indicated. (From Pate, 1975)*

storage proteins of two globulins, that is, salt-soluble proteins known as vicilin and legumin, which comprise 80 per cent of the storage protein in seeds of *Pisum sativum* (Evans *et al.*, 1979). From polysomes that will make a defined protein in this way, it is a relatively direct step to the isolation of the mRNAs for these proteins from the polysomes that are synthesizing them. The purified mRNAs can then be treated with the enzyme reverse transcriptase to make the corresponding DNA, i.e. the legumin and vicilin genes.

Incorporation of this DNA into a bacterial plasmid should, with reasonable luck, afford a means of 'cloning', i.e. multiplying up the legumin and vicilin genes. Now two interesting possibilities arise: on the one hand, if the host bacterium could be persuaded to transcribe and translate the legumin and vicilin genes, the way might be opened to synthesizing them in bulk microbial cultures so that there would no longer be any need for the legumes—at least in so far as their storage proteins are concerned. However, experience with considerably simpler proteins suggests that the bacteria may draw the line at accommodating man's needs in quite so simple a fashion so that, if the genes for the storage proteins can be multiplied but not transcribed or translated in the bacteria, then the emphasis will go on trying to return the cloned genes back into the host whence they came. Plasmids of the plant tumour bacterium *Agrobacterium* suggest one possible route, currently under vigorous attack; the DNA viruses might afford another. This is where the wheel begins to turn full circle; the molecular biologist has to carry out DNA hybridization studies—probably at the cytological level in order to see how many copies of the cloned storage protein genes have been incorporated—and the breeder must now resume an interest in order to explore the stability and inheritance of the enhanced genome. It is here that the basic physiology is needed again, for against the initial description of the seed development one can begin to measure changes due to the increased gene dosage; if we have put in ten more copies of the vicilin gene per genome, does this affect the timing of developmental events such as the phases of RNA or globulin synthesis? Does it affect the amounts of these constituents? Do more copies of the gene mean more mRNA production and does this lead to more legumin and vicilin protein molecules being made, or does the cell resent this intrusion on its economy to the extent of degrading or failing to translate the extra mRNA— if such were found to be produced? This multiplicity of subsequent questions implies a major role for the further involvement of the physiologist at this stage of the enterprise.

Conclusions

I have discussed, perhaps somewhat glibly, some tasks of enormous complexity—and touched in a matter of pages on topics which will fill whole chapters later in this volume; some, indeed, could form the substance of whole symposia in themselves. I have done this in order to try and set the scene for some of our deliberations in the course of this meeting, but I also have a deeper motive. I feel that the time is at hand when for many of our crops the limits to improvement by selection are soon to be reached and I have been trying to grapple with the question of how we might go through this barrier—into the pit of complex underlying processes, arranging our studies in a manner and style such that they can be at once analytic and yet serve the cause of increased crop production. I have tried to show how the physiology should be pressed from the description down to the level of metabolite fluxes and thence to the controlling enzymes, membrane constituents and their determinants so that the breeder can consider specific attributes; but I wish to conclude by reference to another aspect of the problem, which I deem to be crucial if these enterprises are to succeed. I refer to the matter of organization. The physicists learned long ago that the only way to make progress when they wanted to get into the nature of matter or the structure of our universe, was to organize in teams and to apportion the work; I should like to suggest that the time has come when crop scientists must adopt analogous procedures.

Let us consider briefly the major crops in the UK: (a) forage grasses (including forage maize); (b) cereals; (c) potatoes; (d) legumes; (e) brassicas (including oil seed rape). Under these five headings we need all the levels of attack to which I have alluded above in order to ensure that each of the five is being sufficiently and systematically investigated. The Agricultural Research Council (ARC) has made some progress towards such organization, by allocating certain crops to particular centres; but this is not enough. We have an ARC modelling group, but no specialist groups around these several crops. It seems to me that we should now be moving towards the establishment of standing groups around each of these five categories of crops; each group should be monitoring and encouraging integrated programmes for its own crop, ensuring coordinated university and research service work aimed at comprehensive coverage of the many different angles of enquiry which are needed. They would hold annual workshops to exchange ideas, information, align their programmes and reassess imperatives; and at longer intervals, of say five years or so, they would hold review symposia and take a broader look at progress. Perhaps such gatherings could form exciting programmes for future Easter Schools.

References

ALBERSHEIM, P. and ANDERSON-PROUTY, J. (1975). *Ann. Rev. Pl. Physiol.*, **26**, 31

AUSTIN, R. B. (1978). *ADAS Q. Rev.*, **29**, 76

AUSTIN, R. B., EDRICH, J. A., FORD, M. A. and BLACKWELL, R. D. (1977). *Ann. Bot.*, **41**, 1309

AUSTIN, R. B., FORD, M. A., EDRICH, J. A. and HOOPER, B. E. (1976). *Ann. appl. Biol.*, **83**, 425

BENTRUP, F. (1978). *Prog. Bot.*, **40**, 84

BOULTER, D., ELLIS, R. J. and YARWOOD, A. (1972). *Biol. Rev.*, **47**, 113

BROUGHAM, R. W. (1959). *N.Z. J. agric. Res.* **1**, 707

CALLOW, J. A., CALLOW, M. E. and WOOLHOUSE, H. W. (1972). *Cell Diff.*, **1**, 79

COOPER, J. P. (1964). *J. appl. Ecol.*, **1**, 45

CORRENS, C. (1909). *Z. Indukt. Abstamm. Vererbingsl.*, **1**, 291

COWAN, I. R. (1977). *Adv. Bot. Res.*, **4**, 117

DAY, P. R. (1974). *Genetics of Host-Parasite Interaction*. Freeman, San Francisco

ELLIS, R. J. (1975). In *Membrane Biogenesis*, p. 247. Ed by A. Tzagoloff. Plenum Press, New York

EPSTEIN, E. (1972). *Mineral Nutrition of Plants: Principles and Perspectives*. Wiley, New York

EVANS, I. M., CROY, R. R. D., HUTCHINSON, P., BOULTER, D., PAYNE, P. I. and GORDON, M. E. (1979). *Planta*, **144**, 455

EVANS, L. T., WARDLAW, I. F. and FISCHER, R. A. (1975). In *Crop Physiology*, p. 101. Ed by L. T. Evans. Cambridge University Press, Cambridge

GRAY, J. C. (1978). *Eur. J. Biochem.*, **82**, 133

HARDWICK, K., WOOD, M. E. and WOOLHOUSE, H. W. (1968). *New Phytol.*, **67**, 79–86

HIGGINBOTHAM, N. and ANDERSON, W. H. (1974). *Can. J. Bot.*, **52**, 1011

HODGES, T. K. (1976). In *Encyclopedia of Plant Physiology*. New Series, Vol. 2: *Transport in Plants*, Part A, p. 260. Ed. by U. Lüttge and M. Pitman. Springer, Berlin

HOOBER, J. K. and STEGMAN, W. J. (1975). In *Genetics and Biogenesis of Mitochondria and Chloroplasts*, p. 225. Ed. by C. W. Birky, P. S. Perlman and T. J. Byers. Ohio State University Press, Columbus

JENKINS, G. and WOOLHOUSE, H. W. (1981). *J. exp. Bot.*, in press

KOMOR, E. and TANNER, W. (1974). *J. gen. Physiol.*, **64**, 568

MANNING, J. E., WOLSTENHOLME, D. R., RYAN, R. S., HUNTER, J. A. and RICHARDS, O. C. (1971). *Proc. Natl Acad. Sci. USA*, **68**, 1169

MONTEITH, J. L. and ELSTON, J. F. (1971). In *Potential Crop Production*, p. 23. Ed. by P. F. Wareing and J. P. Cooper. Heinemann, London

NESS, P. J. and WOOLHOUSE, H. W. (1980a). *J. exp. Bot.*, **31**, 223

NESS, P. J. and WOOLHOUSE, H. W. (1980b). *J. exp. Bot.*, **31**, 235

OGREN, W. L. and BOWES, G. (1971). *Nature, New Biol.*, **230**, 159

PATE, J. S. (1975). In *Crop Physiology*, p. 191. Ed. by L. T. Evans. Cambridge University Press, Cambridge

POOLE, R. J. (1978). *Ann. Rev. Pl. Physiol.*, **29**, 437

RAISON, J. K. (1973). *Symp. Soc. exp. Biol.*, **27**, 485

RAVEN, J. A. and SMITH, F. A. S. (1974). *Can. J. Bot.*, **52**, 1035

REED, P. W. and LARDY, H. A. (1972). *J. biol. Chem.*, **247**, 6970

ROBSON, M. J. (1967). *J. appl. Ecol.*, **4**, 475

TINKER, P. B. (1975). In *Endomycorrhizas*, p. 353. Ed. by F. E. T. Sanders, B. Mosse and P. B. Tinker. Academic Press, London

TOLBERT, N. E. (1971). *Ann. Rev. Pl. Physiol.*, **22**, 45

WAREING, P. F. (1971). In *Potential Crop Production*, p. 362. Ed. by P. F. Wareing and J. P. Cooper. Heinemann, London

WAREING, P. F., KHALIFA, M. M. and TREHARNE, K. J. (1968). *Nature*, **220**, 453

WARREN-WILSON, J. (1972). In *Crop Processes in Controlled Environments*, p. 7. Ed. by A. R. Rees, K. E. Cockshull, D. W. Hand and R. G. Head. Academic Press, London

WHITTINGHAM, C. P. (1972). In *Crop Processes in Controlled Environments*, p. 175. Ed. by A. R. Rees, K. E. Cockshull, D. W. Hand and R. G. Head. Academic Press, London

WOOLHOUSE, H. W. (1967). *Symp. Soc. exp. Biol.*, **27**, 179

WOOLHOUSE, H. W. (1974). *Sci. Prog. Oxf.*, **61**, 123

WOOLHOUSE, H. W. (1978). In *Endeavour*, New Series, Vol. 2, No. 1. Pergamon Press, Oxford

WOOLHOUSE, H. W. and BATT, T. (1976). In *Perspectives in Experimental Biology*, Vol. 2, p. 163. Ed. by N. Sunderland. Pergamon Press, Oxford

ZELITCH, I. (1975). *Science*, **188**, 626

2

DOES LIGHT LIMIT CROP PRODUCTION?

J. L. MONTEITH
Department of Physiology and Environmental Studies, University of
Nottingham School of Agriculture, UK

Historical Evidence

To plant physiologists, the title of this paper may seem somewhat eccentric. As the energy provided by sunlight plays a central role in the metabolism of green plants, there is a sense in which plant production must always be limited by the availability of light. However, fluctuations in the yield of a crop from year to year can rarely be correlated with the income of radiation because yield is influenced by many environmental variables other than light, in particular, by rainfall, temperature and the incidence of disease. An agronomist might therefore argue that light does *not* limit crop production.

The literature of growth analysis contains many negative conclusions about the influence of sunshine and radiation on growth. When Briggs, Kidd and West (1920) re-examined Kreusler's measurements of maize growth in terms of a unit leaf rate (or net assimilation rate, NAR), they concluded that 'the real assimilation of the plant is not governed by light'. This surprising statement was based on a poor correlation ($r^2 = 0.14$) between weekly hours of sunshine and corresponding values of NAR corrected for respiration to obtain 'real' assimilation. Gregory (1926) attempted a similar analysis for barley grown at Rothamsted where one of the first radiation records in the UK was available from a Callendar recorder. Weekly values of NAR were again poorly correlated with radiation ($r^2 = 0.18$) and relative growth rate (RGR) was negatively correlated. Gregory's work was extended by Watson (1947), who analysed the growth of wheat, barley, potatoes and mangolds in even greater detail but found 'no significant correlation between NAR and mean daily radiation'.

Blackman and Black (1959) were able to obtain a significant positive correlation between the relative growth rate of sunflower and insolation. Not surprisingly, plants grew faster in Adelaide and Brisbane than in Slough and Invergowrie! They concluded that 'under conditions where growth is not restricted by temperature or by supplies of nutrients or water, maximum production of dry matter per unit area will be limited by leaf area index and the amount of solar radiation'. However, Japanese workers who extended Blackman's techniques to other crops and climates found that neither NAR nor RGR was well correlated with solar radiation (Kamiyama and Horie, 1975).

In contrast to these conflicting but mainly unsuccessful attempts to correlate assimilation rate with irradiance, it has been shown that the rate at

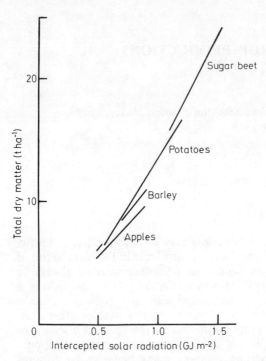

Figure 2.1   *Relationship between total dry matter production of four crops grown in England and intercepted (total) solar radiation. (From Monteith, 1977)*

which arable crops accumulate dry matter during early growth is proportional to the rate at which radiant energy is absorbed by the canopy (Warren Wilson, 1967; Monteith, 1977; Gallagher and Biscoe, 1978; Sibma, 1977). Moreover, the efficiency of growth, defined as the energy equivalent of dry matter per unit of absorbed energy, has a value of about 2.0–2.5 per cent for different (C_3) crops (*Figure 2.1*).

Further relevant evidence comes from field studies of photosynthesis which take three forms: measurements of the CO_2 exchange of single leaves exposed in small chambers (e.g. Marshall and Biscoe, 1977); similar measurements on a number of adjacent plants in large enclosures (e.g. Alberda, 1977); and measurements of the vertical CO_2 flux over whole stands (e.g. Biscoe, Scott and Monteith, 1974). Such studies have clearly established that the rate at which a field crop assimilates CO_2 changes minute by minute in response to changes of irradiance, and that the change of assimilation rate with irradiance is much greater in weak light than in strong light. In extreme cases, as represented by *Figure 2.2*, the response is better represented by two straight lines, following Blackman (1905), than by the rectangular hyperbola commonly used by a later generation of physiologists. In fact, many sets of measurements to which hyperbolas have been fitted could equally well be represented by straight lines that intersect at a limiting value of irradiance (Marshall, 1978).

The evidence so far reviewed suggests that (a) light does not limit yield nor NAR; but (b) the growth rate of a crop is almost proportional to the radiation

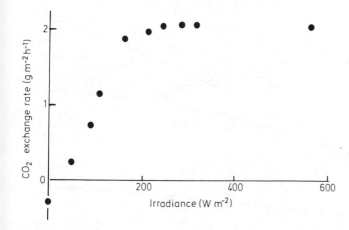

Figure 2.2 *Assimilation of CO_2 by wheat leaf as function of solar irradiance. (From Marshall, 1978)*

intercepted by its canopy; and (c) the rate of photosynthesis increases with irradiance up to a saturating irradiance beyond which it is nearly constant. In an attempt to reconcile the apparent inconsistency of these conclusions, we shall try to predict how the growth rate of a crop is likely to depend on the radiation it receives and on the photosynthetic response of individual leaves. Details of a simple model of crop growth are given in an Appendix at the end of this chapter. A more complex model, which leads to similar conclusions, was recently described by Charles-Edwards (1978).

Sensitivity of Crop Growth Rate to Parameters of the Model

IRRADIANCE AND MAXIMUM LEAF PHOTOSYNTHESIS RATE

The relationship between daily totals of dry matter production for a complete crop canopy and the corresponding income of solar radiation (S) was calculated from the model for values of S between 10 and 20 MJ m^{-2} and for three values of maximum leaf photosynthesis rate (P_m; *Figure 2.3*). The change of photosynthesis rate with irradiance in weak light (ε_m) and the canopy transmission coefficient (K) were set at standard values described in the Appendix. The standard value of $\varepsilon_m = 0.02$ g CH_2O m^{-2} h^{-1} per W m^{-2} (PAR) corresponds to a quantum efficiency of 0.05 mol E^{-1}.

The maximum value of crop growth over much of the UK and parts of western Europe is about 20–25 g m^{-2} d^{-1} (Sibma, 1968; Greenwood *et al.*, 1977), consistent with the curve for $P_m = 2$ g CH_2O m^{-2} h^{-1}. Unless otherwise specified, this value will be used in the calculations which follow.

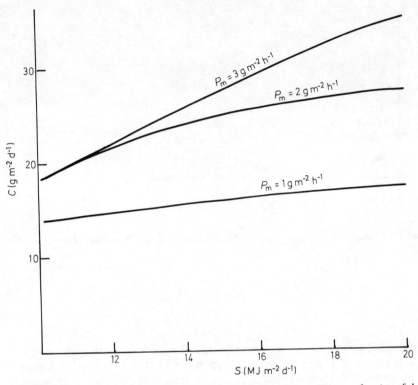

Figure 2.3   *Estimated maximum rate of crop dry matter production (C) as a function of daily insolation (S) and maximum leaf photosynthesis rate (P_m)*

Table 2.1   MONTHLY MEAN VALUES AND STANDARD DEVIATIONS OF SOLAR RADIATION (WHOLE SPECTRUM) RECORDED AT SUTTON BONINGTON, 1967–1976

Month	$MJ\,m^{-2}$
April	11.0 ± 1.3
May	14.5 ± 1.8
June	17.5 ± 2.2
July	15.3 ± 1.7
August	12.9 ± 1.7
September	9.3 ± 1.2
May–July	15.8 ± 1.0
July–September	13.9 ± 1.0

For the observed range of radiation from year to year (*Table 2.1*), the predicted range of growth rate is relatively small, about ± 5 per cent over periods of a month and only ± 3 per cent over a whole growing season. Comparison of the curves for $P_m = 1$ and $2\,g\,m^{-2}\,h^{-1}$ shows that differences of insolation should have even less effect on crop growth when leaf photosynthesis is restricted by, for example, nutrient shortage or disease.

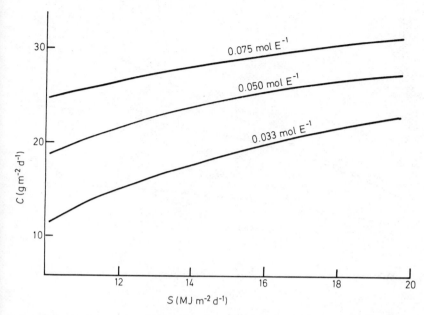

Figure 2.4   *As Figure 2.3 where C is a function of S and of maximum quantum efficiency for a single leaf (ε_m)*

IRRADIANCE AND QUANTUM EFFICIENCY

Figure 2.4 shows the effect of a change in quantum efficiency on the rate of photosynthesis when all other parameters are standard. If the efficiency increased by a factor of 3/2 from 0.05 to 0.75 mol E^{-1}, the predicted rate of photosynthesis would increase by only 17 per cent at 14 MJ m^{-2} d^{-1} and 14 per cent at 18 MJ m^{-2} d^{-1}. On the other hand, a decrease of efficiency by a factor of 2/3 would decrease photosynthesis by 40 per cent at 14 MJ m^{-2} d^{-1}.

Taken together, *Figures 2.3* and *2.4* imply that the combination of a quantum efficiency of 0.05 mol E^{-1} and a maximum assimilation rate of 2 g CH_2O m^{-2} h^{-1} defines a system balanced between light saturation and light dependence, at least when exposed to the levels of radiation which are characteristic of a temperate summer. The lesson for the physiologist and breeder seems to be that crop growth cannot be increased much by increasing quantum efficiency without also increasing P_m and vice versa. Do plants insure against adverse weather by working at a photosynthetic rate which is not completely limited either by levels of irradiance or by physiological factors such as stomatal resistance, which partly determine the value of P_m?

IRRADIANCE AND LEAF EMERGENCE

Figure 2.5 shows a predicted dependence of photosynthesis rate on leaf orientation as represented by three values of the transmission coefficient (K) for crops with vertical leaves (0.3), with very horizontal leaves (0.9), or with

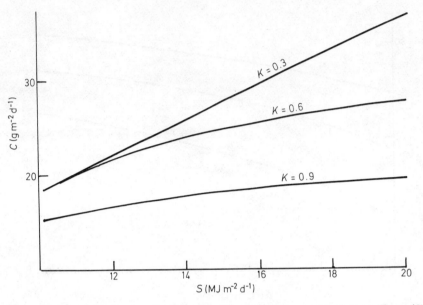

Figure 2.5   *As Figure 2.3 where C is a function of S and of the canopy transmission coefficient (K)*

a standard intermediate arrangement (0.6). Here again, extreme values of K represent contrasting systems: when $K=0.3$, the canopy photosynthesis rate is light-dependent over the whole range of mean insolation encountered during a temperate summer whereas at $K=0.9$ it is almost continuously light-saturated.

Figure 2.5 appears to show a marked advantage in terms of photosynthesis rate for an erect canopy—a consequence of more uniform light distribution—but in a real crop, the gain would be offset by poor light interception during early growth. To examine the influence of canopy structure on growth over a whole growing season, standing dry weight was estimated from equation (2.15). The straight lines in *Figure 2.6* represent the relationship between dry weight and time, extrapolated to zero weight to show how much time is 'lost' during the period when the canopy is incomplete. The disadvantage of vertical leaves (K small) is clear during early growth. Over the period from 80 to 100 days, predicted dry weight is insensitive to the value of K and it is only when the growing season is longer that consistently erect foliage is likely to produce more dry matter. In many real crops such as grasses and cereals, however, foliage which is initially erect becomes more prostrate with age. *Figure 2.6* suggests that a more productive strategy would be horizontal leaves during early growth, which become more vertical as the canopy closes.

Figure 2.6 was calculated for a constant leaf weight ratio (leaf area per total plant weight). According to equation (2.15), the amount of lost time is inversely proportional to this quantity when maximum growth rate is constant. In a similar context, however, Charles-Edwards (1978) has pointed out that maximum photosynthesis rate (P_m) is negatively correlated with the ratio of leaf area to leaf weight. This correlation implies that the apparent advantage of a large leaf weight ratio for intercepting light during the formation of a canopy would be at least partly offset by a slower maximum rate of growth when light interception became complete (or reached a constant maximum value).

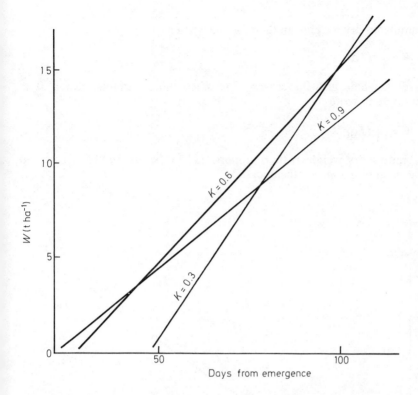

Figure 2.6 *Standing dry weight (W) of a crop as a function of days from emergence and canopy transmission coefficient K (from equation (2.15))*

GROWTH AND INTERCEPTED RADIATION

The fraction of incident light that is transmitted through a canopy with a leaf area index of L is $\exp(-KL)$ so the fraction of absorbed plus reflected radiation is $[1 - \exp(-KL)]$. Since the reflection of photosynthetically active radiation (PAR) is usually less than 10 per cent, the absorbed radiation alone is almost proportional to $[1 - \exp(-KL)]$.

From the model, the area of the directly lit leaves mainly responsible for carbohydrate production is

$$\int_0^L \exp(-KL/\alpha)\, dL = [1 - \alpha \exp(-KL/\alpha)]/K$$

Since α is close to unity for PAR, the area of foliage that contributes to photosynthesis, as estimated in the model, increases with L in much the same way as does the fraction f of intercepted radiation. The model is therefore consistent with the observations summarized in *Figure 2.1*. Photosynthetic rate must also be proportional to irradiance (I) of sunlight and to the efficiency ε $(\leqslant \varepsilon_m)$ with which radiant energy is stored in carbohydrate.

Accumulated dry weight can therefore be written as

$$W = \int \varepsilon f I \, dt$$

If, over a whole growing season, f is much more variable than ε, it is appropriate to write

$$W = \bar{\varepsilon} \int f I \, dt$$

i.e. standing dry weight is then proportional to the integral of intercepted radiation and to a mean efficiency, $\bar{\varepsilon}$.

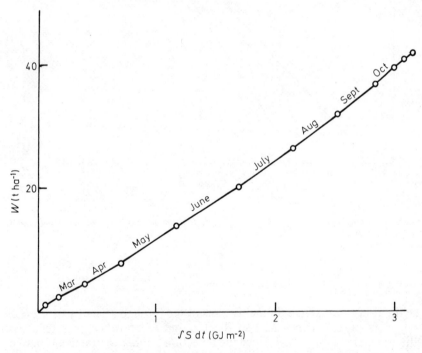

Figure 2.7   *Increase of standing dry matter (W) throughout a year as a function of accumulated solar radiation $\int S \, dt$*

Figure 2.7 reveals the extent to which $\bar{\varepsilon}$ may change over the course of a year. It was constructed by calculating monthly values of photosynthesis from long-term averages of insolation at Sutton Bonington. Maximum leaf photosynthesis rate was set at $2 \text{ g m}^{-2} \text{ h}^{-1}$ when mean monthly temperature $(\bar{T})$ exceeded $10°C$ but was reduced by a factor $\bar{T}/10$ for $\bar{T} < 10°C$ as an arbitrary allowance for the effect of low temperature on CO_2 assimilation. Fixed standard values were assigned to K and to ε (though quantum efficiency is known to be somewhat dependent on temperature).

The predicted accumulation of dry matter from an arbitrary origin at the beginning of the year was nearly proportional to accumulated radiation, which shows that $\bar{\varepsilon}$ did not deviate much from an annual mean value of about

1.3 g MJ^{-1}; but it tended to increase in the autumn when leaves spend more time in a light-dependent state. Observations on sugar beet support this trend (*see Figure 2.1*). During the period of comparable irradiance in spring, $\bar{\varepsilon}$ was smaller than in autumn because the value adopted for P_m was depressed by low temperature.

The increase of dry matter with intercepted radiation is consistent with figures for record crop production at about 20 t ha^{-1} of dry matter for a growing season of about three months from the formation of a complete canopy to harvest. Maximum figures for annual grass production—about 30 t ha^{-1}—are also consistent with *Figure 2.7* if the loss of intercepted radiation by cutting is about 25 per cent of the annual income.

Net Assimilation Rate and Irradiance

When a crop produces dry matter at a rate $\dot{W}$ which is proportional to intercepted radiation, equation (2.5) is valid and the exponential function it contains can be expanded to give

$$\dot{W} = C[1 - \exp(-KL)]$$
$$= C[1 - (1 - KL + K^2L^2/2! - K^3L^3/3! + \cdots)] \qquad (2.1)$$

where C is the maximum crop growth rate which would be achieved if all incident light were intercepted. Terms beyond $K^2L^2/2$ can be neglected when KL is small compared with unity, implying $L < 1$. Then net assimilation rate can be expressed as

$$\dot{W}/L \simeq KC(1 - KL/2) \qquad (2.2)$$

So, for early growth, equation (2.2) implies that NAR ($\dot{W}/L$) should decrease linearly with increasing leaf area from a maximum initial value of KC.

Equation (2.2) is consistent with measurements reported by Watson (1958) but not with the conclusions which he drew from them. He derived empirically an equation which was formerly identical to equation (2.2) and differentiated $\dot{W}$ with respect to L to give an expression which was equivalent to

$$d\dot{W}/dL = KC(1 - KL) \qquad (2.3)$$

On this basis, the maximum growth rate is given by the condition $d\dot{W}/dL = 0$, which implies from equation (2.2) that $KL = 1$, $L = 1/K$ and $\dot{W}/L = 0.5KC$. For larger values of L, $\dot{W}$ decreases and reaches zero when $L = 2/K$. This implausible result is a consequence of extrapolating equation (2.2) beyond the range of L for which it is valid. When equation (2.2) is treated as a limiting case of equation (1.1), valid only when L is small, crop growth rate increases from an initial value of KLC to a maximum value of C effectively independent of L, and this behaviour is consistent with the growth curve of most crops. The decrease of growth rate usually observed at the end of the main period of growth is a consequence of senescence and is irrelevant to this analysis.

When a crop grows at a rate which changes systematically with irradiance (or with any other environmental factor), corresponding changes of NAR tend to be masked by contemporary changes of leaf area index, which always increases during early growth and often decreases towards maturity. Consequently, attempts to correlate NAR with light rarely succeed except during the very early stages of growth before mutual shading of leaves begins. The technique of classic growth analysis is valid for isolated plants with few leaves but it is not an appropriate tool for crop ecologists and attempts to refine its mathematical basis seem irrelevant to agricultural science.

A more rational basis for growth analysis is provided by two variables: the fraction of radiation which a canopy intercepts and the efficiency with which absorbed radiation is used for photosynthesis (Monteith, 1977). Japanese workers (e.g. Murata, 1975) have attempted to modify the algebra of growth analysis to allow for differences in intercepted light at different stages of growth but this exercise seems unnecessarily circuitous.

So is Light a Limiting Factor?

A number of loose ends can now be drawn together to answer the question that heads this paper. Crops cannot grow without radiant energy so it is appropriate to describe light as a *determinant* of production. However, differences of yield from year to year are seldom correlated with light. This is partly because leaves tend to be saturated with light in bright sunshine and partly because the amount of radiation received over a specified length of growing season does not vary much from year to year. In this sense, light does not behave as a *discriminant* of production.

The last statement is open to misinterpretation unless it is carefully qualified. It is the amount of light *incident* on a field that is not a discriminant of yield. The amount of light *intercepted* by a crop is a major discriminant, as

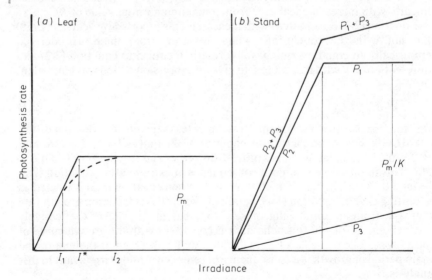

Figure 2.8   *Photosynthesis rate of (a) single leaf and (b) leaf canopy. For definition of symbols, see text*

Watson (1952) pointed out. At a meeting where much attention has been given to the biochemistry of photosynthesis and other aspects of metabolism, it is relevant to stress the simple fact that crop yields depend on the length of the growing season which, in turn, determines the maximum amount of light that a canopy can intercept. The potential yield of a crop is therefore correlated with the rate at which leaves grow and the timing of their senescence. Actual yield is less than potential yield when the photosynthetic efficiency of leaves is limited by drought, disease or other environmental factors.

Finally, the rate of crop growth per unit of intercepted radiation does not change much during the growing season for a given type of crop and is surprisingly similar for a number of crops of which the growth has been analysed in this way. The fraction of intercepted radiation is not proportional to leaf area index nor to standing dry weight except during very early growth. Consequently, the rate of growth per unit of leaf area (NAR) or per unit of weight (RGR) changes systematically during the growing season in a way which usually obscures the correlation with light and other environmental factors.

Appendix: Components of a Simple Growth Model

RESPONSES OF LEAF PHOTOSYNTHESIS TO LIGHT

To explore the proposition that crop growth may be limited by light, it is both convenient and appropriate to represent the response of photosynthetic rate to irradiance by two straight lines (*Figure 2.8*)—a device adopted by De Wit (1959) in one of his early models. Because the 'shoulder' section of the response is neglected, photosynthesis is somewhat overestimated when the irradiance is between the limits shown as I_1 and I_2 in *Figure 2.8a* but the error in estimating daily totals of photosynthesis is usually trivial.

When photosynthesis, P, is limited by irradiance $I(<I_1)$, the photosynthetic efficiency defined as $\varepsilon = P/I$ has a constant maximum value, ε_m, related to the apparent quantum efficiency of the process. On the basis of measurements by Ehleringer and Bjorkman (1977) and others, a round figure of 0.05 mole per absorbed Einstein is adopted here as a standard value for C_3 plants. The true quantum efficiency increases somewhat with decreasing temperature and, at 20°C, may be a few per cent higher than this figure (Charles-Edwards, 1978). Assuming that the quantum content of daylight is 4.6×10^{-6} E per Joule of photosynthetically active radiation (McCartney, 1978) and that 85 per cent of radiation incident on a leaf is absorbed, the amount of CO_2 assimilated per unit of absorbed energy is

$$0.05 \times 4.6 \times 10^{-6} \times 0.85 = 1.96 \times 10^{-7} \text{ mol } CO_2 \text{ J}^{-1}$$

Assuming, further, that the absorption of 1 mole of CO_2 is accompanied by the production of 30 g of carbohydrate, ε_m can be expressed in working units as 0.021 g CH_2O m^{-2} h^{-1} per W m^{-2}, where irradiance (PAR) is expressed in W m^{-2}.

The light-saturated rate of photosynthesis, P_m, obtained when $I > I_2$ is inversely proportional to the total resistance which governs the diffusion of

CO_2 to chloroplasts from the air surrounding leaves. The values of component resistances have been discussed elsewhere (e.g. Jarvis, 1971) and can be used in this type of analysis if necessary. The objectives of the present study can be satisfied by assuming a standard value for P_m to be chosen later.

The intersection of the two linear functions in *Figure 2.8a* occurs at an irradiance I^* defined by the relationship

$$P_m = \varepsilon_m I^* \tag{2.4}$$

With standard values of $\varepsilon_m \simeq 0.02 \, \text{g m}^{-2} \, \text{h}^{-1}$ per W m^{-2} and $P_m = 2 \, \text{g m}^{-2} \, \text{h}^{-1}$, I^* is 100 W m^{-2} (PAR).

DISTRIBUTION OF LIGHT IN SPACE

Most measurements of irradiance in crop canopies are consistent with the relationship

$$I(L) = I(0) \exp(-KL) \tag{2.5}$$

where $I(L)$ is the irradiance on a horizontal surface below a leaf area index of L, and K is an extinction coefficient which depends on the geometry and radiative properties of the foliage. Appropriate values of K can be estimated from the statistics of leaf distribution and orientation but, as these are rarely available, K is usually determined experimentally from measurements of irradiance and leaf area index at different heights.

For the purposes of estimating the photosynthesis of a canopy, equation (2.1) can be interpreted in two simple but constrasting ways. First, $I(L)$ can be defined as a uniform horizontal irradiance below a leaf area index of L. Then, if $I'(L)$ is the mean irradiance of a leaf below a leaf area of L, and α is the fraction of radiation absorbed by a leaf, the radiation dI intercepted by a layer dL is

$$dI = -\alpha I'(L) \, dL \tag{2.6}$$

It follows that the mean irradiance of leaves below an index L is

$$I'(L) = -(dI/dL)/\alpha = (K/\alpha)I(L) = (K/\alpha)I(0) \exp(-KL) \tag{2.7}$$

from equation (2.1). This form of expression was used by Saeki (1963) and his colleagues, and has since been developed by others including Thornley (1976), whose equation (3.29) is a slightly more exact form of our equation (2.7).

Alternatively, the transmission of light may be treated as a process in which rays of light pass through a canopy without being deflected in the manner of a point quadrat (Duncan *et al.*, 1967). It can be shown that below a leaf area index of L, the fractional area which receives light directly from the sun and sky is $\exp(-K'L)$. The coefficient K' depends on leaf geometry and on the angle of incidence of the ray; it is approximately equal to K/α. When the sun is shining, $\exp(-K'L)$ is the fractional area of sunflecks and it can be shown by integration that when L is large, the total leaf area index of sunlit leaves has a limiting value of $1/K'$. Since in this model the total amount of intercepted light $I(0)$ is distributed uniformly over an area $1/K'$, the mean irradiance is $K'I(0)$.

Because the fraction of radiation which is reflected and transmitted by leaves is usually about 10 per cent, it is worth taking account of photosynthesis by leaves exposed to rays which have been intercepted once, but not more than once. The area of such leaves also tends to a value of $1/K'$ at large values of L and the corresponding irradiance is approximately $[(1-\alpha)/2]\,K'I(0)$ where reflection and transmission coefficients of leaves are assumed to have the same value of $(1-\alpha)/2$. The two classes of leaves will be referred to as 'directly' and 'indirectly' lit.

Both models of light distribution contain unrealistic features. The first model takes no account of the pattern of light and shade in a canopy, significant in calculations of photosynthesis when the response to irradiance is not linear. The second model does not distinguish differences between the directional properties of incident light and scattered light. More rigorous models are available (*see* Ross, 1975, for example), but for estimating canopy photosynthesis, usefulness tends to diminish as complexity expands. The second model discussed above was used for the present analysis.

Measured values of K range from about 1.0 for crops with predominantly horizontal leaves to less than 0.3 for exceptionally vertical foliage (Monteith, 1975). Many cereals have a value between 0.5 and 0.7. Since a representative figure for α is 0.85, an appropriate standard value for $K'=K/\alpha$ is 0.6 and values of 0.3 and 0.9 will be used to investigate the behaviour of extreme types of canopy geometry.

DISTRIBUTION OF LIGHT IN TIME

To estimate daily totals of radiation, irradiance was expressed as a simple sinusoidal function of time, that is:

$$I(t)=(\pi S/2T)\,\sin(\pi t/T) \tag{2.8}$$

where T is daylength, t is time after sunrise and $S=\int_0^T I(t)\,dt$ is daily insolation available from climatological records as the income of radiation over the whole spectrum. The ratio of PAR to whole spectrum irradiance was assumed to be 0.5 (Szeicz, 1974).

When $I(t)$ exceeds the saturation irradiance I^* for at least part of a day, the time t^* at which all the directly lit leaves become light-saturated is given by $K'I(t)=I^*$ so that

$$t^*=(T/\pi)\,\sin^{-1}(2TI^*/K'\pi S) \tag{2.9}$$

which has a maximum value of $T/2$ when $I^*=K'\pi S/2T$. The amount of radiation received when $I\leqslant I^*$ is given by

$$2\int_0^{t^*} I(t)\,dt = S[1-\cos(\pi t^*/T)] \qquad t^*\leqslant T/2 \tag{2.10}$$

CANOPY PHOTOSYNTHESIS

Daily totals of photosynthesis were calculated for a canopy with a large leaf area index so that the area of directly lit leaves could be taken as $1/K'$. For an hour in which the irradiance was $I(t)$ (derived from equation (2.8)), two components of the foliage were assumed to contribute to photosynthesis:

(1) directly lit foliage irradiated *above* the saturation limit I^*:

$$P_1 = P_m/K' \qquad K'I(t) > I^*$$

or (2) directly lit foliage irradiated below the limit I^*:

$$P_2 = \varepsilon I(t) \qquad K'I(t) < I^*$$

and (3) indirectly lit foliage which is always irradiated below I^* to give

$$P_3 = [(1-\alpha)/2]\varepsilon I(t)$$

Figure 2.8a illustrates the difference between the photosynthesis of a single leaf and canopy photosynthesis determined in this way. Note that because of the contribution from the third component, light saturation does not occur but there is a break in the relationship between P and I at the point where $P_2 = P_1$ or

$$I(t) = P_m/(\varepsilon K') \tag{2.11}$$

For standard values of P_m, ε and K', $I(t)$ is approximately 160 W m^{-2} (PAR).

CROP GROWTH RATE

The rate at which uniform crop stands produce dry matter during early growth is proportional to the rate at which radiant energy is intercepted by foliage. During the main period of growth, most arable crops intercept 90–95 per cent of incident PAR so that an estimate of maximum crop growth rate can be obtained from figures of photosynthesis with an appropriate reduction for respiration. In terms of the symbols already used, maximum daily crop growth rate can be estimated as

$$C = 0.6 \sum (P_{1,2} + P_3) \tag{2.12}$$

when respiration is assumed to be 40 per cent of gross photosynthesis and $\sum$ implies summation over a day. Recent measurements by Yamaguchi (1978) show that the ratio of C to gross photosynthesis is much more constant than values predicted from the widely used formula of McRee (1970).

When the leaf area index is L, the rate of dry matter production can be written as

$$\dot{W} = C[1 - \exp(-KL)] \tag{2.13}$$

Strictly, the term in square brackets is the fraction of incident radiation *not* transmitted by a canopy, i.e. it is the sum of absorbed and reflected components. But as the reflectivity of foliage rarely exceeds 0.1 in the visible spectrum, $1 - \exp(-KL)$ is a good approximation to absorbed radiation.

Equation (2.13) can be integrated if the leaf weight ratio L/W is assumed to have a constant value D. The integral is

$$W(t) = W(0) + Ct + \ln\left\{\frac{[1 - \exp(-KL_0)]}{1 - \exp(-KL)}\right\} \Big/ DK \tag{2.14}$$

where t is time from emergence and L_0 is the leaf area index at $t=0$. Since most arable crops intercept little light at emergence, $\{1 - \exp(-KL_0)\} \simeq KL_0$; and when interception is complete, $\{1 - \exp(-KL)\} \simeq 1$. Equation (2.14) can therefore be approximated by the straight line equation:

$$W(t) = W(0) + Ct + (\ln KL_0)/DK$$
$$= W(0) + C[t + (\ln KL_0)/CDK] \qquad (2.15)$$

The term $(\ln KL_0)/CDK$ is negative and can be interpreted as the time lost by a crop while it is establishing the complete canopy that is needed to achieve a maximum growth rate of C (*see Figure 2.6*).

References

ALBERDA, TH. (1977). Crop photosynthesis: methods and compilation of data obtained with a mobile field equipment. Agriculture Research Reports, Wageningen

BISCOE, P. V., SCOTT, R. K. and MONTEITH, J. L. (1974). *J. appl. Ecol.*, **12**, 269–293

BLACKMAN, F. F. (1905). *Ann. Bot.*, **19**, 281–295

BLACKMAN, G. E. and BLACK, J. N. (1959). *Ann. Bot.*, **23**, 131–145

BRIGGS, G. E., KIDD, F. and WEST, C. (1920). *Ann. appl. Biol.*, **7**, 202–223

CHARLES-EDWARDS, D. A. (1978). *Ann. Bot.*, **42**, 717–731

DE WIT, C. T. (1959). *Neth. J. agric. Sci.*, **7**, 141–149

DUNCAN, W. G., LOOMIS, R. S., WILLIAMS, W. A. and HANAU, R. (1967). *Hilgardia*, **38**, 181–205

EHLERINGER, J. and BJORKMAN, O. (1977). *Plant Physiol.*, **59**, 86–90

GALLAGHER, J. N. and BISCOE, P. V. (1978). *J. agric. Sci.*, **91**, 47–60

GREENWOOD, D. J., CLEAVER, T. J., LOQUENS, S. H. M. and NIENDORF, K. B. (1977). *Ann. Bot.*, **41**, 987–997

GREGORY, F. G. (1926). *Ann. Bot.*, **40**, 1–26

JARVIS, P. G. (1971). In *Plant Photosynthetic Production*. Ed. by Z. Sestak, J. Catsky and P. G. Jarvis. W. Junk, The Hague

KAMIYAMA, K. and HORIE, M. (1975). In *Crop Productivity and Solar Energy Utilization in Various Climates in Japan (JIBP)*. Ed. by Y. Murata. University of Tokyo Press, Tokyo

MARSHALL, B. (1978). PhD Thesis, University of Nottingham

MARSHALL B. and BISCOE, P. V. (1977). *J. exp. Bot.*, **28**, 1008–1017

MCCARTNEY, H. A. (1978). *Quart. J. roy. met. Soc.*, **104**, 911–926

MCCREE, K. J. (1970). In *Prediction and Measurement of Photosynthetic Productivity*. Ed. by I. Setlik. Pudoc, Wageningen

MONTEITH, J. L. (1975). *Principles of Environmental Physics*. Edward Arnold, London

MONTEITH, J. L. (1977). *Phil. Trans. roy. Soc. B*, **281**, 277–294

MURATA, Y. (1975). In *Crop Productivity and Solar Energy Utilization in Various Climates in Japan (JIBP)*. Ed. by Y. Murata. University of Tokyo Press, Tokyo

ROSS, J. (1975). In *Vegetation and the Atmosphere*, Vol. I. Ed. by J. L. Monteith. Academic Press, London

SAEKI, T. (1963). In *Environmental Control of Plant Growth*, pp. 79–94. Ed. by L. T. Evans. Academic Press, London

SIBMA, L. (1968). *Neth. J. agric. Sci.*, **16**, 211–216
SIBMA, L. (1977). *Neth. J. agric. Sci.*, **25**, 278–287
SZEICZ, G. (1974). *J. appl. Ecol.*, **11**, 617–636
THORNLEY, J. (1976). *Mathematical Models in Plant Physiology*. Academic Press, London
WARREN WILSON, J. (1967). In *The Collection and Processing of Field Data*. Ed. by E. F. Bradley and O. T. Denmead. Interscience, New York
WATSON, D. J. (1947). *Ann. Bot.*, **11**, 42–76
WATSON, D. J. (1952). *Adv. Agron.*, **4**, 101–145
WATSON, D. J. (1958). *Ann. Bot.*, **22**, 37–54
YAMAGUCHI, J. (1978) *J. Fac. Agric. Hokkaido Univ.*, **59**, 59–129

3

EFFICIENCY OF WATER, SOLAR ENERGY AND FOSSIL FUEL USE IN CROP PRODUCTION

G. STANHILL
Institute of Soils and Water, Agricultural Research Organization, The Volcani Center, Israel

Introduction

The necessary, but not the sufficient, condition for maximizing crop production is the attainment of a dry matter production rate limited by the physical and biochemical environment of the crop. Maximum rates of carbon fixation by a crop necessitate minimum diffusive resistances to the influx of CO_2 at the crop surface. Because of their common air layer and stomatal diffusion pathways, this implies that the crop's resistance to the efflux of water vapour will be similarly low, and hence its transpiration flux correspondingly high. Both the chemical fixation of CO_2 and the change of state of water within the crop canopy are energy-intensive processes, powered by solar radiation absorbed by the crop canopy. Therefore, high rates of dry matter production and water loss require that the size and duration of the crop canopy will be sufficient to intercept all the incident solar radiation during the growing season of the crop.

Crop physiologists have established that the high yield levels attained in modern crop production by the cultivars and cultivation practices used, can largely be explained by the increased solar radiation interception achieved by larger and longer-living crop canopies, the development of which is in phase with that of incident solar energy.

To ensure the rapid establishment and maintenance of a large crop canopy requires correctly timed cultivation, elimination of water and nutrient deficiencies, and protection from pests and diseases, while canopies which function in a productive state for a prolonged period require the elimination of any increase in gas-exchanged resistance due to water deficiency. In industrialized crop production systems, these cultural practices depend on production and distribution processes powered by fossil fuel.

Thus the inputs of water, solar energy and fossil fuel needed to maximize crop yields through dry matter production are closely related and their efficiencies are interdependent. The purpose of this paper is to examine these efficiencies and their interrelationships and to compare them for crop production systems practised under the two major constraints: water and light deficiencies.

Water Use Efficiency

The essential role of water in plant growth, established by Van Helmont almost 400 years ago, was first studied quantitatively in western Europe 100 years ago using isolated plants and soil. Similar investigations, carried out by Briggs and his coworkers in the semiarid central regions of the USA during the first three decades of this century, established clear linear relationships between increase in dry matter content (C) and transpiration loss (T) for a wide variety of crop plants. After eliminating differences due to soil and cultural factors, the wide range in the slopes (α) of the relationship

$$C = \alpha T \tag{3.1}$$

was found to be due to specific crop differences and seasonal or regional differences in the climatologically determined rate of water loss, E_0. For example, the water use efficiency, i.e. the weight of total dry matter production per unit water transpired, was twice as large for winter wheat grown in the temperate climate of southern England as in the semiarid climate of Colorado (Russell, 1950).

De Wit's (1958) analysis of isolated plant container measurements, and later of field crop measurements, showed that all the results could be represented by a relationship of the form

$$C = \alpha T E_0^{-k} \tag{3.2}$$

where the exponent k varied from 1 in the arid, water-limiting conditions of central USA to 0 in the light-limiting conditions of north-western Europe.

Sibma (1968) has shown that in the Netherlands maximum rates of total dry matter production by closed canopies of actively growing field crops are remarkably similar, $C \simeq 250 \text{ kg ha}^{-1} \text{ d}^{-1}$. Assuming that $T = E_0$ for the summer cropping period when $E_0 = 4 \text{ mm d}^{-1}$ (estimated by Penman's combined heat balance and aerodynamic equation), gives a figure for the water use efficiency of field crops of $\alpha = 6.3 \text{ g kg}^{-1}$ or $63 \text{ kg ha}^{-1} \text{ mm}^{-1}$.

In arid zones, where $C = \alpha T E_0^{-1}$, values of α and C are in the same units and water use efficiency, CT^{-1}, equals α divided by the climatologically determined evaporation rate, E_0. The ratio TE_0^{-1} varies considerably in arid zones with crop, growth stage and irrigation treatment (Shalhevet et al., 1976), but Shalhevet and Bielorai (1978) have shown that for three very different irrigated crops maximum yields may be obtained when seasonal crop water losses, E_T, are 0.6 of the loss from adjacent screened class A evaporation pans. With values of E_0 estimated by the Penman equation, the corresponding seasonal crop water use ratio for maximum yields was $E_T E_0^{-1} = 1$.

Monteith's (1978) reassessment of maximum crop growth rates shows maximum cropping season averages of $C = 220 \pm 36 \text{ kg ha}^{-1} \text{ d}^{-1}$ for field stands of C_4 crop plants in regions of high insolation, values similar to those reported by Sibma (1968) for temperate region crops in the Netherlands. No measurements of E_t or E_0 were given for the arid zone crops; in their absence, an average seasonal value of $E_t = E_0 = 6.8 \text{ mm d}^{-1}$ has been taken from measurements in Israel in a region of similarly high insolation, $30 \text{ MJ m}^{-2} \text{ d}^{-1}$. These values and assumptions yield a water use efficiency of $WT^{-1} = 32 \text{ kg ha}^{-1} \text{ mm}^{-1}$, half of the corresponding value for temperate zone crops.

This comparison of water use efficiencies of crops which attain their maximum growth rates under different climatic conditions is, of course, only illustrative. Apart from the fact that the values of C and E_0 used were taken from different sources, the assumption that $T = E_t$ is not strictly valid for

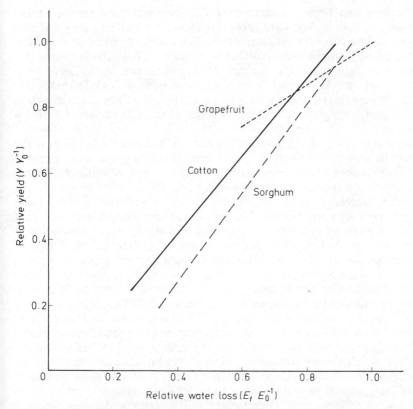

Figure 3.1   *Relative yield, Y Y_0^{-1}, of three irrigated arid zone crops as a function of relative water loss, E_t E_0^{-1} (after Shalhevet and Bielorai, 1978).*

Cotton: $Y = Y_0$ (1.20 E_t E_0^{-1} −0.06) $r = 0.94$, $n = 59$
Sorghum: $Y = Y_0$ (1.36 E_t E_0^{-1} −0.29) $r = 0.94$, $n = 29$
Grapefruit: $Y = Y_0$ (0.61 E_t E_0^{-1} −0.38) $r = 0.78$, $n = 27$

either temperate or arid zone crops even when closed canopies ensure that water loss by direct evaporation from the soil represents only a small component of the total crop water loss. In addition, the complex relationships that exist between yield and dry matter production render calculations of water use efficiency, of the type presented, insufficiently accurate to be used quantitatively for the application of irrigation water.

The importance of the partitioning of both E_t and C to practical problems of water allocation in arid zone cropping is well illustrated by Shalhevet and Bielorai's (1978) analysis of the results of 20 water requirement experiments carried out in Israel with three different crops, which were grown under semicommercial conditions in a wide range of soil and climatic regions. The results were analysed in the form

$$Y Y_0^{-1} = aE_t E_0^{-1} - b \qquad (3.3)$$

with yields Y, normalized to the maximum obtained in each experiment Y_0, and crop water loss E_t normalized originally to screened class A pan evaporation and here to estimates of E_0, open water evaporation using Penman's equation. The data shown in *Figure 3.1* indicate very different slopes and intercepts for the three crops. The sensitivity of relative yield of the two annual summer crops, cotton and sorghum, are similar but twice that of the perennial tree crop, grapefruit. These differences, as well as those in the intercepts, are interpreted by Shalhevet and Bielorai (1978) as being due to the different fraction of evaporation from soil in the three crops as well as the different timing and sensitivity to water stress of their yield formation processes.

The significance of the crop differences in water use efficiency of the magnitude illustrated in *Figure 3.1* for water allocation decisions in irrigated agriculture is obvious but, after more than half a century of research, it is not clear how crop-specific these differences are. In a recent review of the literature, largely of container experiments, Fischer and Turner (1978) suggest that there are important differences between the water use efficiencies of plants with crassulacean acid metabolism (CAM), C_4 and C_3 metabolisms, in that order of efficiency, but that within each group there is little difference between plants of the same life form exposed to similar E_0 conditions. Doorenbos and Kassam (1979), in a general synopsis, have assigned 27 crops to four groups according to their water use efficiencies; using a relationship of the type described by equation (3.3), the values of the slope constant a ranged from 1.25 in the most sensitive group to 0.75 in the least sensitive.

The effect of the stage of crop development on water use efficiency is also unclear. The success of the use of E_t as a parameter for C, and in some cases Y, suggests that there is little effect. Most of this evidence is from temperate zone crops or container experiments, but Hanks *et al.* (1978) have recently shown—by subjecting field stands of maize growing in the arid environment of Utah to a wide range of irrigation and salinity treatments—that the large reductions in C caused by water stress at flowering were accompanied by a proportional reduction in E_t, so that the water use efficiency was unaffected.

In contrast, Doorenbos and Kassam's (1979) general synopsis of crop yield response to water suggests that there is an important effect of crop growth stage. They indicate that values of a, the slope in equation (3.3), vary from a maximum of 1.2 during the flowering stage to a minimum of 0.25 during ripening.

The inverse relationship between CT^{-1} and E_0 in arid zones, implied by equation (3.2), indicates that even local climatic differences may have an important effect on crop water use efficiency and there is much field evidence to support this. In Israel, Shalhevet *et al.* (1976) have shown that the yield response to irrigation of several crops, as parameterized by the slope of the relationship between relative yield and net water application, was inversely related to values of E_0 during the irrigation season. In *Table 3.1* values of the slope, intercept, coefficient of determination and mean value of E_0 are given, based on 20 water requirement experiments with cotton crops carried out in three regions of differing evaporative demands. When the crop water loss was normalized to the evaporative demand, all the experimental data were adequately represented by the single relationship given in *Figure 3.1*.

For the reasons given above, a seasonal variation in crop water use efficiency may be expected to occur in arid regions, so that efficiency of water use is least at the time of greatest atmospheric demand. The ratio of monthly

Table 3.1 PARAMETERS OF THE RELATIONSHIP $Y Y_0^{-1}=aI+b$, BETWEEN RELATIVE YIELD, $Y Y_0^{-1}$, AND NET WATER APPLICATION, $I\,mm$; DATA FROM 20 COTTON IRRIGATION EXPERIMENTS IN ISRAEL. (FROM SHALHEVET *ET AL.*, 1976)

Parameter		Region	
	Coastal plain	Inland valleys	Jordan rift
Slope, a	0.150	0.138	0.103
Intercept, b (per cent)	43	21	9
Coefficient of determination, r^2	0.90	0.89	0.90
Evaporative demand in cropping season, E_0 mm d^{-1}	7.2	8.2	9.5

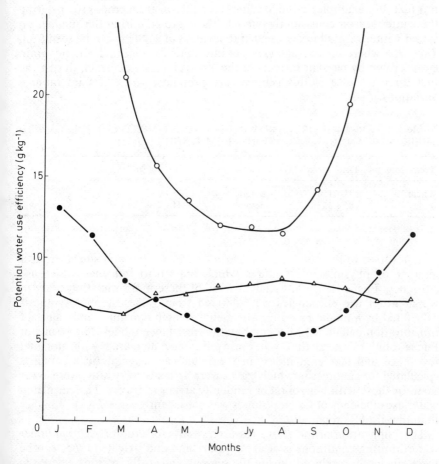

Figure 3.2 *Seasonal trends in potential water use efficiency, CT^{-1}, in three climates. C calculated from De Wit's equation (1965), and T is equal to E_0 calculated from Penman's equation. Monthly means in g kg^{-1}.*

Temperate climate (○–○)	*central England*	*2°W, 52°N*
Semiarid climate (●–●)	*central Israel*	*35°E, 32°N*
Tropical climate (△–△)	*southern Nigeria*	*3°E, 7°N*

values of potential photosynthesis, calculated by De Wit's (1965) equation, to those of potential evaporation calculated by Penman's equation, is presented in *Figure 3.2* to show the seasonal variation to be expected theoretically in arid, temperate and tropical climates.

Experimental confirmation of the large seasonal range in the arid climate, where it is of most importance, has been reported by Fischer and Turner (1978), who cite the results of measurements made in the Mediterranean climate of South Australia from a series of monthly sowings of six wheat cultivars. Water use efficiencies were found to vary seasonally from a maximum of 8.5 g kg^{-1} in midwinter to 2.5 g kg^{-1} in midsummer.

When crop production is water-limited, these seasonal changes in water use efficiency are of utmost importance for dryland agriculture, and the traditional crop rotations of dryland farming in semiarid regions is based on this fact. For example, in the Mediterranean basin grain crops grown during the winter season consistently outyield those grown during the summer on stored winter rainfall under comparable values of E_t. This can be seen from *Table* 3.2, where the average yields of unirrigated winter and summer grain crops grown by peasant farmers in the Yizre'el Valley of Israel, during the first three decades of this century, are presented with their interannual variations.

Table 3.2 AVERAGE DRYLAND WINTER AND SUMMER GRAIN YIELDS WITH THEIR INTERANNUAL VARIATION. (FROM STANHILL, 1978)

Winter grain	Yield (kg ha^{-1})	Summer grain	Yield (kg ha^{-1})
Wheat	531 ± 156	Sorghum	237 ± 236
Barley	785 ± 233	Sesame	244 ± 294

Data collected by Elazari-Volcani over 15 years for 675 ha in the Yizre'el Valley during the first three decades of this century.

In addition to the crop species, growth stage, season and site, there is a number of physiological features which appear to influence water use efficiency. Fuchs (1975) has studied some of these in a simulation study in which water use efficiencies were derived from values of photosynthesis calculated by a model based on Michaelis–Menten dynamics and values of transpiration calculated by a combination resistance model. The results of Fuchs' study showed that water use efficiency increased with stomatal resistance and leaf area index, in the latter case asymptotically. Values calculated for canopies in which leaves were arranged vertically, were lower than for those with horizontal or randomly arranged leaves. The calculated water use efficiency of C_4 crop stands was consistently some 7 g kg^{-1} higher than that for C_3 crops.

The implication of the above research results for crop production under water-limiting conditions is clear: both dryland and irrigated crops can be most efficiently produced during the seasons and in the regions where the atmospheric demand for water is minimal. Where there is sufficient water for a crop under conditions of high atmospheric demand, then the size and shape of the irrigated areas and the irrigation regime practised should be such as to lead to the maximum microclimate modification and hence to a decrease in E_0 and an increase in water use efficiencies.

The results of Fuchs' simulation study also suggest that those crop factors that increase the rate of photosynthesis, and hence solar energy conversion efficiency, will lead to an increase in water use efficiency.

However, in irrigation farming, water use efficiency is rarely the major factor to influence the choice of crop, season or site, and the major advances that have been achieved in increasing this efficiency can hardly be attributed to the application of the principles outlined above. For an example on a national scale, in Israel the average irrigation water application per irrigated hectare fell from 7545 m^3 ha^{-1} in 1956 to 6771 m^3 ha^{-1} in 1976, an average reduction of 0.5 per cent per annum; during the same 20 year period, yields of the major irrigated crops increased by an average of 2.2 per cent per annum (Anon., 1978). During this same period the proportion of irrigation water diverted to the arid southern areas and to summer crops increased, so that the large increase in water use efficiency in the field can scarcely be explained physiologically.

The major reason was undoubtedly the increased proportion of irrigation water productively transpired through crop plants and the reduction in the proportion unproductively lost through deep seepage or by evaporation from open water, and bare soil surfaces, or transpiration of phreatophytic vegetation.

The increase in the efficiency of transporting and applying water to irrigated crops has been achieved through technical advances which allow greater control and precision in the timing and placement of water in the field. In traditional gravity-fed, surface irrigation systems the timing of irrigation is often fixed on a district level and the evenness of water application in the field is very limited. In such cases, the proportion of irrigation water that is transpired is often less than one quarter. At the other extreme, an automated trickle irrigation system can ensure that more than 90 per cent of the water that enters the irrigation system is transpired by virtually eliminating losses in conveyance, evaporation from bare soil and deep seepage.

This high water use efficiency is achieved through a pressurized supply of water distributed through concrete, metal and plastic tubes and controlled by electronic equipment, all of which degrade or sequester energy obtained from fossil fuel sources.

The different inputs of non-solar energy needed by various systems of irrigation are considered in the next section.

Fossil Fuel Efficiency in Irrigated Crop Production

The energy needed in irrigation depends on the sources of water used and the methods of conveyance and application.

Surface sources, such as large rivers, require the least direct energy input per unit water. In many cases, because the river flow is out of phase with crop water requirements, major storage structures are needed. Sometimes, as in the Aswan High Dam in Egypt or the Snowy Mountain scheme of south-eastern Australia, the indirect energy input into constructing such storage facilities is less than the hydroelectric power produced, and the stored water can be considered as energetically free. In other major irrigation schemes, e.g. the National Water Carrier in Israel and the Salt River Irrigation Project in central Arizona, a significant energy cost is involved in exploiting stored water for irrigation.

An important side-effect of increasing surface water storage for irrigation purposes is the enhanced evaporation loss, which reduces the proportion of water available for crop transpiration.

Evaporative losses during storage are minimal underground, but the energy cost of lifting water from the aquifer to the surface can be considerable. Averaged over the 17 irrigated western states of the USA, the gross energy requirement for lifting water one metre was 65.5 KJ m^{-3} (Dvoskin, Nicol and Heady, 1975).

The same source lists estimates of the total direct energy requirements, i.e. the gross energy requirement for both distributing and applying water for irrigation, in the 105 major irrigation districts of the USA. The average value, 2.84 MJ m^{-3}, varied from a maximum of 7.60 MJ m^{-3} in the High Plains districts of the Texas Panhandle, where 91 per cent of the water is pumped to an average height of 100 m from the water table, to a minimum of 0.13 MJ m^{-3} for a river valley in Arizona. For those districts in which 99 per cent or more of the water was taken from surface sources, the total direct energy requirement was 0.529 MJ m^{-3}. The average energy requirement for the districts in which ground water resources formed more than 99 per cent of the irrigation water supply, was ten times greater, 5.604 MJ m^{-3}.

The overexploitation of low energy water resources has stimulated interest in new sources of water for irrigation. The results of Roberts and Hagan's

Table 3.3 TOTAL ENERGY REQUIREMENTS FOR NEW SOURCES OF IRRIGATION WATER. (FROM ROBERTS AND HAGAN, 1975)

Water source	Total dissolved salt concentration $(kg\,m^{-3})$	Treatment	Energy requirement $(MJ\,m^{-3})$
Sea	35	Desalination to 50 g m^{-3} by multistage flash distillation	
		(a) single-purpose plant	61.9
		(b) dual-purpose power and water plant	49.2
		by vapour compression distillation	58–73
		by vacuum freezing	29–58
Brackish ground water	1–6	Desalination to 500 g m^{-3}	
		by electrodialysis	5–31
		by reverse osmosis	9.5–11.5
Agricultural waste water	3	Desalination to 500 g m^{-3} in dual electrodialysis and reverse osmosis plant	11.5
Municipal waste water	1	Secondary treatment (activated sludge, sludge, digestion, landfill) plus electrodialytic desalination to 500 g m^{-3}	10.5
Antarctic icebergs	0	Transport to coast plus melting and overland conveyance	4.4

(1975) survey of energy requirements of alternative irrigation water supply for California are presented in *Table 3.3*. They show that nearly all the new sources suggested have energy requirements greater by one order of magnitude than those now used. For comparison with these new sources it may be noted that in 1974 the average energy requirement for irrigation water in California was 0.97 MJ m^{-3}. Expressed as an annual requirement per unit crop land, this equalled 4.75 GJ ha^{-1} per annum, constituting 18 per cent of the total fossil fuel energy requirements for crop production (Federal Energy Administration, 1976).

The surface, gravity-flow water application methods used today over most of the earth's irrigated area require a minimal additional energy input for earth levelling and furrow construction. Even the energy input to reduce—by lining—seepage and salination in the main distribution canals is comparatively small. Rawlins (1977) has estimated that the energy cost of lining 1 km of water distribution canal in California would be 49 GJ per annum.

By contrast, the new and more water-efficient, pressurized on-demand irrigation application systems have much higher energy requirements. They include sprinklers—both single line and central pivot—rain guns and travelling booms as well as drip and trickle systems. Apart from the energy needed to distribute the water under pressure, such irrigation systems sequester energy by the amortization of the energy-intensive components of the distribution system, such as iron, aluminium and plastic.

Preliminary estimates for irrigation in Israel show that the amortization component is 0.29 GJ ha^{-1} per annum for irrigation with small alloy sprinklers on portable aluminium pipes and 50 times as much, 14 GJ ha^{-1} per annum, for plastic solid set trickle irrigation systems. The lengths of life assumed for the two systems were generous—15 and 7 years, respectively.

Even more efficient water application is possible by eliminating waste due to human errors and mechanical breakdown with electronically controlled, computer-managed, solid set irrigation systems. Water savings of up to 15 per cent have been claimed on installing such controls even in comparison to well-managed sprinkler irrigation systems.

An example of the very high total energy requirements for modern, water-efficient irrigation systems is provided by Israel. In 1970 the direct energy input into distributing water through the pressurized National Water Carrier averaged 11.2 MJ m^{-3}, and the fossil fuel input into irrigation amounted to 63 per cent of the total used in crop production (Stanhill, 1974).

Portable sprinkler systems used on 90 per cent of the irrigated area required, on average, 60 GJ ha^{-1} per annum of fossil fuel energy input for irrigation. Solid set trickle irrigation, used for the remaining 10 per cent, was estimated to degrade 75 GJ ha^{-1} per annum. Fully automated computer control is currently used in Israel to irrigate approximately 600 ha of crops. After allowing for 15 per cent water savings, the fossil fuel requirement for this system was estimated to be 85 GJ ha^{-1} per annum.

Even with modern irrigation systems the energy requirements will vary considerably with water source, topography, crop and irrigation regime. Nevertheless, it seems justifiable to conclude that the three-fold increase in field water application efficiency that commonly separates modern from traditional irrigation systems is usually accompanied by a more than commensurate increase in fossil fuel inputs. Thus, to adapt Odum's striking phrase, modern irrigation cropping is based on substituting oil for water. This substitution of inputs limits the crops that can be economically produced using modern irrigation methods. The following simple calculation illustrates this point.

Assuming very high efficiencies of both crop water use (i.e. $CT^{-1} = 5 \text{ g kg}^{-1}$) and field water application (i.e. $I = T$), then in California where each 1 m^3 of irrigation water requires 1 MJ fossil fuel energy, the energy equivalent of the total dry matter produced would be 70 times that used in irrigation. In Israel, where each 1 m^3 of irrigation water requires 11.5 MJ, the energy fixed in dry matter would be six times that dissipated in irrigation. As the proportions of both the total dry matter production harvested and of the total energy input into crop production used in irrigation varies between one fifth and one half, it seems clear that even the most efficient irrigated crop production in Israel cannot be energetically profitable, and that in California a similar situation will prevail if new and more energy-intensive water sources have to be used.

Confirmation of the conclusion is available for the wheat crop. In California it is still energetically profitable to irrigate this crop (Avlani and Chancellor, 1977). In Israel this is not so even with the relatively high marginal water use efficiency; thus the marginal response obtained in ten wheat irrigation experiments was 1.16 kg grain per 1 m^3 irrigation application (Shalhevet *et al.*, 1976).

Interrelationships between Water, Solar Energy and Fossil Fuel Use

The inverse relationship between the efficiency of water and solar energy use in crop production and that of fossil fuel appears to apply to whole agroecosystems and national food production systems. To illustrate these interrelationships, the energy balances and efficiencies of four systems have been contrasted in *Figure 3.3* and *Table 3.4*.

The two national agricultural systems contrasted—those of the UK and Egypt—represent opposite extremes with respect to their use of irrigation water and human labour. Similar contrasts are provided by the two contrasted agroecosystems—one from the past-traditional peasant dry cropping in the

Table 3.4 ANNUAL INPUTS OF LAND, SOLAR ENERGY, IRRIGATION WATER, HUMAN LABOUR AND FOSSIL FUEL ENERGY TO FEED ONE PERSON ONE MILLION K CAL PER ANNUM

Production system input	UK 1968 (National System)	Egypt 1972–1974 (National System)	Yizre'el Valley 1914–1923 (dryland peasant)	Sinai Coast 1985 (nuclear desalination)
Land (m²)	3066	890	17 000	520
Solar energy (TJ)	9.99	6.69	122.76	3.91
Irrigation water (m³)	0	1600	0	410
Labour (man years)	0.02	0.16	0.88	0.003
Fossil fuel energy (GJ)	6.0	1.9	0	15.2
Reference	Leach, 1975	Stanhill, 1979	Stanhill, 1978	Stanhill, 1979

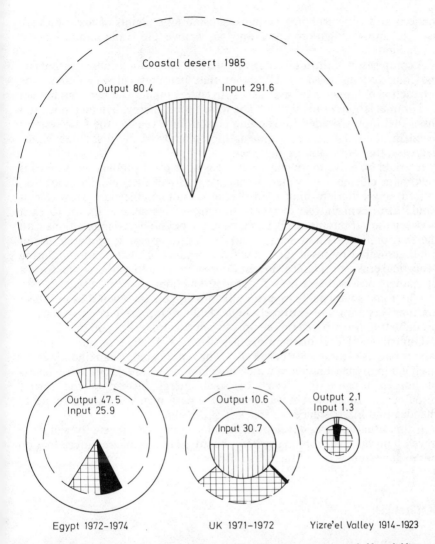

Figure 3.3 *Energy inputs and outputs in four agricultural systems. Areas encircled by solid lines are proportional to food energy output—partitioned into primary crop production (unhatched) and secondary animal production (vertically hatched). Areas encircled by broken lines proportional to non-solar energy input—partitioned into fossil fuel sources (unhatched), nuclear energy (diagonally hatched), human labour (solid), and imported plant produce for animal food production or animal work (cross-hatched). Scale: edible food outputs and inputs in GJ per cultivated hectare per annum*

Mediterranean basin, and the second a hypothetical one from the future-industrialized double cropping in a coastal desert region based on irrigation provided by nuclear desalination.

A comparison of the total energy inputs into the UK example of an industrialized, temperate zone agroecosystem shows that the fossil fuel used to substitute for low levels of incident solar energy, e.g. glasshouse heating and crop drying, form a considerable fraction of the total input into crop production (Joint Consultative Organisation, 1975). As shown previously, in modern arid zone cropping systems, substantial amounts of fossil fuel are used to apply irrigation water and so remove the water limitations of production.

A comparison of the total energy inputs and edible food energy outputs by the four systems (*Figure 3.3*) shows that per unit cultivated area, the production of human food increases with energy input, whether from human and animal labour or fossil fuels. The water, solar energy, human labour and fossil fuel inputs needed to feed one person for a year by the four systems listed in *Table 3.4*, indicate that as the amounts of the first three inputs decrease, that of the fourth increases.

It would be naive to imagine that the goal of plant production systems is maximum efficiency of water, solar energy, fossil fuel, human labour or, indeed, any criterion other than that of economic efficiency. However, it would also seem unwise to ignore the strong interactions shown to exist between the various input efficiencies and the physiological interactions with the environment. It now seems unlikely that the present, profligate use of inputs powered by fossil fuel—itself the product of photosynthesis during geological eras—to enhance current rates of crop photosynthesis, can form a permanent or worldwide basis for crop production.

Water and solar energy inputs, unlike fossil fuel, are renewable and there are many examples of extremely high yielding crop production systems in preindustrial times based on the use of such resources (Kropotkin, 1899).

Unfashionably, all of them are associated with large inputs of human labour but, as Georgescu-Roegen (1971) has speculated, 'Mankind may find itself again in the situation in which it will find it advantageous to use beasts of burden because they work on solar energy instead of the earth's resources . . . In a different way than in the past, man will have to return to the idea that his existence is a free gift of the sun.'

If this situation does develop, the challenge to man's ingenuity will be to develop productive cropping systems based on renewable resources that do not include him as one of the beasts of burden.

References

ANON. (1978). Statistical Abstract of Israel, No. 29. Central Bureau of Statistics, Jerusalem. [In Hebrew, with 121 pp. English summary]

AVLANI, P. K. and CHANCELLOR, W. J. (1977). *Trans. Am. Soc. agric. Eng.*, **20**, 429–437

DE WIT, C. T. (1958). Transpiration and crop yields. Mededeling 59, No. 64.6—S-Gravenhage. Landbouwkundige Onderzoekingen

DE WIT, C. T. (1965) Photosynthesis of leaf canopies. Agricultural Research Report 663, Mededeling 274. Wageningen

DOORENBOS, J. and KASSAM, A. H. (1979). Yield response to water. FAO Irrigation and Drainage Paper No. 33. Food and Agricultural Organization, Rome

DVOSKIN, D., NICOL, K. and HEADY, E. O. (1975). Energy use for irrigation in the seventeen western states. Special Report Center for Agricultural and Rural Development, Iowa State University

FEDERAL ENERGY ADMINISTRATION (1976). Energy and US Agriculture: 1974 Data Base, Vol. 1. FEA/D-76/459. US Govt Print. Off., Washington DC

FISCHER, R. A. and TURNER, N. C. (1978). *Ann. Rev. Pl. Physiol.* **29**, 277–317

FUCHS, M. (1975). In *Arid Zone Development: Potentialities and Problems*, pp. 139–155. Ed. by Y. Mundlak and S. F. Singer. Bollinger, Cambridge, Mass.

GEORGESCU-ROEGEN, N. (1971). *The Entropy Law and the Economic Process.* Harvard University Press, Cambridge, Mass.

HANKS, R. J., ASHCROFT, G. L., RASMUSSEN, V. P. and WILSON, G. P. (1978). *Irrig. Sci.*, **1**, 47–59

JOINT CONSULTATIVE ORGANISATION (1975). Report of the Energy Working Party, No. 1. London

KROPOTKIN, P. (1899). *Fields, Factories and Workshops.* Hutchinson, London

LEACH, G. (1975). *Energy and Food Production.* International Institute of Environment and Development, London

MONTEITH, J. L. (1978). *Exp. Agric.*, **14**, 1–5

RAWLINS, S. L. (1977). In *Agriculture and Energy*, pp. 131–147. Academic Press, New York

ROBERTS, E. B. and HAGAN, R. M. (1975). Energy requirements of alternatives in water supply, use, and conservation: a preliminary report, No. 155. Californian Water Resources Center, University of California, Davis

RUSSELL, E. W. (1950). *Soil Condition and Plant Growth*, 8th edn. Longman, London

SHALHEVET, J. and BIELORAI, H. (1978). *Soil Sci.*, **125**, 240–247

SHALHEVET, J., MANTELL, A., BIELORAI, H. and SHIMSHI, D. (1976). Irrigation of field and orchard crops under semi-arid conditions. International Irrigation Information Centre Publication No. 1. Bet Dagan, Israel

SIBMA, L. (1968). *Neth. J. agric. Sci.*, **16**, 211–214

STANHILL, G. (1974). *Agro-ecosyst.*, **1**, 204–217

STANHILL, G. (1978). *Agro-ecosyst.*, **4**, 433–488

STANHILL, G. (1979). *Agro-ecosyst.*, **5**, 213–230

4

CHLOROPLAST STRUCTURE AND PHOTOSYNTHETIC EFFICIENCY

GUNNAR ÖQUIST
Department of Plant Physiology, University of Umeå, Sweden

Introduction

The objective of this communication is to give a brief review of the relationship between chloroplast structure and function. Firstly, a basic model for the chloroplast and thylakoid organization will be presented; secondly, the photochemical and structural properties of developing and mature chloroplasts, isolated from plants exposed to various environmental conditions, will be compared. I will concentrate on conifers and other C_3 plants, but results obtained from other photosynthetic organisms will sometimes be used in order to illustrate basic properties.

For more extensive studies of chloroplast structure and function, the reader is referred to the following review articles: Boardman, 1970; Park and Sane, 1971; Bishop, 1974; Anderson, 1975; Arntzen and Briantais, 1975; Murakami, Torres-Pereira and Packer, 1975; Boardman, 1977a; Mühlethaler, 1977; Sane, 1977; Arntzen, 1978; Boardman, Anderson and Goodchild, 1978.

Abbreviations used in this chapter are as follows:

ALA	D-aminolevulinic acid
ATP	adenosine 5'-triphosphate
Chl	chlorophyll
DCMU	3,-(3,4-dichlorophenyl)-1,1-dimethylurea
DPC	diphenylcarbazide
DNA	deoxyribonucleid acid
DPIP	2,6-dichlorophenol-indophenol
LH-CP$a/b$	light-harvesting chlorophyll a/b-protein complex
NADP	nicotinamide adenine dinucleotide phosphate
P700-CPa	P700-chlorophyll a-protein complex
PSII-CPa	photosystem II-chlorophyll a-protein complex
PSU	photosynthetic unit
RuBPc	ribulose-1,5-bisphosphate carboxylase
SDS-PAGE	sodium dodecyl sulphate-polyacrylamide gel electrophoresis.

53

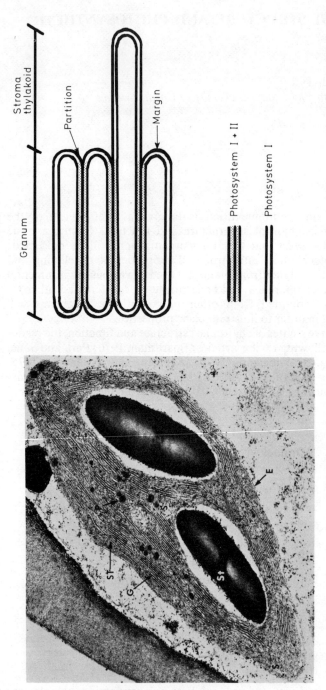

Figure 4.1 *Mature chloroplast of* Pinus silvestris. *E: Double membrane surrounding the chloroplast. G: Granum. P: Plastoglobuli. S: Stroma. ST: Stroma thylakoid. St: Starch. For preparation of samples for electron microscopy, needles were fixed in glutaraldehyde and formaldehyde and postfixed in osmium tetroxide. Cut sections were stained with uranyl acetate and lead citrate (Martin and Öquist, unpublished). Sketch after Park and Sane (1971)*

Chloroplast and Thylakoid Organization

ULTRASTRUCTURE OF CHLOROPLASTS

Mature chloroplasts of C_3 plants have the overall morphological organization shown in *Figure 4.1*. The photochemical reactions and the antennae pigments are located in the internal membrane system, the thylakoids (Menke, 1962). These sack-like vesicles can be more or less aggregated in stacks called grana (Park and Sane, 1971). The outer surface of the thylakoids is exposed to the proteinaceous matrix, the stroma, which contains the Benson–Calvin cycle enzymes (Kelly, Latzko and Gibbs, 1976). The inner surface of the thylakoids encloses the intrathylakoid space or loculus, which seems to be one single anastomosing chamber that links stacked and unstacked thylakoids. Those parts of a granum which are exposed to the stroma are called margins, whereas regions of tightly compressed thylakoids in grana are called partitions. The unstacked thylakoids are called stroma thylakoids or frets. The chloroplasts also contain ribosomes and strands of DNA, which take part in the synthesis of some chloroplast proteins (Ellis, 1976). The black dots seen in osmium-fixed chloroplasts (*Figure 4.1*) have no limiting membranes; they are called plastoglobuli and are mainly composed of quinones (Lichtenthaler and Sprey, 1966).

The chloroplasts are bounded by envelopes comprised of two membranes. Large-scale transport of various metabolites into and out of the chloroplast occurs across the envelope (Heber, 1974; Ellis, 1976). The inner membrane contains specific carriers for the transport of primary energy-rich metabolites in photosynthesis. The inner membrane is also thought to play a role in the formation of thylakoids (Menke, 1962). Direct connections between the inner membrane of the envelope and the thylakoids are not likely but Joyard and Douce (1976) have shown that the inner membrane is the regulatory site of galactosyl lipid biosynthesis for the thylakoids.

When isolated chloroplasts are fragmented with mechanical methods or detergents, a heavy fraction of grana thylakoids can be separated from a light fraction of stroma thylakoids (Boardman, 1970; Park and Sane, 1971; Öquist, 1975). Analyses of these two fractions have shown that the heavy grana fraction contains both photosystems I and II, whereas the light stroma thylakoid fraction contains only photosystem I. A model was suggested by Jacobi and Lehmann (1968) and Park and Sane (1971) in which photosystem II is restricted to the partition regions of the grana and photosystem I is present in both stroma and grana thylakoids (*Figure 4.1*). Additional fragmentation of isolated grana fractions with detergents, followed by sucrose-density gradient centrifugation, resulted in a final separation of the two photosystems (Vernon *et al.*, 1971; Wessels and Borchert, 1975). Thus, there seem to be two kinds of photosystem I: one physically close to photosystem II in the grana and one physically separated from photosystem II in the stroma thylakoids or margins. These two units of photosystem I are most likely identical (Wessels and Borchert, 1975). However, there is not unanimous agreement on the model given in *Figure 4.1* Some 20 per cent of photosystem II may be present in stroma thylakoids and margins (Armond and Arntzen, 1977). It has also been suggested that the partition regions may contain predominantly photosystem II, whereas photosystem I is restricted to margins and stroma thylakoids (Åkerlund, Andersson and Albertsson, 1976). In the latter case some kind of diffusible electron carrier must interconnect the two photosystems. However, the existence of a mechanism which regulates the distribution of excitation energy between the two photosystems (*see* below) predicts that the photosystems must be close together, e.g. in the partition regions of grana.

Exceptions from this general feature of C_3 plant chloroplasts are found in algae and in the bundle sheath cell chloroplasts of some C_4 plants (Coombs and Greenwood, 1976).

MOLECULAR ORGANIZATION OF THYLAKOIDS

Membrane model

The molecular organization of thylakoids is considered from the standpoint of the fluid mosaic membrane model (Singer and Nicolson, 1972; Anderson, 1975). According to this model, the thylakoids are composed of a lipid bilayer with their polar heads exposed to the stroma. Globular proteins are more or less embedded in the lipid bilayer. Large proteins may extend through the membrane. The model allows changes in the interaction between membrane constituents, as exemplified by the observation that proteins as well as lipids can move laterally in the thylakoids (Ojakian and Satir, 1974; Staehelin, 1976). This dynamic structural organization is essential to explain conformational and functional changes within the thylakoids (*see* below).

Electron transport

Biophysical studies by Witt and coworkers (Witt, 1975) have suggested a vectorial electron flow in the thylakoid, which yields a positive charge on the inside and a negative charge on the outside. Oxidation of water by photosystem II is supposed to occur at the internal surface of the thylakoid, whereas the reduction of the primary (Q) and secondary (plastoquinone) electron acceptor of photosystem II occurs at the outer surface. When plastiquinone becomes reduced it picks up two protons from the stroma. A pool of plastoquinone then transfers the reduced equivalents towards the inner surface of the thylakoid, where plastoquinone is oxidized by photosystem I, via the 'Rieske' iron–sulphur centre (White, Chain and Malkin, 1978), cytochrome f and plastocyanin, and protons are released into the intrathylakoid space. The reaction of photosystem I reduces the primary electron acceptor X, located in the external surface of the thylakoid. Subsequently, electrons are transferred via ferredoxin and $NADP^+$-reductase to $NADP^+$ which, on reduction, picks up protons from the stroma. The proton gradient created by the electron transport is, according to Mitchell (1966), the driving force for ATP synthesis. Both ATPase and $NADP^+$-reductase are thought to be localized on those external surfaces of the thylakoids which are exposed to the stroma (Sane, 1977).

The asymmetrical model of Witt and coworkers was established primarily on a functional basis but subsequent biochemical analyses have given it strong support (Trebst, 1974; Sane, 1977; Andersson and Åkerlund, 1978).

It must, however, be remembered that the presented model of the asymmetric organization of the thylakoids is still a hypothesis and the identification and assembling of its components to a functional photochemical apparatus is only at its beginning.

Chemical composition

Chemical analyses have shown that thylakoids are made up of approximately 50 per cent lipids and 50 per cent proteins (Lichtenthaler and Park, 1963). Quantitatively, the most important lipids are the neutral glycolipids, e.g. mono- and digalactosyl diglyceride, which together make up more than 50 per cent of the acyl lipids in *Pinus silvestris* (*Table 4.1*). The glycolipids have

Table 4.1 ACYL LIPIDS AND THEIR MAJOR FATTY ACIDS OF THYLAKOIDS ISOLATED FROM CHLOROPLASTS OF SECONDARY NEEDLES OF GLASSHOUSE-GROWN *PINUS SILVERSTRIS* (ÖQUIST, UNPUBLISHED OBSERVATIONS)

	mol % of lipids	mol % of acids							
		16:0	16:1	16:3	18:0	18:1	18:2	18:3	>18
Monogalactosyl diglyceride	27.5	4.8	tr	8.5	tr	tr	9.9	76.8	tr
Digalactosyl diglyceride	28.5	17.2	tr	3.1	2.6	2.2	10.3	64.6	tr
Sulpholipid	13.0	41.6	tr	tr	3.4	6.1	19.2	29.7	—
Phospholipids	31.0	25.7	6.6	4.7	1.9	5.6	18.0	31.5	6.1

The method described by Selldén and Selstam (1976) was used. tr = Trace.

a very high amount of the polyunsaturated linolenic acid (18:3). The lipid and fatty-acid composition of pine thylakoids is typical for higher plants but the content of phospholipids seems to be relatively high. The neutral glycolipids are thought to comprise the fluid matrix in the lipid bilayer, whereas the ionic phospholipids (mainly phosphatidylglycerol) and the sulpholipid are believed to function as boundary lipids between proteins and the fluid lipid matrix (Anderson, 1975). Other lipids which occur in significant amounts are chlorophylls, carotenoids and quinones (Lichtenthaler and Park, 1963). In addition to their functional role in light harvesting and electron transport, Anderson (1975) gives them a structural function as boundary lipids, e.g. in chlorophyll-protein complexes. The lipid and fatty-acid composition in the chloroplast envelope differs from that of the thylakoids (Douce, Holtz and Benson, 1973).

Sodium dodecyl sulphate-polyacrylamide gel electrophoresis (SDS-PAGE) has been widely used to identify the proteins in the thylakoids. By use of this and chromatographic techniques, two well-characterized chlorophyll-protein complexes have been isolated (for a more detailed description *see* reviews by Thornber, 1975; Boardman, Anderson and Goodchild, 1978). One of the complexes has an apparent molecular weight of about 90 000, it has only chlorophyll *a* and some β-carotene and it contains the reaction centre molecule P700 of photosystem I. It is therefore considered to represent the antenna of photosystem I and it is denoted P700-chlorophyll *a*-protein complex (P700-CPa). The apoprotein of the complex has a molecular weight of about 70 000, but it may be further solubilized in two or more subunits.

The other complex has an apparent molecular weight of about 30 000. It has a chlorophyll *a/b* ratio of 1 or slightly higher and it is denoted light-harvesting chlorophyll *a/b* protein (LH-CP*a/b*) because of its lack of a photochemical reaction centre. On chlorophyll extraction, two polypeptides with molecular weights of about 24 000 and 27 000 are obtained. This complex was originally considered as a photosystem II complex but it is now clear that it is not necessary for photosystem II activity (Thornber, 1975).

Recently, a minor chlorophyll-protein complex with slightly lower mobility in the SDS-PAGE than LH-CP*a/b* has been identified (Hayden and Hopkins, 1977; Remy, Hoarau and Leclerc, 1977; Boardman and Anderson, 1978; Henriques and Park, 1978). Wessels and Borchert (1978) have studied a similar complex with an apparent molecular weight of about 40 500. When chlorophyll was extracted from the complex it converted to a polypeptide with a molecular weight of 47 000. As this polypeptide is also found after electrophoresis of purified photosystem II particles, Wessels and Borchert conclude that the complex is derived from the photosystem II reaction centre. This polypeptide is most likely identical with the reaction centre polypeptide of photosystem II described earlier (Kretzer, Ohad and Bennoun, 1976; Machold and Høyer-Hansen, 1976). Furthermore, Bishop and Öquist (1980) have correlated this chlorophyll-protein complex with the occurrence of photosystem II in *Scenedesmus oblicuus*. The complex contains only chlorophyll *a* and some carotenoids. As no P680 has yet been shown in the complex, I denote it photosystem II-chlorophyll *a*-protein complex (PSII-CP*a*). This complex should not be confused with the dimer of LH-CP*a/b*, which has an apparent molecular weight of around 45 000 (Hiller, Genge and Pilger, 1974; Wessels and Borchert, 1978).

Other polypeptides identified by SDS-PAGE are the subunits of the coupling factor and the RuBPc, ferrodoxin, ferrodoxin NAPD$^+$-reductase, cytochrome f, cytochrome b_6 and plastocyanin (Klein and Vernon, 1974; Wessels and Borchert, 1978).

Although as many as 43 polypeptides have been separated by SDS-PAGE of mature barley chloroplasts (Høyer-Hansen and Simpson, 1977), only about 15 major polypeptides have been identified so far. It must be remembered that these polypeptides are denatured monomers, which are thought to be aggregated into native proteins in the functional thylakoid (Machold, 1975).

Organization of chlorophyll-protein complexes

Thornber and coworkers have postulated that the free chlorophyll, which is solubilized during SDS-PAGE, is derived from the reaction centre antenna of photosystem II and from an unidentified light-harvesting chlorophyll *a* protein (Thornber *et al.*, 1976). As the anionic detergent SDS disrupts hydrophobic bonds, it can be suggested that the SDS-solubilized chlorophyll is derived from fractions of hydrophobically associated chlorophylls. Deroche and Briantais (1974) have shown that the apolar solvent, petroleum ether, primarily extracts *β*-carotene and some chlorophyll *a* from lyophilized wheat chloroplasts, whereas more chlorophyll *a* and then chlorophyll *b* were extracted when the polarity of the petroleum ether was increased by addition of ethanol.

In order to investigate the localization of hydrophobically bound chlorophylls we have verified the experiments by Deroche and Briantais (1974) with lyophylized chloroplasts from *Pisum sativum* and *Scenedesmus obliquus*, and the chlorophyll proteins of the extracted thylakoids have been separated with SDS-PAGE (Öquist and Samuelsson, 1980). The PSII-CPa in lyophilized chloroplasts of Scenedesmus could not be isolated by SDS-PAGE when carotenoids were first extracted with petroleum ether (Bishop and Öquist, unpublished data). P700-CPa and LH-CPa/b were not affected by this treatment. By extracting lyophilized chloroplast thylakoids of pea with petroleum ether of increasing polarity (petroleum ether with 0–1 per cent ethanol) prior to SDS-PAGE we found that chlorophyll a of P700-CPa was extracted by less polar solvents than were needed to extract chlorophyll a and b from LH-CPa/b (Öquist and Samuelsson, 1980). Petroleum ether with 0.1 per cent ethanol extracted carotenoids, approximately 10 per cent of chlorophyll a and traces of chlorophyll b. P700-CPa isolated from such thylakoids contained no β-carotene but the amount of P700-CPa obtained was only slightly decreased relative to LH/CPa/b. Concomitant with the extraction of β-carotene from P700-CPa there was a shift of the chlorophyll a absorption peak from 676 to 670 nm and the low-temperature fluorescence peak of the complex at 720 nm disappeared in favour of a 680 nm peak. Extraction of carotenoids and chlorophylls from lyophilized thylakoids did not affect the relative absorption or low-temperature fluorescence properties of LH-CPa/b. When the polarity of the solvent (petroleum ether with approximately 0.5 per cent ethanol) reached the level where P700-CPa was almost totally extracted, the a/b ratio of the SDS-solubilized free chlorophyll decreased, which indicated that an increasing proportion of the free chlorophyll was stripped of LH-CPa/b.

These results are interpreted to show that the chlorophyll a in P700-CPa is in a more hydrophobic environment than chlorophyll a and b in LH-CPa/b. The long-wavelength absorbing chlorophyll a form of P700-CPa, which emits at 720 nm, is thought to be in a particularly hydrophobic environment where β-carotene is a structurally important lipid. The findings that PSII-CPa is easily destroyed in SDS-PAGE and that P700-CPa is less resistant than LH-CPa/b to high SDS-concentrations (Öquist et al., 1978b) also support the idea that the chlorophylls of PSII-CPa and P700-CPa are more hydrophobically associated *in vivo* than the chlorophylls of LH-CPa/b. Both P700-CPa and LH-CPa/b are known to have high contents of hydrophobic amino-acid residues (Thornber, 1975). It is possible that the chlorophyll a-protein complexes of the two photosystems are relatively more associated with the neutral glycolipids in the matrix of the lipid bilayer than is LH-CPa/b, which is thought to span the membrane, thereby exposing part of the protein to the hydrophilic stroma.

In order to explain the sequential extraction of chlorophyll from the pigment-protein complexes with increasing polarity of the solvent, we favour a model in which chlorophylls as boundary lipids can range from being primarily hydrophilically bound to being primarily hydrophobically bound (Anderson, 1975; Liljenberg and Selstam, 1978). With improved techniques it will probably be shown that practically all chlorophyll can be isolated bound to proteins (Boardman and Anderson, 1978). *In vivo*, however, we believe that the fraction of hydrophobically bound chlorophyll, which is easily extracted with organic solvents or SDS, is stabilized by association with lipids. It would therefore be more appropriate to talk about lipid-chlorophyll-protein complexes. According to this view there is no reason to postulate a fourth unknown chlorophyll *a* protein in order to explain the front chlorophyll seen in the SDS-PAGE. Based on the data obtained by Thornber and others (Thornber, 1975; Boardman, Anderson and Goodchild, 1978) and by us, we propose a model (*Figure 4.2*) for the *in vivo* chlorophyll organization based on only three chlorophyll-protein complexes: P700-CP*a*, PSII-CP*a* and LH-CP*a/b*. The relative areas of the shaded portions illustrate the amounts of chlorophylls which are preferentially hydrophobically associated. As more or less of the hydrophobically bound chlorophyll can be stripped off according to the SDS-PAGE method used, chlorophyll-protein complexes of different spectral properties or aggregations are obtained.

This model is also consistent with the modified functional tripartite model proposed by Butler and coworkers in order to explain mathematically the energy distribution in the photosynthetic apparatus (Butler and Kitajuma, 1975; Butler, 1978). In their model the low-temperature fluorescence bands of higher plant chloroplasts, F685, F695 and F735, have been ascribed to LH-CP*a/b*, PSII-CP*a* and P700-CP*a*, respectively (*Figure 4.2*). It has recently been shown that isolated purified subchloroplast particles which contain the chlorophyll *a/b*-protein or the separate reaction centre antennae have fluorescence properties which basically agree with the model (Shutilova and Kutyurin, 1976; Satoh and Butler, 1978). We have also shown that F695 is absent in Scenedesmus mutant PS-11, which lacks photosystem II activity and PSII-CP*a* (Bishop and Öquist, 1980).

Morphological expressions of chlorophyll-protein complexes

The freeze-fracturing technique of membranes has made it possible to examine internal structural details of membrane organization (Park and Sane, 1971; Mühlethaler, 1977). The fracture plane goes through the central hydrophobic middle layer of the membrane to leave two fracture faces. In stacked thylakoids the inner leaflet (the EF-face) has relatively large particles, whereas the outer leaflet (the PF-face) has smaller tightly packed particles (Staehelin, 1976). Stroma thylakoids lack large EF-particles. Armond, Staehelin and Arntzen (1977) have followed the changes of size and distribution of particles on these two faces during greening of pea seedlings. They have correlated the accumulation of LH-CP*a/b* with the formation of the large EF-particles in the grana stacks. They and others (Miller, Miller and McIntyre, 1976) have presented a structural model for the association of the *in vivo* chlorophylls and photosystems, which basically agrees with the more functional biochemical model given in *Figure 4.2*. According to their model, the large EF-particles are the morphological expressions of an association between PSII-CP*a* and LH-CP*a/b*, whereas at least some of the small PF-particles, present in both stroma and grana thylakoids, are morphological equivalents of P700-CP*a* (Arntzen, 1978). Furthermore, freeze-fracture studies of tobacco and barley mutants which lack photosystem II have shown that the large EF-particles fail to assemble even though LH-CP*a/b* is present (Miller, 1978). This suggests that the large EF-particles cannot be formed in the absence of PSII-CP*a*.

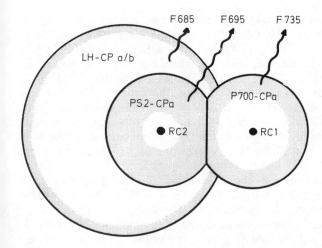

Figure 4.2 *Functional and biochemical model for* in vivo *organization of chlorophyll in a photosynthetic unit. Shaded areas represent chlorophylls primarily hydrophobically bound to polypepetides or lipids. White areas represent chlorophylls primarily hydrophilically bound to polypeptides. F685, F695 and F735 are the fluorescence emission bands at − 196°C. RC: Reaction centre*

IONIC REGULATION OF PHOTOSYNTHESIS

The chloroplast can be considered as a two-compartment system composed of the stroma and the intrathylakoid space. As stated above, there is a light-driven pumping of protons across the thylakoids, which results in an alkalization of the stroma and an acidification of the intrathylakoid lumen (Trebst, 1974; Jagendorf, 1975). The net uptake of protons into the intrathylakoid compartment must be balanced by a counter movement of cations or a comovement of anions in order to maintain electrical neutrality. Depending on various experimental conditions, there are some conflicting results on the species of co-ions for the proton pump but, as has been pointed out in a recent review of this field by Barber (1976), Mg^{2+} is most likely to be a very important co-ion. The light-driven counter exchange of protons and Mg^{2+} causes a shrinkage of the thylakoids, which eventually, by influencing the osmotic properties, affects the whole morphology of the chloroplast (Murakami, Torres-Pereira and Packer, 1975). The initial change of the thylakoid conformation is thought to be coupled to the protonation of negatively charged groups in the intrathylakoid space (Murakami and Packer, 1970).

Analyses by Murata, Bonaventura, Myers and others (cf. Myers, 1971; Butler, 1978) have shown that photosystem II is preferentially excited when dark-adapted chloroplasts or cells are illuminated. This state has been called state 1 and it is characterized by a high fluorescence yield from photosystem II (Govindjee and Papageorgiou, 1971). After a few minutes' illumination, the fluorescence yield decreases due to an increased distribution of excitation energy into photosystem I at the expense of photosystem II. This is called state 2. The transition from state 1 to state 2 can also be controlled by divalent cations, e.g. Mg^{2+} (Homann, 1969; Murata, 1969; Bennoun, 1974), and strong evidence is emerging that state 1 is characterized by binding of Mg^{2+} to the thylakoids, whereas state 2 is characterized by a removal of Mg^{2+} from the thylakoids (Barber, 1976; Arntzen, 1978; Butler, 1978). Gross and coworkers (Davis and Gross, 1975) have shown that LH-CP$a/b$ has cation binding sites. Burke, Ditto and Arntzen (1978) have recently given direct evidence for the involvement of LH-CP$a/b$ in a cation-regulated redistribution of excitation energy between the two photosystems. We thus conclude that state 1 exists in the dark because of a maximal Mg^{2+} binding to LH-CP$a/b$, and that the transition from state 1 to state 2 is coupled to the conformational change caused by the displacement of Mg^{2+} from its sites by the light-driven proton pump (Barber, 1976). The state 1 to state 2 transition represents an important regulatory process in order to equalize the excitation of the two photosystems, thus optimizing the photosynthetic efficiency. It furthermore explains why the quantum yield of photosynthesis is essentially wavelength independent, except in blue-violet where light is absorbed by the photosynthetically inactive carotenoids as well as by the photosynthetic pigments (McCree, 1972). The transition may be of great ecological importance in maintaining a high quantum yield of photosynthesis in leaf canopies, where both the flux density and the wavelength distribution of quanta can be highly variable.

Two mechanisms for the light- and Mg^{2+}-controlled redistribution of excitation energy have long been discussed (Myers, 1971; Butler, 1978): (a) regulation of the energy distribution from the light-harvesting chlorophyll into the two photosystems; or (b) spillover of excitation energy from photosystem II to photosystem I. The significance of these two mechanisms is still under debate but Butler and Kitajima (1975) include both in their tripartite model. Finally, it must be mentioned that the light-mediated counter exchange of protons and Mg^{2+}, i.e. alkalization and increase of Mg^{2+} in the stroma, also activates RuBPc and several other enzymes in the Benson–Calvin cycle (Kelly, Latzko and Gibbs, 1976; Walker, 1976). Light-induced modulation of enzyme activities is also mediated by the ATP/ADP ratio and reduced ferredoxin and NADPH. Thus, light-driven electron transport in the thylakoids not only provides energy for CO_2 reduction, but also optimizes its efficiency by conformational adjustments and simultaneously regulates the activities of specific photosynthetic enzymes.

Development of Chloroplast Structure and Function

In studies of correlations between chloroplast structure and photosynthetic function an important approach has been to examine changes in structure, chemical composition and function of chloroplasts during their maturation. Most of these studies have been performed on angiosperms and algae. For a more extensive study of the chloroplast development of these plant groups, the reader is referred to Kirk (1971), Park and Sane (1971), Rosinski and Rosen (1972), Arntzen and Briantais (1975), Boardman (1977a), Boardman, Anderson and Goodchild (1978), Treffry (1978), Sundqvist, Björn and Virgin (1979). However, gymnosperms, which can synthesize chlorophyll in the dark (Boardman, 1966), are very little considered in this respect. In this section, after a short description of the chloroplast development in angiosperms, I will discuss the chloroplast development of *Pinis silvestris* and *Picea abies*.

CHLOROPLAST DEVELOPMENT IN ANGIOSPERMS

Dark-grown angiosperm seedlings develop etioplasts which, on subsequent illumination, develop into chloroplasts. Etioplasts contain small amounts of protochlorophyll(ide) and no chlorophyll. Each etioplast possesses one or more three-dimensional tubular structures, prolamellar bodies, from which single perforated membranes may extend. The protochlorophyll(ide) holo-chrome is considered to be associated with the prolamellar bodies (Kahn, 1968). Chemically, the prolamellar bodies and the membranes are not identical structures (Lütz, 1978). The membranes contain lipids and proteins, whereas the prolamellar bodies contain mainly two steroids which, according to Kesselmeier and Ruppel (1978), are saponins. The membranes may be the direct precursors of the mature thylakoids, since the lipid composition of the two structures is basically similar (Lütz, 1978) and relatively small changes in the amount of lipids occur during the first hours of greening (Roughan and Boardman, 1972; Selldén and Selstam, 1976) before grana stacks start to develop. SDS-PAGE of etioplasts does not reveal any apoprotein to the chlorophyll-protein complexes typical for mature chloroplasts (*see* previous section), although the subunits of the coupling factors are present in etioplasts together with some ten major unidentified polypeptides (Høyer-Hansen, Machold and Kahn, 1976). The photosystems are not functional in etioplasts but according to Bradbeer *et al.* (1974) etioplasts of bean do contain the Benson–Calvin cycle enzymes. On illumination of etiolated angiosperm seedlings the following general changes occur during the first hours: (a) photoconversion of protochlorophyll(ide) to chlorophyll(ide); (b) phytolation of chlorophyllide; (c) spectral shifts in the red chlorophyll *a* absorption band; and (d) disappearance of the prolamellar bodies and formation of thylakoids. It is generally agreed that the development of photosystem activities is a stepwise process in which photosystem I develops before photosystem II. The time-course of all these changes varies with species of plant, age of seedlings and growing and experimental conditions. For a more extensive discussion of these changes *see* the above references.

CHLOROPLAST DEVELOPMENT IN CONIFERS

Development of structure

A feature common to most gymnosperms is that in darkness they form etioplasts, which are more or less differentiated to mature chloroplasts, sometimes called etiochloroplasts (von Wettstein, 1958; Laudi, 1964; Walles, 1967; Nicolić and Bogdanović, 1972; Laudi and Bonatti, 1973; Michel-Wolwertz and Bronchart, 1974; Walles and Hudák, 1975; Lewandowska and Öquist, 1980b). These organelles, called chloroplasts here, usually contain prolamellar bodies, developed thylakoids which are more or less appressed in grana-like structures, and large starch grains. Analyses of chlorophyll-protein complexes with SDS-PAGE show that cotyledons of dark-grown *Pinus silvestris*, unlike angiosperms, contain LH-CP*a/b*, P700-CP*a* (Król, 1978) and PSII-CP*a* (Lewandowska and Öquist, 1980b). Michel-Wolwertz (1977) has shown that *Pinus jeffreyi* cotyledons accumulate protochlorophyll(ide) at 10°C but not at 23°C. She has also found that protochlorophyll(ide) accumulated at 10°C or after ALA treatment needs light to be photoreduced (Wolwertz, 1978). This suggests two alternative pathways for chlorophyll synthesis in pine, one proceeding in darkness and one that is light dependent. The presence of prolamellar bodies in *Pinus jeffreyi* is not correlated with the occurrence of protochlorophyll(ide) or chlorophyll(ide) (Michel-Wolwertz and Bronchart, 1974). Such a correlation has been suggested for angiosperms (Henningsen and Boynton, 1970; Treffrey; 1970). The prolamellar bodies in pine may represent accumulations of molecules, which cannot be used up if the assembling of thylakoids is the rate-limiting step in gymnosperm chloroplast development in darkness.

Walles and Hudák (1975) have reported that dark-formed chloroplasts of *Pinus silvestris* contain grana of 2–4 thylakoids, whereas corresponding chloroplasts of *Picea abies* have large grana which contain up to ten thylakoids. It is known that the amount of LH-CP*a/b* is positively correlated with the size of the grana stacks (Anderson, 1975), thus explaining why dark-grown cotyledons of *Pinus silvestris* show higher chlorophyll *a/b* ratios (Król, 1978; Lewandowska and Öquist, 1980b) than corresponding cotyledons of *Picea abies* (Oku, Hayashi and Tomita, 1975).

Development of function

Chloroplasts prepared from dark-grown seedlings of *Pinus silvestris* (Oku, Sugahara and Tomita, 1974; Król, 1978; Lewandowska and Öquist, 1980a), *Pinus nigra* (Tyszkiewicz, Popovic and Roux, 1977) and *Picea abies* (Oku, Hayashi and Tomita, 1975) have active photosystem I. Calculated on a chlorophyll basis, the photosystem I activities increased only slightly on subsequent illumination (*Table 4.2*). The results obtained in studies of photosystem II activities in dark-grown seedlings of pine and spruce are, however, more variable. We have shown that chloroplasts isolated from dark-grown seedlings of *Pinus silvestris* have a substantial electron transport from water to NADP (*Table 4.2*). The rate was limited on the water-splitting side of photosystem II, since the artificial electron donor diphenylcarbazide increased the electron transport rate about two-fold. Fluorescence kinetic studies also showed that the electron acceptor side of photosystem II was fully organized and active. Our finding of an active photosystem II in chloroplasts isolated from dark-grown pine seedlings disagree with other reports that show no photosystem II activity, even in the presence of diphenylcarbazide (Oku, Sugahara and Tomita, 1974; Król, 1978). This discrepancy may be explained by the observation that the presence of polyethylene glycol in the preparation medium causes inactivation of the water-splitting system of chloroplasts isolated from dark-grown pine cotyledons (Lewandowska and Öquist, 1980a). Both Oku and coworkers and Król used polyethylene glycol in their preparation medium, as this compound

Table 4.2 PHOTOSYSTEM ACTIVITIES MEASURED IN CHLOROPLASTS ISOLATED FROM DARK-GROWN COTYLEDONS OF *PINUS SILVESTRIS* AND AFTER SIMILARLY TREATED PLANTS HAD BEEN ILLUMINATED FOR THE TIME INDICATED. VALUES ARE IN µMOL REDUCED ELECTRON ACCEPTOR PER mg CHLOROPHYLL PER HOUR. (FROM LEWAMDOWSKA AND ÖQUIST, 1980a)

Period of illumination (hours)	Photosystems II + I	Photosystem II − DPC	+ DPC	Photosystem I
0	14	33	78	34
1/60	—	42	81	—
5/60	—	68	96	—
10/60	—	65	84	—
30/60	—	153	185	—
1	—	121	124	—
2	57	103	96	51
4	56	100	102	40
8	79	124	155	41
16	50	113	132	35
24	40	113	140	41
48	63	111	121	49

Photosystems II + I: $H_2O \rightarrow NADP$; photosystem II: $H_2O \rightarrow DPIP$; photosystem I: $Asc/DPIP \rightarrow NADP$.

is known to be a powerful protective agent when preparing photoactive chloroplasts from mature pine needles (Oku and Tomita, 1971; Öquist, Martin and Mårtensson, 1974). Chloroplasts isolated from dark-grown *Pinus jeffreyi* (Farineau and Popovic, 1975) and *Picea abies* (Oku and Tomita, 1976) were also prepared in the presence of polyethylene glycol. However, these chloroplasts showed some photosystem II activities, which became pronounced in the presence of diphenylcarbazide. This again shows that the water-splitting side of photosystem II limits the electron transport of dark-grown conifer seedlings.

Common to all these studies is that the activity of photosystem II is greatly enhanced if the cotyledons are exposed to light before chloroplast preparation (Farineau and Popovic, 1975; Oku and Tomita, 1976; Król, 1978; Lewandowska and Öquist, 1980a). Jeske and Senger (1978) have also observed photoactivation of oxygen evolution from dark-grown seedlings of *Pinus silvestris*. We found a significant stimulation of the activity of photosystem II when the seedlings were illuminated for only a few minutes prior to chloroplast isolation (*Table 4.2*). Maximal photosystem II activities obtained within 30 minutes of illumination of the seedlings. The activity did not change during a 48 hour illumination period. However, Oku and Tomita (1976) have reported that the photoactivation of photosystem II in seedlings of *Picea abies* shows a lag phase of 10–20 hours after a rapid activation during the first minutes of illumination. They have ascribed the increased activity after the lag phase to structural changes in the chloroplasts. The rapid photoactivation of photosystem II activities indicates that there is a light-induced rearrangement of already existing molecules, possibly an incorporation of manganese into existing polypeptides (Farineau and Popovic, 1975; Inoue *et al.*, 1976). This interpretation is supported by the finding that during the first two hours of illumination of dark-grown seedlings of *Pinus silvestris*, the prolamellar bodies disappear and the thylakoids appear well differentiated and more or less appressed (Lewandowska and Öquist, 1980b). It should also be remarked that we found no net accumulation of chlorophyll or change in the a/b ratio during photoactivation of photosystem II and the observed structural rearrangements of the chloroplasts.

The structural and functional properties of dark-grown conifer chloroplasts resemble in many respects angiosperm chloroplasts greened under intermittent, far-red or weak light (Sironval *et al.*, 1968; De Greef, Butler and Roth, 1971; Michel and Sironval, 1972; Remy, 1973; Ogawa and Shibata, 1973; Akoyunoglou and Michelinaki-Maneta, 1975). Like conifer chloroplasts, these angiosperm chloroplasts generally have (a) prolamellar bodies, (b) well-developed thylakoids and no or weakly developed grana, (c) an incomplete water-splitting system, (d) functional reaction centres of both photosystem I and photosystem II, and (e) a developed electron transport path between the two photosystems. As in dark-grown conifer seedlings, it is possible to photoactivate the water-splitting system with continuous white light for a few minutes, probably due to photoactivation of manganese (Cheniae and Martin, 1973; Inoue *et al.*, 1975). We thus conclude that these angiosperm chloroplasts are in the same physiological phase of development as chloroplasts in dark-grown conifers (Lewandowska and Öquist, 1980a, 1980b).

On the assembling of photosynthetic units

As discussed by Herron and Mauzerall (1972) and by Baker and Hardwick (1976), there are two basic models for the development of the photosynthetic units (PSU) in greening plants. One model suggests that the reaction centres are first synthesized and the antennae chlorophyll is then added to the reaction centres. Angiosperm chloroplasts seem to develop according to this model (Kirk and Goodchild, 1972; Henningsen and Boardman, 1973; Egnéus, Selldén and Andersson, 1976). Typical for this model is that the light-harvesting apparatus becomes increasingly effective during greening. The other model suggests that the synthesis of PSUs occurs in a sequential manner so that one PSU is completely synthesized before the next one. Baker and Hardwick (1976) have found that the development of PSUs in cocoa leaves follows this model. It is characterized by a constant efficiency of the light-harvesting apparatus during greening.

We do not know the model for PSU development in dark-grown cotyledons of conifers. However, the constant chlorophyll a/b ratio of about 3 during rapid chlorophyll synthesis in dark-grown *Pinus nigra* (Bogdanović, 1973) is consistent with the sequential model. Chloroplasts isolated from dark-grown *Pinus silvestris* have a chlorophyll/P700 ratio of about 600 and the ratio remains constant as more chlorophyll accumulates in light (Lewandowska and Öquist, 1980b). This constant chlorophyll/P700 ratio as well as the constant light-saturated electron transport rate per unit chlorophyll and time during the course of chlorophyll accumulation (Lewandowska and Öquist, 1980a, 1980b) also provide strong evidence for a sequential development of the PSUs of greening pine cotyledons. However, these cotyledons show a decreasing chlorophyll a/b ratio during greening in light because of a preferential accumulation of LH-CP$a/b$ (Lewandowska and Öquist, 1980b).

According to Alberte, McClure and Thornber (1976) a decreasing a/b ratio is consistent with an increasing size of the PSUs. Our results indicate that a decreasing a/b ratio may not necessarily mean that the size of the PSUs is affected. Instead, the relative proportion between the three *in vivo* chlorophyll proteins shown in *Figure 4.2* may vary relative to each other, thus keeping the total amount of chlorophyll constant per PSU.

Finally, it must be assumed that the developmental pattern of structure and function of gymnosperm chloroplasts, like other plants, is dependent on viability of seeds, age of seedlings, moisture and temperature during growth of the seedlings.

Environmental Effects on Chloroplast Structure and Function

Having discussed the physiological significance of grana, we will concentrate in this section on the effects of seasonal climatic variations on the photosynthetic apparatus of conifers. For a more extensive study of the effects of contrasting light and temperature regimes on chloroplast organization and function, the reader is referred to the following representative articles: Taylor and Craig (1971), Björkman (1973), Murata and Fork (1976), Armond, Schreiber and Björkman (1977), Boardman (1977b), Nolan and Smillie (1977) and Lewandowska and Jarvis (1978).

PHYSIOLOGICAL SIGNIFICANCE OF GRANA

It is well known that shade plants have more developed grana structures than sun plants (Björkman *et al.*, 1972; Goodchild, Björkman and Pyliotis, 1972) and it is assumed that the extent of grana stacking is the means by which the plant regulates the utilization of solar energy. Today it is generally agreed that grana are not obligate for photosystem II activity but that the onset of photosystem II activity, grana stacking, formation of LH-CP$a/b$ and the appearance of large freeze-fracture particles on the EF-fracture faces may be closely linked on a timescale in greening seedlings (Anderson, 1975; Arntzen and Briantais, 1975). As mentioned above, the large EF particles are restricted to the partition regions in the grana and they are believed to be the morphological expression of an association between PSII-CPa and LH-CP$a/b$. Although the presence of LH-CP$a/b$ is not necessary for thylakoid appression to occur, it is most likely that LH-CP$a/b$ is normally involved in the mechanism for grana formation in green algae and higher plants (Anderson, 1975). This explains why shade plants with large grana stacks usually show lower chlorophyll a/b ratios than exposed plants with small stacks (Anderson, Goodchild and Boardman, 1973).

Essentially, all photosystem II activity is localized to the partition regions of the grana (*Figure 4.1*), whereas photosystem I is present in both grana and stroma thylakoids. Arntzen and coworkers (Armond *et al.*, 1976; Davis *et al.*, 1976) have shown that the onset of cation regulation of excitation energy distribution between the two photosystems is causally linked to the appearance of LH-CP$a/b$ and grana formation in greening pea. They have recently given direct evidence for the involvement of LH-CP$a/b$ in the cation regulation of energy redistribution between the two photosystems (Burke,

Ditto and Arntzen, 1978). Furthermore, it has been demonstrated that the quantum yield of whole-chain electron transport increases when LH-CP*a*/*b* is formed or grana appear in greening Scenedesmus or pea (Senger *et al.*, 1975; Armond *et al.*, 1976). It can be suggested, therefore, that the partitions in grana are necessary structures for linking the two photosystems close enough together in order to make it possible for regulating an equal excitation of the two photosystems by conformational changes, thus optimizing the photosynthetic quantum yield. Such a mechanism is probably of great ecological importance in shaded habitats because, when light is limiting, an efficient use of absorbed light should be an important factor for plant survival and concurrence.

It is often assumed that increased grana formation and low chlorophyll *a*/*b* ratios are expressions of large PSUs, thereby increasing the light-collecting capacity by a high content of LH-CP*a*/*b* (Alberte, McClure and Thornber, 1976). By using the ratio chlorophyll/P700 as an index of the PSU size, a very small or insignificant increase of the ratio is usually found when plants are grown in shade, which indicates that the size of PSU is not very much affected by the light conditions (Björkman *et al.*, 1972; Goodchild, Björkman and Pyliotis, 1972; Anderson, Goodchild and Boardman, 1973). Also typical for plants adapted to various irradiances is an equal quantum yield of photosynthesis at rate-limiting quantum flux densities (Björkman *et al.*, 1972; Brunes, Öquist and Eliasson, 1980). This indicates that when a high density of light-collecting pigments accumulates in shade, the photosystems and the antennae must be assembled without lowering the photosynthetic quantum yield, i.e. in grana partitions where an equal redistribution of quanta between the two photosystems is controlled by conformational changes. The chlorophyll *a*/*b* ratio may therefore be used as an approximate index of the ratio between the length of stacked and unstacked thylakoids, or as an index of the relative content of LH-CP*a*/*b*, rather than as an index of the PSU size.

SEASONAL EFFECTS ON THE PHOTOSYNTHETIC FUNCTION OF CONIFER CHLOROPLASTS

It is established that there are seasonal variations in apparent photosynthesis in herbaceous and deciduous plants (De Puit and Caldwell, 1973; Schaedle, 1975; Taylor and Pearcy, 1976) and in evergreen conifers (Freeland, 1944; Bourdeau, 1959; Parker, 1961; Schulze, Mooney and Dunn, 1967; Troeng and Linder, 1977). Conifers are of particular interest in this respect as they retain their needles during the winter, even at subzero temperatures. Short days and low temperatures above zero during the autumn induce frost hardening, which prepares the plants to survive the low temperature stress during the winter. For a discussion of the possible mechanisms involved in frost hardening *see* Levitt (1972) and Heber and Santarius (1973). Winter depression of apparent photosynthesis in conifers is a well-known phenomenon, although the severity of the depression may vary widely from climate to climate. In areas with 'mild' winters, the winter depression in the rate of apparent photosynthesis may depend primarily on the environmental conditions (Neilson, Ludlow and Jarvis, 1972), whereas more 'severe' winter climates, with prolonged periods of temperature below zero, cause a winter depression which needs several days or weeks at above zero temperatures to recover (Parker, 1961; Tranquillini, 1964; Pharis, Hellmers and Schuurmans, 1970). A winter depression which needs such a long time to recover must be associated with changes of the photosynthetic mechanism at the biochemical level (Zelawski and Kucharska, 1967). In order to study the mechanism of winter depression and the recovery of photosynthesis in conifers, photoactive chloroplasts must be utilized. Due to the high content of polyphenols and resins in the needles, chloropla st separation and purification are impossible with conventional isolation media.

In 1971 Oku and Tomita overcame, at least partly, the problem of isolating photoactive chloroplasts from conifers by adding polyethylene glycol to the isolation medium (Oku and Tomita, 1971). Their method has since been adopted and modified by others, and the reaction conditions for studying photosynthetic electron transport and photophosphorylation have been optimized for pine and spruce chloroplasts (Öquist, Martin and Mårtensson, 1974, Öquist and Martin, 1975; Lewandowska, Hart and Jarvis, 1976; Nicolić *et al.*, 1977; Senser and Beck, 1977, 1978). Gezélius (1975) and Gezélius and Hallén (1980) have developed additional methods for both extraction and assay of RuPBc from *Pinus silvestris*.

Seasonal effects on the electron transport properties of chloroplasts isolated from *Pinus silvestris* (Martin, Mårtensson and Öquist, 1978b) and *Picea abies* (Senser and Beck, 1977, 1978) have been analysed. By isolating chloroplasts from a 20-year-old pine tree growing at a site near the campus of Umeå University, we showed in 1976–1977 that the rate of light-saturated electron transport from water to NADP decreased during the autumn and the whole-chain electron transport was already totally inhibited in October. This inhibition lasted until May of the following year, when the electron transport activity became fully recovered. As the partial reactions of both photosystems I and II showed low but detectable activities during the winter, we conclude that the site for winter inhibition is somewhere in the electron transport chain that links the two photosystems. Artificial frost hardening and prolonged winter stress at $-5°C$ in climate chambers have given similar results (Martin, Mårtensson and Öquist, 1978a). Later analyses have revealed that the time for winter inhibition of photosynthesis can occur as late as March. Such differences are thought to depend on climatic variations from year to year.

Senser and Beck (1977, 1978) have reported basically similar seasonal changes of the ferricyanide-dependent oxygen evolution in chloroplasts isolated from *Picea abies*. However, they never noted total winter inhibition. They have also shown that the non-cyclic photophosphorylation decreases in parallel with the Hill activity and that there is a concomitant marked increase of the phenazine methosulphate-mediated cyclic photophosphorylation during the winter. They suggested that the seasonal changes in cyclic photophosphorylation reflect an alteration of the composition of the thylakoids during frost hardening/dehardening. Although we have not analysed cyclic electron transport or photophosphorylation, we have found that frost hardening causes an increase in the content of cytochrome b-563 and that part of cytochrome b-559HP is converted to cytochrome b-559LP (Öquist and Hellgren, 1976). As cytochrome b-563 is thought to function in a cyclic electron transport around photosystem I (Cramer and Whitmarsh, 1977), an increased content of this cytochrome may indicate an increased potential for cyclic electron transport in frost-hardened plants, as suggested by Senser and Beck (1977). During favourable temperature conditions the existence of such a mechanism may support additional ATP during the winter, although there may be a partial or total inhibition of non-cyclic electron transport.

In order to investigate if our results on the climatic effects on the photosynthetic properties of chloroplast or subchloroplast levels are consistent with the photosynthetic function of the intact plant structure, we have made a comparative study of how prolonged artificial frost hardening (short day, $+3°C$) and winter stress (short day, $-5°C$) of pine seedlings affect apparent photosynthesis, electron transport properties of isolated chloroplasts and the activity of extracted RuBPc (Öquist *et al.*, 1980). Frost hardening did not affect the quantum yields of apparent photosynthesis or photosynthetic electron transport (2,6-dichlorophenol-indophenol as electron acceptor) when light was rate limiting. At light saturation, however, there were significantly lower rates of both apparent photosynthesis and electron transport in the frost-hardened compared to the control seedlings (*Table 4.3*). Winter stress induced a parallel decay of the rates of both apparent photosynthesis and electron transport. Although both apparent photosynthesis and electron transport were severely inhibited in the experiment, the activity of extracted RuBPc was only slightly affected. This is consistent with the substantial RuBPc activity of *Pinus silvestris* in midwinter when there is no whole-chain electron transport (Gezélius and Hallén, 1980). We conclude, therefore, that the decay and ultimate inhibition of apparent photosynthesis in a winter climate are coupled to the electron transport properties of the thylakoids.

The findings that winter stress of pine causes an inhibition of overall electron transport from water to NADP while the partial photosystems remain active, show that the site of the winter inhibition must be in the electron transport chain between the two photosystems (Martin, Mårtensson and Öquist, 1978a, 1978b). Fluorescence kinetic measurements of isolated pine chloroplasts have shown that frost hardening reduces the pool of the electron acceptors on the reducing side of photosystem II, probably plastoquinone (Öquist and Hellgren, 1976). Further analyses with the same technique have revealed that the winter inhibition of photosynthesis must be coupled to the function of plastoquinone since winter stress and DCMU cause the same type of inhibition of the reoxidation of the primary electron acceptor Q of photosystem II (Öquist and Martin, 1980). The inability of plastoquinone to oxidize Q in winter-inhibited chloroplasts may be due to conformational changes in the thylakoids and/or to a destruction of plastoquinone (Martin, Mårtensson and Öquist, 1978b; Öquist and Martin, 1980). Winter inhibition caused no electron transport inhibition between cytochrome f and P700.

SEASONAL EFFECTS ON CHLOROPHYLL-PROTEIN COMPLEXES OF PINE

Usually, conifers have a higher chlorophyll content in summer than in winter (Linder, 1972; Lewandowska and Jarvis, 1977); in northern Scandinavia the

Table 4.3 A COMPARISON OF THE RATES OF LIGHT-SATURATED PHOTO-REACTIONS OF ISOLATED PINE CHLOROPLASTS, SATURATED APPARENT PHOTOSYNTHESIS OF PINE SEEDLINGS AND ACTIVITIES OF EXTRACTED RuBPc. (CALCULATED FROM ÖQUIST *ET AL.*, 1980)

	Photoreactions (μmol reduced acceptor per mg chlorophyll per hour)			Apparent photosynthesis (μmol CO_2 per mg chlorophyll per hour)		RuBPc activity (μmol CO_2 per mg chlorophyll per hour)
	Photo-systems II+I	Photo-system II	Photo-system I	2.0 kPa O_2	21.2 kPa O_2	
Control seedlings	49	137	50	126	100	248
Frost-hardened seedlings	36	103	32	86	65	239
Seedlings winter-stressed for:						
3 days	22	68	40	48	36	—
9 days	—	—	—	—	—	250
12 days	15	38	40	20	12	—
30 days	3	15	8	—	—	—
42 days	—	—	—	—	—	201

Photosystems II + I: $H_2O \rightarrow$ NADP; photosystem II: $H_2O \rightarrow$ DPIP; photosystem I: Asc/CPIP $\rightarrow$ NADP

chlorophyll content of pine needles reaches a minimum in late winter (Öquist *et al.*, 1978a). With reference to the three chlorophyll-protein complexes shown in *Figure 4.2*, we have found by SDS-PAGE and low-temperature fluorescence emission measurements that the winter destruction of chlorophyll in pine is more pronounced in the reaction centre antennae of chlorophyll *a* than in LH-CP*a/b* (Öquist *et al.*, 1978a). This is also reflected by a lower chlorophyll *a/b* ratio of pine needles in winter than in summer. Model experiments in climate chambers have also shown that the winter destruction of the reaction centre antennae is more pronounced in high than in low irradiance. When P700-CP*a* is bleached during winter its apoprotein, with a molecular weight of 70 000, appears in the polypeptide scan profile (Öquist *et al.*, 1978b). Except for this change, winter stress does not induce any major changes among the polypeptides solubilized from the thylakoids. However, experiments with artificially frost-hardened pine seedlings have revealed that when the temperature is lowered below the frost killing point there is a gradual destruction of the polypeptides after thawing of the needles (Öquist *et al.*, 1978b). This is again consistent with the observation by Senser and Beck (1977) that frost killing of spruce chloroplasts is probably not caused by a direct effect on the thylakoids by the low temperature as such but due to toxic compounds or hydrolytic enzymes released when the intracellular compartmentation is destroyed by freezing injury.

The winter destruction of P700-CP*a* in pine is accompanied by an increased ratio of chlorophyll/P700; in midwinter a ratio of 2000–3000 has been measured (Martin, Mårtensson and Öquist, 1978b). This indicates an increased size of the PSUs if the ratio is taken as an index of the PSU size. In contrast, time course studies of P700 photo-oxidation in chloroplasts isolated from winter-stressed and active pine have shown that the real PSU sizes must be about equal (approximately 700) in the two types of chloroplast (Öquist and Martin, 1980). The high chlorophyll/P700 ratio in winter-stressed pine is due to the formation of photosynthetically inactive chlorophyll. Model experiments in climate chambers have shown that destruction of chlorophyll-protein complexes and increased chlorophyll/P700 ratios are not observed before the temperature is lowered below 0°C and the changes are more pronounced in high than in low irradiances (Martin, Mårtensson and Öquist, 1978a; Öquist *et al.*, 1978a). Thus, no destruction is observed in seedlings frost-hardened at +3°C, which is consistent with the unaffected photosynthetic quantum yield of frost-hardened seedlings as compared with a control (Öquist *et al.*, 1980).

Correlations between Seasonal Effects on Function and Structure of Pine Chloroplasts

For a more extensive study of seasonal effects on the ultrastructure of conifers, *see* Schmidt (1936), Parker and Philpott (1961, 1963), Campbell (1972), Walles, Nyman and Aldén (1973), Chabot and Chabot (1975), Senser, Schötz and Beck (1975) and Martin and Öquist (1979). Reported seasonally induced structural changes seem to follow a general pattern, although the changes can be more or less pronounced depending on the climatic conditions. In the summer we found symmetrically shaped chloroplasts spread around the cell walls in pine. The chloroplasts were heavily loaded with starch. The thylakoids were well differentiated into grana and stroma thylakoids. In the autumn, during frost hardening, the starch content was reduced and the chloroplasts assumed an amoeboid appearance with pronounced regions which lacked thylakoids. The chloroplasts also became swollen and aggregated in one part of the cell. During winter the stroma thylakoids were the first to be reduced in number, then the number of thylakoids per granum was reduced and in late winter the chloroplasts contained mostly single and disorganized thylakoids. No envelope could be seen surrounding the chloroplasts in the late winter. This may be real, or the envelope may have become destroyed during the preparation of the sample. Another possibility is that the double bonds of the membrane acyl lipds were destroyed in late winter so that the osmium tetroxide could not bind and give contrast to the membranes (Kopp, 1972). In spring and early summer the chloroplasts recovered their typical summer structure as they migrated to the proximity of the cell walls. We observed basically the same kind of structural changes both in naturally grown pine and in artificially frost-hardened, winter-stressed and dehardened pine seedlings. We were not able to correlate the inhibition of whole-chain electron transport in October to any visible changes in the structure of grana and stroma thylakoids, although the chloroplasts became swollen and clumped together during this time. The loss of stroma thylakoids in the winter may be related to the destruction of P700-CP*a*, thus indicating the structural importance of this chlorophyll-protein complex for the maintenance of intact stroma thylakoids (*see* above). In April the ratio between LH-CP*a*/*b* and the remaining P700-CP*a* decreased sharply a few weeks before chlorophyll synthesis started (Öquist *et al.*, 1978a). This is thought to be due to a predominant loss of LH-CP*a*/*b* and it may be related to a pronounced, but not complete, loss of grana stacking, thus reflecting the assumed importance of this chlorophyll-protein in the stacking process (Anderson, 1975).

Mechanism for Winter Inhibition of Photosynthesis and Chloroplast Disorganization in Pine

Gerold (1959) has shown that chlorophyll degradation in winter is more pronounced in exposed than in shaded needles. Additionally, we have shown that the inhibition of photosynthesis and the destruction of chlorophyll-protein complexes are more pronounced in high than in low irradiance (Martin, Mårtensson and Öquist, 1978a; Öquist *et al.*, 1978a). The effects of strong light on the chlorophyll-protein complexes in isolated pine chloroplasts are the same as those observed in stressing winter climates. However, when the chloroplasts are irradiated in nitrogen atmosphere, only minor chlorophyll destruction is observed. This has led us to suggest that winter destruction in the photosynthetic apparatus of pine is primarily caused by photo-oxidation (Öquist *et al.*, 1978a). When the temperature falls in the autumn, stomata close and absorbed excitation energy cannot be used for CO_2 reduction. Under such circumstances photorespiration may be an important mechanism (Krause and Heber, 1976; Powles and Osmond, 1978), together with carotenoids (Krinsky, 1968, 1971) and superoxide dimutase (Fridovitch, 1976) for the de-excitation of the photosynthetic apparatus, thus preventing photo-oxidation. At subzero temperatures these protective mechanisms may not be of any substantial importance due to their more or less enzymatic character. When chlorophyll cannot be de-excited by a physiological mechanism it might be converted to the metastable triplet state (Krinsky, 1968, 1971). Such chlorophyll has the potential for exciting oxygen to its singlet state or for producing superoxide radicals. These species of oxygen are very strong oxidizing agents, capable of destruction of thylakoid constituents (Foote, 1976). The chloroplast destruction is most pronounced in late winter (February to April) when day length, solar height and reflection from the snow cause a very intense light exposure of the needles. As the plants at this time of the year also suffer from water stress due to a frozen ground and repeated frequent cycles of freezing and thawing, the pressure for photodynamic damage should be most pronounced.

We conclude that the effects of autumn and winter climates on photosynthesis in conifers can be divided into two phases, which we denote as frost hardening and winter stress (Öquist *et al.*, 1980). Frost hardening is induced by short day and low temperature conditions above zero (Levitt, 1972). During these conditions the efficiency of photosynthesis is maintained at a high level and the suppression of the rate of light-saturated photosynthesis may be compensated by a shift of the temperature optimum to the prevailing temperature conditions (Neilson, Ludlow and Jarvis, 1972). No disorganization of thylakoids or destruction of chlorophyll-protein complexes occurs during this phase and the chlorophyll/P700 ratio is not affected. The next phase, winter stress, which occurs in light exposed areas with temperatures below zero for a prolonged time, is characterized by a reduced efficiency of photosynthesis which ultimately becomes totally inhibited. Chlorophyll bleaching, destruction of chlorophyll-protein complexes and reaction centres, disorganization of chloroplast structures and decreased activity of RuBPc are additional typical symptoms. The primary mechanism involved is most likely to be photo-oxidation. According to this view, frost hardening is the result of an adaptive physiological response whereas winter stress is characterized by physical and chemical destruction of the photosynthetic apparatus. However, frost hardening makes it possible for plants to survive the winter stress phase and the photosynthetic apparatus can be repaired during the following spring.

Summary

Chloroplast thylakoids are asymmetric structures. On illumination, electron transport mediates an uptake of protons from the stroma into the intrathylakoid space, and a counter movement of cations. These processes give reduced NADP, cause ATP synthesis and activate the Benson–Calvin cycle enzymes. They also cause conformational changes in the thylakoid, which regulates the partitioning of excitation energy between the two photosystems, thus optimizing the photosynthetic efficiency. A model based on biochemical and functional data for the *in vivo* organization of chlorophyll is given.

When angiosperm seedlings germinate in darkness they form chlorophyll-free etioplasts which, on illumination, develop to mature chloroplasts. Most conifers develop chlorophyll-containing chloroplasts in darkness. These chloroplasts contain prolamellar bodies but in other aspects they show properties typical for mature chloroplasts. Chloroplasts from dark-grown pine have a substantial electron transport activity, which is limited on the water-splitting side of photosystem II. The electron transport rate is optimized within 30 minutes of illumination. No significant changes in pigmentation occur during the first 20 hours in light but only two hours' exposure causes the prolamellar body to disappear and well-defined grana and stroma thylakoids are formed. Grana formation in higher plants is most likely necessary in order to optimize the photosynthetic efficiency.

It is well documented that the photosynthetic rate of plants changes with the state of development as well as with the environmental conditions. The effect of autumn and winter climates on photosynthesis in pine can be divided into two phases: frost hardening and winter stress. During frost hardening, which is induced by short days and temperatures slightly above zero, the photosynthetic efficiency is maintained at a high level. Winter stress occurs at subzero temperatures and is more pronounced in high than in low irradiances. This phase is characterized by a reduced efficiency of photosynthesis, which ultimately becomes inhibited at the site of plasto-quinone. Typical for this phase is also chlorophyll bleaching, destruction of reaction centres and chlorophyll-protein complexes, disorganization of chloroplast structures and decreased levels of RuBPc. Frost hardening is considered as an adaptive physiological response, while winter stress is a physically and chemically induced destruction (probably by photo-oxidation) of the photosynthetic apparatus. The damage caused by winter stress is fully repaired during the spring.

Acknowledgements

The author wishes to thank the colleagues at the Department of Plant Physiology, University of Umeå, Professor N. I. Bishop, Oregon State

University, Dr B. Andersson, University of Lund and Dr Conny Liljenberg, University of Gothenburg, for suggestions in preparing the manuscript.

References

ÅKERLUND, H.-E., ANDERSSON, B. and ALBERTSSON, P.-Å. (1976). *Biochim. Biophys. Acta*, **449**, 525–535
AKOYUNOGLOU, G. and MICHELINAKI-MANETA, M. (1975). In *Proceedings of the Third International Congress on Photosynthesis*, Vol. III, pp. 1885–1896. Ed. by M. Avon. Elsevier, Amsterdam
ALBERTE, R. S., MCCLURE, P. R. and THORNBER, J. P. (1976). *Pl. Physiol.*, **58**, 341–344
ANDERSON, J. M. (1975). *Biochim. Biophys. Acta*, **416**, 191–235
ANDERSON, J. M., GOODCHILD, D. J. and BOARDMAN, N. K. (1973). *Biochim. Biophys. Acta*, **325**, 573–585
ANDERSSON, B. and ÅKERLUND, H.-E. (1978). *Biochim. Biophys. Acta*, **503**, 462–472
ARMOND, P. A. and ARNTZEN, C. J. (1977). *Pl. Physiol.*, **59**, 398–404
ARMOND, P. A., ARNTZEN, C. J., BRIANTAIS, J.-M. and VERNOTTE, C. (1976). *Arch. Biochim. Biophys.*, **175**, 54–63
ARMOND, P. A., SCHREIBER, U. and BJÖRKMAN, O. (1977). In *Carnegie Institution Year Book*, Vol. 76, pp. 335–341
ARMOND, P. A., STAEHELIN, L. A. and ARNTZEN, C. J. (1977). *J. Cell Biol.*, **73**, 400–418
ARNTZEN, C. J. (1978). In *Current Topics in Bioenergetics. 8. Photosynthesis*, Part B, pp. 111–160. Ed. by D. R. Sanadi and L. P. Vernon. Academic Press, New York
ARNTZEN, C. J. and BRIANTAIS, J.-M. (1975). In *Bioenergetics of Photosynthesis*, pp. 51–113. Ed. by Govindjee. Academic Press, New York
BAKER, N. R. and HARDWICK, K. (1976). *Photosynthetica*, **10**, 361–366
BARBER, J. (1976). In *Topics in Photosynthesis. 1. The Intact Chloroplast*, pp. 89–134. Ed. by J. Barber. Elsevier, Amsterdam
BENNOUN, P. (1974). *Biochim. Biophys. Acta*, **368**, 141–147
BISHOP, D. G. (1974). *Photochem. Photobiol.*, **20**, 281–299
BISHOP, N. I. and ÖQUIST, G. (1980). *Physiol. Plant.*, **49**, 477–486.
BJÖRKMAN, O. (1973). In *Photophysiology*, Vol. 8, pp. 1–63. Ed. by A. C. Gise. Academic Press, New York
BJÖRKMAN, O., BOARDMAN, N. K., ANDERSON, J. M., THORNE, S. W., GOODCHILD, D. J. and PYLIOTIS, N. A. (1972). In *Carnegie Institution Year Book*, Vol. 71, pp. 115–135
BOARDMAN, N. K. (1966). In *The Chlorophylls*, pp. 437–479. Ed. by L. P. Vernon and G. R. Seely. Academic Press, New York
BOARDMAN, N. K. (1970). *Ann. Rev. Pl. Physiol.*, **21**, 115–140
BOARDMAN, N. K. (1977a). In *Encyclopedia of Plant Physiology*, New Series 5, *Photosynthesis I. Photosynthetic Electron Transport and Photophosphorylation*, pp. 583–600. Ed. by A. Trebst and M. Avron. Springer-Verlag, Berlin
BOARDMAN, N. K. (1977b). *Ann. Rev. Pl. Physiol.*, **28**, 355–377
BOARDMAN, N. K. and ANDERSON, J. M. (1978). In *Proceedings of the International Symposium in Chloroplast Development*, pp. 1–14. Ed. by G. Akoyunoglou and J. H. Argyroudi-Akoyunoglou. Elsevier, Amsterdam

BOARDMAN, N. K., ANDERSON, J. M. and GOODCHILD, D. J. (1978). In *Current Topics of Bioenergetics*, 8. *Photosynthesis*, Part B, pp. 35–109. Ed. by D. R. Sanadi and L. P. Vernon. Academic Press, New York
BOGDANOVIĆ, M. (1973). *Physiol. Plant.*, **29**, 17–18
BOURDEAU, P. F. (1959). *Ecology*, **40**, 63–67
BRADBEER, J. W., IRELAND, H. M. M., SMITH, J. W., REST, J. and EDGE, H. J. W. (1974). *New Phytol.*, **73**, 263–270
BRUNES, L., ÖQUIST, G. and ELIASSON, L. (1980). *Plant. Physiol.*, **66**, 940–944
BURKE, J. J., DITTO, C. L. and ARNTZEN, C. J. (1978). *Arch. Biochem. Biophys.*, **187**, 252–263
BUTLER, W. L. (1978). *Ann. Rev. Pl. Physiol.*, **29**, 345–378
BUTTER, W. L. and KITAJIMA, M. (1975). *Biochim. Biophys. Acta*, **396**, 72–85
CAMPBELL, R. (1972). *Ann. Bot.*, **36**, 711–720
CHABOT, J. F. and CHABOT, B. F. (1975). *Can. J. Bot.*, **53**, 295–304
CHENIAE, G. M. and MARTIN, I. F. (1973). *Photochem. Photobiol.*, **17**, 441–459
COOMBS, J. and GREENWOOD, A. D. (1976). In *Topics in Photosynthesis*. 1. *The Intact Chloroplast*, pp. 1–51. Ed. by J. Barber. Elsevier, Amsterdam
CRAMER, W. A. and WHITMARSH, J. (1977). *Ann. Rev. Pl. Physiol.*, **28**, 133–172
DAVIS, D. J., ARMOND, P. A., GROSS, E. L. and ARNTZEN, C. J. (1976). *Arch. Biochem. Biophys.*, **175**, 64–70
DAVIS, D. J. and GROSS, E. L. (1975). *Biochim. Biophys. Acta*, **387**, 557–567
DE GREEF, J., BUTLER, W. L. and ROTH, T. F. (1971). *Pl. Physiol.*, **47**, 457–464
DE PUIT, E. J. and CALDWELL, M. M. (1973). *Am. J. Bot.*, **60**, 426–435
DEROCHE, M. E. and BRIANTAIS, J. M. (1974). *Photochem. Photobiol.*, **19**, 233–240
DOUCE, R., HOLTZ, R. B. and BENSON, A. A. (1973). *J. biol. Chem.*, **248**, 7215–7222
EGNÉUS, H. SELLDÉN, G. and ANDERSSON, L. (1976). *Planta*, **133**, 47–52
ELLIS, R. J. (1976). In *Topics in Photosynthesis*. 1. *The Intact Chloroplast*, pp. 335–364. Ed. by J. Barber. Elsevier, Amsterdam
FARINEAU, J. and POPOVIC, R. (1975). *Comptes Rendus Hebdom. Séances L'acad. Sci., Paris (Série D, t.)*, **281**, 1317–1320
FOOTE, C. S. (1976). In *Free Radicals in Biology*, Vol. II, pp. 85–133. Ed. by W. A. Pryor. Academic Press, New York
FREELAND, R. O. (1944). *Pl. Physiol.*, **19**, 179–185
FRIDOVITCH, I. (1976). In *Free Radicals in Biology*, Vol. I, pp. 239–277. Ed. by W. A. Pryor. Academic Press, New York
GEROLD, H. D. (1959). *Forest Sci.*, **5**, 333–343
GEZÉLIUS, K. (1975). *Photosynthetica*, **9**, 192–200
GEZÉLIUS, K. and HALLÉN, M. (1980). *Physiol. Plant.*, **48**, 88–98
GOODCHILD, D. J., BJÖRKMAN, O. and PYLIOTIS, N. A. (1972). In *Carnegie Institution Year Book*, Vol. 71, pp. 102–107
GOVINDJEE and PAPAGEORGIOU, G. (1971). *Photophysiology*, Vol. 6, pp. 1–46. Ed. by A. C. Giese. Academic Press, New York
HAYDEN, D. B. and HOPKINS, W. G. (1977). *Can. J. Bot.*, **55**, 2525–2529
HEBER, U. (1974). *Ann. Rev. Pl. Physiol.*, **25**, 393–421
HEBER, U. and SANTARIUS, K. H. (1973). In *Temperature and Life*, pp. 232–263. Ed. by H. Precht, J. Christophensen, H. Hensel and W. Larcher. Springer-Verlag, Berlin
HENNINGSEN, K. W. and BOARDMAN, N. K. (1973). *Pl. Physiol.*, **51**, 1117–1126

HENNINGSEN, K. W. and BOYNTON, J. E. (1970). *J. Cell Biol.*, **44**, 290–304
HENRIQUES, F. and PARK, R. B. (1978). *Biochem. biophys. Res. Commun.*, **81**, 1113–1118
HERRON, H. A. and MAUZERALL, D. (1972). *Pl. Physiol.*, **50**, 141–148
HILLER, R. G., GENGE, S. and PILGER, D. (1974). *Pl. Sci. Lett.*, **2**, 239–242
HOMANN, P. H. (1969). *Pl. Physiol.*, **44**, 932–936
HØYER-HANSEN, G., MACHOLD, O. and KAHN, A. (1976). *Carlsberg. Res. Commun.*, **41**, 349–357
HØYER-HANSEN, G., and SIMPSON, D. J. (1977). *Carlsberg. Res. Commun.*, **42**, 379–389
INOUE, Y., FURUTA, S., OKU, T. and SHIBATA, K. (1976). *Biochim. Biophys. Acta*, **449**, 357–367
INOUE, Y., KOBAYASHI, Y., SAKAMOTO, E. and SHIBATA, K. (1975). *Pl. Cell Physiol.*, **16**, 327–336
JACOBI, G. and LEHMANN, H. (1968). *Zeitschr. Pfl.-physiol.*, **59**, 457–476
JAGENDORF, A. T. (1975). In *Bioenergetics of Photosynthesis*, pp. 413–492. Ed. by Govindjee. Academic Press, New York
JESKE, C. and SENGER, H. (1978). In *Proceedings of the International Symposium of Chloroplast Development*, pp. 475–480. Ed. by G. Akoyunoglou and J. H. Argyróudi-Akoyunoglou. Elsevier, Amsterdam
JOYARD, J. and DOUCE, R. (1976). *Biochim. biophys. Acta*, **424**, 125–131
KAHN, A. (1968). *Pl. Physiol.*, **43**, 1769–1780
KELLY, G. J., LATZKO, E. and GIBBS, M. (1976). *Ann. Rev. Pl. Physiol.*, **27**, 181–205
KESSELMEIER, J. and RUPPEL, H. G. (1978). In *Recent Advances in the Biochemistry and Physiology of Plant Lipids*. Abstracts of the symposium held at Göteborg, 28–30 August 1978
KIRK, J. T. O. (1971). *Ann. Rev. Biochem.*, **40**, 161–196
KIRK, J. T. O. and GOODCHILD, D. J. (1972). *Austr. J. biol. Sci.*, **25**, 215–241
KLEIN, S. M. and VERNON, L. P. (1974). *Photochem. Photobiol.*, **19**, 43–49
KOPP, F. (1972). *Cytobiol.*, **6**, 287–317
KRAUSE, G. H. and HEBER, U. (1976). In *Topics in Photosynthesis 1. The Intact Chloroplast*, pp. 171–214. Ed. by J. Barber. Elsevier, Amsterdam
KRETZER, F., OHAD, I. and BENNOUN, P. (1976). In *Genetics and Biogenesis of Chloroplasts and Mitochondria*, pp. 25–32. Ed. by Th. Bücher, W. Neupert, W. Sebald and S. Werner. Elsevier, Amsterdam
KRINSKY, N. I. (1968). In *Photophysiology*, pp. 123–195. Ed. by A. C. Giese. Academic Press, New York
KRINSKY, N. J. (1971). In *Carotenoids*, pp. 669–716. Ed. by O. Isler. Birkhäuser Verlag, Basel
KRÓL, M. (1978). *Zeitschr. Pfl.-physiol.*, **86**, 379–387
LAUDI, G. (1964). *Giorna. Bot. Ital.*, **71**, 177–182
LAUDI, G. and BONATTI, P. M. (1973). *Caryologia*, **26**, 107–114
LEVITT, J. (1972). In *Physiological Ecology*. Ed. by T. T. Kozlowski. Academic Press, New York
LEWANDOWSKA, M., HART, J. W. and JARVIS, P. G. (1976). *Physiol. Plant.*, **37**, 269–274
LEWANDOWSKA, M. and JARVIS, P. G. (1977). *New Phytol.*, **79**, 247–256
LEWANDOWSKA, M. and JARVIS, P. G. (1978). *Physiol. Plant.*, **42**, 277–282
LEWANDOWSKA, M. and ÖQUIST, G. (1980a). *Physiol. Plant.*, **48**, 134–138

LEWANDOWSKA, M. and ÖQUIST, G. (1980b). *Physiol. Plant.*, **48**, 39–46
LICHTENTHALER, H. K. and PARK, R. B. (1963). *Nature*, **198**, 1070–1072
LICHTENTHALER, H. K. and SPREY, B. (1966). *Zeitschr. Naturforsch.*, **21b**, 690–697
LILJENBERG, C. and SELSTAM, E. (1978). In *Recent Advances in the Biochemistry and Physiology of Plant Lipids*. Abstracts of the symposium held at Göteborg, 28–30 August 1978
LINDER, S. (1972). *Stud. Forest. Suecica*, **100**, 1–37
LÜTZ, C. (1978). In *Proceedings of the International Symposium on Chloroplast Development*, pp. 481–488. Ed. by G. Akoyunoglou and J. H. Argyroudi-Akoyunoglou. Elsevier, Amsterdam
MACHOLD, O. (1975). *Biochim. Biophys. Acta*, **382**, 494–505
MACHOLD, O. and HØYER-HANSEN, G. (1976). *Carlsberg Res. Commun.*, **41**, 359–366
MARTIN, B., MÅRTENSSON, O. and ÖQUIST, G. (1978a). *Physiol. Plant.*, **43**, 297–305
MARTIN, B., MÅRTENSSON, O. and ÖQUIST, G. (1978b). *Physiol. Plant.*, **44**, 102–109
MARTIN, B. and ÖQUIST, G. (1979). *Physiol. Plant.*, **46**, 42–49
MCCREE, K. J. (1972). *Agric. Meteorol.*, **9**, 191–216
MENKE, W. (1962). *Ann. Rev. Pl. Physiol.*, **13**, 27–44
MICHEL, J.-M. and SIRONVAL, C. (1972). *Febs Lett.*, **27**, 231–234
MICHEL-WOLWERTZ, M. R. (1977). *Pl. Sci. Lett.*, **8**, 125–134
MICHEL-WOLWERTZ, M. R. and BRONCHART, R. (1974). *Pl. Sci. Lett.*, **2**, 45–54
MILLER, K. R. (1978). In *Proceedings of the International Symposium on Chloroplast Development*, pp. 17–30. Ed. by G. Akoyunoglou and J. H. Argyroudi-Akoyunoglou. Elsevier, Amsterdam
MILLER, K. R., MILLER, G. J. AND MCINTYRE, K. R. (1976). *J. Cell Biol.*, **71**, 624–638
MITCHELL, P. (1966). *Bot. Rev.*, **41**, 445–502
MÜHLETHALER, K. (1977). In *Encyclopedia of Plant Physiology. Photosynthesis I. Photosynthetic Electron Transport and Photophosphorylation*, pp. 503–521. Ed. by A. Trebst and M. Avron. Springer-Verlag, Berlin
MURAKAMI, S. and PACKER, L. (1970). *Pl. Physiol.*, **45**, 289–299
MURAKAMI, S., TORRES-PEREIRA, J. and PACKER, L. (1975). In *Bioenergetics of Photosynthesis*, pp. 555–618. Ed. by Govindjee. Academic Press, New York
MURATA, N. (1969). *Biochim. Biophys. Acta*, **189**, 171–181
MURATA, N. and FORK, D. C. (1976). *Biochim. Biophys. Acta*, **461**, 365–378
MYERS, J. (1971). *Ann. Rev. Pl. Physiol.*, **22**, 289–312
NEILSON, R. E., LUDLOW, M. M. and JARVIS, P. G. (1972). *J. appl. Ecol.*, **9**, 721–745
NICOLIĆ, D. and BOGDANOVÍC, M. (1972). *Protoplasma*, **75**, 205–213
NICOLIĆ, D., POPOVIC, R., TYSZKIEWICZ, E. and SARIC, M. (1977). In *4th International Congress on Photosynthesis*. Abstracts, p. 272
NOLAN, W. G. and SMILLIE, R. M. (1977). *Pl. Physiol.*, **59**, 1141–1145
OGAWA, T. and SHIBATA, K. (1973). *Physiol. Plant.*, **29**, 112–117
OJAKIAN, G. K. and SATIR, P. (1974). *Proc. natl Acad. Sci. USA*, **71**, 2052–2056
OKU, T., HAYASHI, H. and TOMITA, G. (1975). *Pl. Cell Physiol.*, **16**, 101–108
OKU, T., SUGAHARA, K. and TOMITA, G. (1974). *Pl. Cell Physiol.*, **15**, 175–178

OKU, T. and TOMITA, G. (1971). *Photosynth.*, **5**, 28–31
OKU, T. and TOMITA, G. (1976). *Physiol. Plant.*, **38**, 181–185
ÖQUIST, G. (1975). *Physiol. Plant.*, **34**, 300–305
ÖQUIST, G., BRUNES, L., HÄLLGREN, J.-E., GEZELIUS, K., HALLÉN, M. and MALMBERG, G. (1980). *Physiol. Plant.*, **48**, 526–531
ÖQUIST, G. and HELLGREN, N.-O. (1976). *Pl. Sci. Lett.*, **7**, 359–369
ÖQUIST, G. and MARTIN, B. (1975). In *Proceedings of the Third International Congress on Photosynthesis*, Vol. I, pp. 729–734. Ed. by M. Avron. Elsevier, Amsterdam
ÖQUIST, G. and MARTIN, B. (1980). *Physiol. Plant.*, **48**, 33–38
ÖQUIST, G., MARTIN, B. and MÅRTENSSON, O. (1974). *Photosynth.*, **8**, 263–271
ÖQUIST, G., MARTIN, B., MÅRTENSSON, O., CHRISTERSSON, L. and MALMBERG, G. (1978b). *Physiol. Plant.*, **44**, 300–306
ÖQUIST, G., MÅRTENSSON, O., MARTIN, B. and MALMBERG, G. (1978a). *Physiol. Plant.*, **44**, 187–192
ÖQUIST, G. and SAMUELSSON, G. (1980). *Physiol. Plant.*, **50**, 57–62
PARK, R. B. and SANE, P. V. (1971). *Ann. Rev. Pl. Physiol.*, **22**, 395–430
PARKER, J. (1961). *Ecol.*, **42**, 372–380
PARKER, J. and PHILPOTT, D. E. (1961). *Protoplasma*, **53**, 575–583
PARKER, J. and PHILPOTT, D. E. (1963). *Protoplasma*, **56**, 355–361
PHARIS, R. P., HELLMERS, H. and SCHUURMANS, E. (1970). *Photosynth.*, **4**, 273–279
POWLES, S. B. and OSMOND, C. B. (1978). *Austr. J. Pl. Physiol.*, **5**, 619–629
REMY, R. (1973). *Photochem. Photobiol.*, **18**, 409–416
REMY, R., HOARAU, J. and LECLERC, J. C. (1977). *Photochem. Photobiol.*, **26**, 151–158
ROSINSKI, J. and ROSEN, W. G. (1972). *Quart. Rev. Biol.*, **47**, 160–191
ROUGHAN, P. G. and BOARDMAN, N. K. (1972). *Pl. Physiol.*, **50**, 31–34
SANE, P. V. (1977). In *Encyclopedia of Plant Physiology. Photosynthesis I. Photosynthetic Electron Transport and Photophosphorylation*, pp. 522–542. Ed. by A. Trebst and M. Avron. Springer-Verlag, Berlin
SATOH, K. and BUTLER, W. L. (1978). *Pl. Physiol.*, **61**, 373–379
SCHAEDLE, M. (1975). *Ann. Rev. Pl. Physiol.*, **26**, 101–115
SCHMIDT, E. (1936). *Thar. Forstl. Jahrb.*, **87**, 1–43
SCHULZE, E. D., MOONEY, H. A. and DUNN, E. L. (1967). *Ecol.*, **48**, 1044–1047
SELLDÉN, G. and SELSTAM, E. (1976). *Physiol. Plant.*, **37**, 35–41
SENGER, H., BISHOP, N. I., WEHRMEYER, W. and KULANDAIVELU, G. (1975). In *Proceedings of the Third International Congress on Photosynthesis*, Vol. III, pp. 1913–1923. Ed. by M. Avron. Elsevier, Amsterdam
SENSER, M. and BECK, E. (1977). *Planta*, **137**, 195–201
SENSER, M. and BECK, E. (1978). *Photosynth.*, **12**, 323–327
SENSER, M., SCHÖTZ, F. and BECK, E. (1975). *Planta*, **126**, 1–10
SHUTILOVA, N. I. and KUTYURIN, V. M. (1976). *Sov. Pl. Physiol.*, **23**, 31–36
SINGER, S. J. and NICOLSON, G. L. (1972). *Science*, **175**, 720–731
SIRONVAL, C., BRONCHART, R., MICHEL, J.-M., BROUERS, M. and KUYPER, Y. (1968). *Bull. Soc. Franc. Physiol. végét.*, **14**, 195–225
STAEHELIN, L. A. (1976). *J. Cell Biol.*, **71**, 136–158
SUNDQVIST, C., BJÖRN, L. O. and VIRGIN, H. I. (1979). In *Results and Problems in Cell Differentiation*. Ed. by W. Beerman, W. Gehryng, J. B. Gurdon, F. C. Kafatos and J. Reinerf. Springer, Berlin

80   *Chloroplast structure and photosynthetic efficiency*

TAYLOR, A. O. and CRAIG, A. S. (1971). *Pl. Physiol.*, **47**, 719–725
TAYLOR, R. J. and PEARCY, R. W. (1976). *Can. J. Bot.*, **54**, 1094–1103
THORNBER, J. P. (1975). *Ann. Rev. Pl. Physiol.*, **26**, 127–158
THORNBER, J. P., ALBERTE, R. S., HUNTER, F. A., SHIOZAWA, J. A. and KAN, K.-S. (1976). *Brookhaven Symp. Biol.*, **28**, 132–148
TRANQUILLINI, W. (1964). In *The Formation of Wood in Forest Trees*, pp. 505–518. Ed. by M. H. Zimmerman. Academic Press, New York
TREBST, A. (1974). *Ann. Rev. Pl. Physiol.*, **25**, 423–458
TREFFRY, T. (1970). *Planta*, **91**, 279–284
TREFFRY, T. (1978). *Int. Rev. Cytol.*, **52**, 159–196
TROENG, E. and LINDER, S. (1977). Swedish Coniferous Forest Project, Internal Report 56
TYSZKIEWICZ, E., POPOVIC, R. and ROUX, E. (1977). *Febs Lett.*, **81**, 65–68
VERNON, L. P., SHAW, E. R., OGAWA, T. and RAVEED, D. (1971). *Photochem. Photobiol.*, **14**, 343–357
VON WETTSTEIN, D. (1958). *Brookhaven Symp. Biol.*, **11**, 138–159
WALKER, D. A. (1976). In *Topics in Photosynthesis* 1. *The Intact Chloroplast*, pp. 235–278. Ed. by J. Barber. Elsevier, Amsterdam
WALLES, B. (1967). Studia Forestalia Suecia 60
WALLES, B. and HUDÁK, J. (1975). Studia Forestalia Suecia 127
WALLES, B., NYMAN, B. and ALDÉN, T. (1973). Studia Forestalia Suecia 106
WESSELS, J. S. C. and BORCHERT, M. T. (1975). In *Proceedings of the Third International Congress on Photosynthesis*, Vol. I, pp. 473–484. Ed. by M. Avron. Elsevier, Amsterdam
WESSELS, J. S. C. and BORCHERT, M. T. (1978). *Biochim. Biophys. Acta*, **503**, 78–93
WHITE, C. C., CHAIN, R. K. and MALKIN, R. (1978). *Biochim. Biophys. Acta*, **502**, 127–137
WITT, H. T. (1975). In *Bioenergetics of Photosynthesis*, pp. 493–554. Ed. by Govindjee. Academic Press, New York
WOLWERTZ, M.-R. (1978). In *Proceedings of the International Symposium on Chloroplast Development*, pp. 111–118. Ed. by G. Akoyunoglou and J. H. Argyroudi-Akoyunoglou. Elsevier, Amsterdam
ZELAWSKI, W. and KUCHARSKA, J. (1967). *Photosynth.*, **1**, 207–213

5

PRODUCTION EFFICIENCY OF CONIFEROUS FOREST IN THE UK

P. G. JARVIS
Department of Forestry and Natural Resources, University of Edinburgh, UK

Introduction

Conifers are crops which have been widely planted in the UK over the past 30 years. During the 1960s and early 1970s planting of new sites was proceeding at a rate of about 40 000 hectares per annum, about half of which was by the Forestry Commission and half by private enterprises. More recently, the rate of planting by private enterprises has dropped off dramatically as a result of the introduction of new forms of taxation, but it is starting to increase again as confidence is restored. The rate of planting by the Forestry Commission has also fallen because sufficient new land is not becoming available for planting. It is estimated that there are about 200 000 hectares of land in England and Wales and 1.5 million hectares in Scotland which could reasonably be afforested over the next 50 years (Forestry Commission, 1977). Over the last five years, 84 per cent of the new planting has been in Scotland. The current areas of productive woodland in the UK are given in *Table 5.1*: 87 per cent of the broadleaves plus coppice are in private woodlands; 58 per cent of the conifers are in Scotland and 12 per cent in Wales. Over half the national forest is in Wales and Scotland and most of the recently planted forest is of coniferous species in Scotland.

Table 5.1 THE STOCKED AREA OF PRODUCTIVE FOREST IN THE UK AT 31 MARCH 1980 (FROM THE FORESTRY COMMISSION, 1980)

Type of forest	Stocked area $(ha \times 10^{-3})$		Total
	Private woodlands	*Forestry Commission*	
Conifers	521	834	1355
Broad leaves	318	49	367
Coppice	26	1	27
Total	865	884	1749

Notwithstanding the sustained efforts to create a productive national forest over the last 30 years, the UK, with the exception of Eire, has the lowest proportion of afforested land area in Europe. We grow only 9 per cent of our current requirements, the annual bill for our imports currently exceeding £3000 million.

In the UK, in general, land is only afforested if it is not required for agriculture. All new plantings are made with the agreement of the Ministry of Agriculture, Fisheries and Foods (MAFF) or the Department of Agriculture and Fisheries for Scotland (DAFS). Nearly all the land planted to conifers in the last 30 years has been grade 4 or 5 (in Scotland grade D) land. That is to say, the land given over to the growth of tree crops has been the poorest available both with regard to soils and altitude. In Scotland a typical pattern of land use is: arable in the valley bottoms, pasture on the lower slopes and forest at the higher elevations. In general, the newly planted forests are on land that is marginal for agriculture and which has been maintained in agricultural production by the help of extensive subsidies. Much of this land is poor hill grazing, which produces only about 2 tonnes (Mg) of herbage dry matter per annum and which carries only about one sheep per hectare with very low utilization efficiency (Eadie and Cunningham, 1971). In view of this, it is difficult to make realistic comparisons between the productivity of tree crops and agricultural crops. However, in this paper I do propose to compare their yields and to analyse the differences.

Actual Production

Because of the very wide range of soils and altitudes on which tree crops are grown in the UK, there is a very wide range of growth rates. In addition, growth rates vary widely between species and also within species. While varieties in an agricultural sense do not yet exist in any quantity, plants raised from seed collected from different sources, i.e. provenances, grow at widely different rates. The species most widely planted at the present time are Sitka spruce (*Picea sitchensis* (Bong.) Carr.) and lodgepole pine (*Pinus contorta* Dougl.), both exotics from the west coast of North America. The provenances of Sitka spruce which do best are from the Queen Charlotte Islands and from northern Washington state. The provenances of lodgepole pine which have been most widely planted are from coastal sites in southern Washington state and Oregon, but it now seems that intermediate provenances (further inlaid but not montane) from the Queen Charlotte Islands and mainland British Colombia do rather better. In addition, there are substantial plantations of Scots pine (*Pinus sylvestris* L.), Norway spruce (*Picea abies* (L.) Karst.), Japanese and hybrid larch (*Larix kaempferi* (Lamb.) Carr. and X. *eurolepis* Henry), and Corsican pine (*Pinus nigra* var. *maritima* (Ait.) Melville) in different parts of the country (Locke, 1978).

Regional and national average rates of production of the harvestable product can be derived from standard Forestry Commission inventory methods, as described by Hamilton and Christie (1971), and from information on the yield class of the growing stock, such as provided by Locke (1978). Average annual rates of production derived from the average yield class of the six major species grown in the UK are listed in *Table 5.2*.

The values given in *Table 5.2*, columns 3 and 6, are averages over a period of time which is approximately the length of the rotation. That is to say, the maximum mean annual dry matter increment (column 6) is $\sum_0^m W / \sum_0^m n$, where W is dry matter and n the number of years to the time m at which $\sum W / \sum n$ is maximum. During that interval the current annual increment (column 7),

Table 5.2 NATIONAL AVERAGE RATES OF FOREST PRODUCTION AS DRY MATTER (d.m.)

Crop	(1) Area stocked ($ha \times 10^{-3}$)	(2) Rotation length (yr)	(3) Average UK yield class ($m^3 ha^{-1} yr^{-1}$)	(4) Bark vol. fraction	(5) Basic density ($kg\ m^{-3}$)	(6) Mean d.m. yield increment ($Mg\ ha^{-1} yr^{-1}$)	(7) Max. current d.m. total increment ($Mg\ ha^{-1} yr^{-1}$)
Sitka spruce	329	60	11	0.09	350	3.5	9.1
Scots pine	104	79	7	0.15	410	2.4	6.6
Lodge pole pine	86	68	7	0.12	390	2.4	6.3
Norway spruce	73	77	11	0.09	340	3.4	8.4
Japanese and hybrid larch	58	45	9	0.16	410	3.1	8.2
Corsican pine	35	59	12	0.18	400	3.9	11.7

(1) From Locke (1978)
(2) Taken as age at which the maximum mean annual rate of volume production of stem is reached.
(3) Mean annual volume increment of stem (> 70 mm diameter) calculated as volume of stem (V) divided by number of years from planting (n) ($\sum V/\sum n$) for the age (m) at which this quotient reaches its maximum. The age is given in column (2). (From Locke, 1978)
(4) Proportion of bark expressed as the average volume fraction of total stem volume. (From Hamilton, 1975)
(5) Dry weight divided by fresh volume of stem wood. (From Hamilton, 1975)
(6) Mean increment in stem wood dry matter calculated from (3), (4) and (5).
(7) Maximum current annual increment (dW/dt) for the tree as a whole assuming a harvest index of 0.65 and the densities in (5). (Original data from Hamilton and Christie, 1971)

dW/dt, rises to a maximum, which is about $1\frac{1}{2}$–2 times the mean annual increment, and falls again. The maximum mean annual dry matter increment ($\sum W/\sum n$; *Table 5.2*, column 6) is about two-thirds of the current annual increment because of the period of several years at the beginning of the rotation before a closed canopy has developed.

The figures in *Table 5.2* can be compared with the average annual rates of production of agricultural crops in the UK listed in *Table 5.3*. Clearly, the actual annual rates of production of the tree crops and the majority of the agricultural crops are very similar. This might be regarded as rather surprising in view of the wide differences in the class of land on which trees and agricultural crops are grown, and because of the big differences in the intensity of management and supply of fertilizers. In addition, both sets of figures indicate surprisingly low average rates of production in comparison to the maximum rates which are known to be possible (e.g. Christie and Lines, 1979, and *see* next section). A rate of dry matter production of 5 Mg ha^{-1} yr^{-1} corresponds to a solar energy conversion efficiency of only 0.3 per cent in northern Europe.

Table 5.3 NATIONAL YIELD OF AGRICULTURAL CROPS AS DRY MATTER (d.m.) IN THE UK AVERAGED OVER PERIODS OF 1–7 YEARS BETWEEN 1967 AND 1975 (DATA MAINLY FROM MAFF, 1974, 1976)

Crop		Crop area ($ha \times 10^{-3}$)	Average d.m. yield ($Mg\ ha^{-1}\ yr^{-1}$)
Grass:	grazing	13 940	7.5
	rough grazing	6 620	~2.0
	hay	2 070	4.0
Barley:	grain	2 360	3.2
	straw		1.5
Wheat:	grain	970	3.2
	straw		1.5
Potatoes:	tubers	270	5.3
Oats:	grain	250	3.0
Sugar beet:	dry root	190	8.0
	sugar		5.6
Turnips:	fodder	110	4.6
Pulses:	seeds	75	2.7
Brassicas:	fodder	60	3.6

Potential Production

Potential production can be estimated using appropriate models if the parameters and functions are adequately known. Monteith (1977) has estimated the potential production 'that could be achieved by a (C_3) crop with a complete canopy throughout the year and with a photosynthesis rate which was not slowed either by low temperature in winter or drought in summer' as 54 Mg ha^{-1} yr^{-1}. The maximum rate of photosynthesis was assumed to be 0.3 g m^{-2} s^{-1} and the ratio of dark respiration rate to photosynthetic rate was assumed to be 0.4; an extinction coefficient of 0.6 was assumed. Other assumptions can be found in Monteith (1977, pp. 281–

283). De Wit (1965) arrived at estimates of potential production in the Netherlands for the six months April to September of 36 Mg ha^{-1}.

While this approach has proved useful for agricultural crops, the necessary parameters to make the calculations are poorly known for forest crops. On the other hand, maximal measured values are probably good approximations to potential values provided that there is a sufficiently large literature of reasonably error-free data available (Gifford, 1974). *Table 5.4* shows that the maximum rates of annual production of some of the agricultural and forest crops listed in *Tables 5.2* and *5.3* are up to an order of magnitude larger than the rates achieved under normal management practice.

Table 5.4   MAXIMAL RATES OF DRY MATTER (d.m.) PRODUCTION OF SOME FOREST AND C$_3$ AGRICULTURAL CROPS IN TEMPERATE CLIMATES

Crop	Rate of d.m. production (Mg ha^{-1} yr^{-1})	Harvest index
Coniferous		
Cryptomeria japonica (L.f.) D. Don	53	(0.65)
Pinus radiata D. Don	46	0.66
Tsuga heterophylla (Raf.) Sarg.	43	(0.65)
Abies sachaliensis Mast.	29	(0.65)
Pseudotsuga menziesii (Mirb.) Franco	28	0.71
Pinus nigra Arnold var. *nigra*	25	0.46
Picea abies (L.) Karst.	22	0.61
Thuja plicata Donn	20	0.68
Agricultural		
Sugar beet	42	0.45
Wheat	30	0.40
Perennial rye grass	26	0.85
Potatoes	22	0.82
Field bean	20	0.31
Barley	18	0.39
Calabrese	12	0.17
Soya beans	10	0.30

The figures for the tree crops are current or short-term (1–5 years) periodic annual productions. Where information on roots and stumps are not provided, 20 per cent has been added.
Data from Minderman (1967), Cooper (1975), Kira (1975), Loomis and Gerakis (1975), Hanley (1976), Kestemont (1977) and Thompson and Taylor (1979).

The maximal rates of production of the agricultural crops are higher in most cases because of more intensive management: most of them were obtained on research plots. Maximum rates of *mean* annual increment for forest crops are less often quoted than high rates of *current* annual increment. Some very high rates of above-ground current annual production of 35–45 Mg ha^{-1} yr^{-1} for some conifers have been reported (Kira, 1975) but there are very few data for the species grown in the UK. Ovington (1962) concluded, from a consideration of harvest data, that the maximum net rate of current annual production of coniferous forest in western Europe might be 22 Mg ha^{-1} yr^{-1} with a corresponding figure for maximum mean net production over the life of the crop of 15 Mg ha^{-1} yr^{-1}. His corresponding figures for the stems alone were 17 and 12 Mg ha^{-1} yr^{-1}, respectively. His figures are consistent with those in *Table 5.4*, where the highest values come from Mediterranean types of climate.

Problem Formulation

These measurements of growth rate pose a major question with respect to the functioning of our coniferous tree crops. How is it that our tree crops have similar values of production to agricultural crops in the UK when they are planted on the poorest soils and managed at a very low intensity? Is it, for example, because they have particular physiological or structural adaptations which result in high assimilation rates by the canopy, or is it largely because they exploit their site more effectively, both with respect to space and time?

I propose, therefore, to examine the photosynthetic capability of conifers in terms of the partial processes of photosynthesis (*Figure 5.1*) beginning with what is known at the level of the chloroplast and working up to consider the functioning of the canopy.

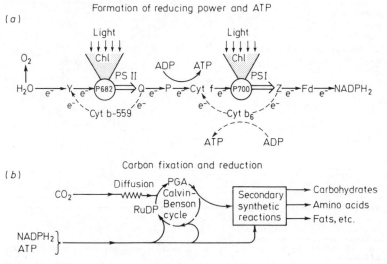

Figure 5.1   *The partial processes of photosynthesis. (a) Light harvesting and electron transport. (b) CO_2 diffusion, carboxylation and carbon cycling. (From Björkman, 1973)*

Comparisons of Photosynthetic Properties of Conifers and Other C_3 Species

THE PHOTOSYNTHETIC APPARATUS

Almost all the evidence indicates that conifers have typical C_3 photosynthesis. One set of observations has been interpreted to suggest the presence of some C_4 characteristics in Japanese larch, *Larix kaempferi* (Lamb.) Carr. (Sieb. & Zuco. Gord.) (Fry and Phillips, 1976) but this is unconfirmed.

The structure of the chloroplasts in conifers is similar to that in other C_3 species (Walles and Hudak, 1975; Soikkelli, 1978; Martin and Öquist, 1979).

The amounts of chlorophyll present and the relative proportions of chlorophyll *a* and *b* are also generally similar to those of other C_3 species (e.g. Gabrielsen, 1948; Linder, 1972; Wood, 1974; Senser, Schötz and Beck, 1975; Lewandowska and Jarvis, 1977). The protein, lipid and fatty-acid composition of the thylakoid membranes of *Tsuga heterophylla* (Raf.) Sarg. has been shown to be qualitatively similar to that of other C_3 species but significant quantitative differences in the proportions of the lipids and fatty acids were found (De Yoe and Brown, 1977).

There are very substantial amounts of carotenoids in coniferous needles (Linder, 1972; Clark and Lister, 1975) but there may be fewer kinds of accessory pigment present than in other species. Using thin-layer chromatography and spectroscopy only one carotene and two xanthophylls were detected in Sitka spruce whereas 11 or 12 were detected in leaves of tomato by the same methods (Lewandowska, unpublished observations).

There seem to be two well-substantiated unusual features of the photosynthetic apparatus of conifers: the photosynthetic unit size, and the development of chlorophyll and electron transport capacity in the dark.

The number of chlorophyll molecules per reaction centre (i.e. photosynthetic unit size) seems to be over twice as large as is generally accepted as the norm, i.e. approximately 350 chlorophylls per P700. Alberte, McClure and Thornber (1976) have found up to 1600 chlorophyll molecules per P700 in several North American conifers and the minimal Chl/P700 ratio seems to be 600–700 in Scots pine (Öquist and Hellgren, 1976; Martin, Mårtensson and Öquist, 1978a, 1978b). It seems, too, that the size of the photosynthetic unit responds to acclimation, a larger number of chlorophyll molecules per P700 being found in shade-acclimated foliage (Alberte, McClure and Thornber, 1976) and in low temperature-acclimated plants (Martin, Mårtensson and Öquist, 1978a). In the latter case, however, the increase in Chl/P700 found after low-temperature hardening was in part at least the result of the destruction of antennae chlorophyll associated with the reaction centre (Öquist *et al.*, 1978a, 1978b).

Secondly, seedlings grown in the dark contain chloroplasts differentiated into prolamellar bodies, grana and stroma thylakoids (Nikolic and Bogdanović, 1972; Michel-Wolwertz and Bronchart, 1974; Walles and Hudak, 1975) and the chloroplasts contain significant amounts of chlorophyll *a* and *b* (Bogdanović, 1973) and possess capability for photosynthetic electron transport prior to illumination. In several species (Scots pine, *Pinus jeffreyi* Murr. and Norway spruce) photosystem I (PSI) activity has been clearly shown in dark-grown seedlings (Oku, Sugahara and Tomita, 1974; Farinaeu and Popovic, 1975; Inoue *et al.*, 1976; Oku and Tomita, 1976; Król, 1978; Lewandowska and Öquist, 1980) but there is some controversy with respect to the presence of photosystem II (PSII) activity: Oku, Sugahara and Tomita (1974), Oku and Tomita (1976) and Król (1978) found negligible PSII activity in dark-grown Scots pine and Norway spruce seedlings, whereas appreciable activity was found by Inoue *et al.* (1976) and by Lewandowska and Öquist (1980). Although the last authors found appreciable PSII activity present in dark-grown seedlings of Scots pine, they also found that illumination for five minutes markedly promoted PSII activity and the activity of overall electron transport from water to nicotinamide adenine dinucleotide phosphate (NADP), whereas there was little effect on the activity of PSI. Fluoresence kinetic measurements confirmed that the capacity for photosynthetic electron transport in dark-grown seedlings was limited on the water-splitting side of PSII, whereas the primary and secondary electron receptors of PSII were fully synthesized and functional in darkness. In Norway spruce, too, full expression of PSII activity in dark-grown seedlings seems to be limited on the water-splitting side and the limitation is removed by a short period of illumination (Oku et al., 1978). Thus, these conifers are capable of producing almost completely functional chloroplasts in darkness and only a short period of illumination is required for the development of full function. The significance of this unique feature of chloroplast development in conifers is not at all clear.

PHOTOCHEMICAL PROCESSES

Measured rates of electron transport through PSI and PSII in conifers are similar to the rates in other C_3 species but generally somewhat lower. In Sitka spruce and Scots pine the maximal uncoupled rates of electron transport through PSII, measured as the reduction of 2,6-dichlorophenol indophenol (DCIP), were about 80 and 100 μmol h^{-1} per mg of chlorophyll, respectively (e.g. Lewandowska, Hart and Jarvis, 1976; Martin, Mårtensson and Öquist, 1978b) in comparison with rates of 100–200 μmol mg^{-1} h^{-1} common in other C_3 species (e.g. Ku et al., 1974). Maximal rates of electron transport through PSI in Sitka spruce and Scots pine, measured as oxygen evolved or NADP reduced, are also low (100 and 35 mol mg^{-1} h^{-1}, respectively) when using ascorbate and DCIP as the electron donor system (Lewandowska, Hart and Jarvis, 1977; Lewandowska and Jarvis, 1978; Martin, Mårtensson and Öquist, 1978b).

Substantial rates of non-cyclic photophosphorylation have been observed in the summer and significant rates of phenazine methosulphate (PMS)-mediated cyclic photophosphorylation in the winter (Senser and Beck, 1978).

CARBOXYLATION

Carbon fixation by isolated chloroplasts from conifers has not been accomplished *in vitro*, as far as the author is aware. However, the properties and activity of the ribulose-1,5-bisphosphate carboxylase (RuBPc) have been established for Scots pine by Gezélius (1975) and for Sitka spruce by Beadle (1977). They found the K_M (CO_2) to be 0.18 and 0.11 mM, the $K_M(HCO_3^-)$ to be 8 and 4.7 mM, and the $K_M(RuBP)$ to be 0.18 and 0.12 mM for Scots pine and Sitka spruce, respectively. These values for the K_M are similar to values for the 'intermediate' form of the carboxylase determined for other C_3 species in similar conditions of pH 8 and high Mg^{2+} concentrations, e.g. a $K_M(HCO_3^-)$ of 2.5–3 mM and a $K_M(CO_2)$ of 0.05–0.25 mM (Bahr and Jensen, 1974; Andrews, Badger and Lorimer, 1975). Since a $K_M(CO_2)$ of 0.11 mM is equivalent to a CO_2 concentration of 3200 cm^3 m^{-3} at the chloroplast (i.e. ten times the normal ambient CO_2 concentration), small differences in the

K_M cannot be regarded as indicative of real differences in CO_2 affinity *in vivo*.

Rates of RuBPc activity, measured using a standard assay *in vitro*, are also similar to rates for other C_3 species but somewhat lower. In Sitka spruce the measured rates lie in the range of 150–300 μmol h^{-1} per mg chlorophyll (Beadle, 1977; Beadle and Jarvis, 1977) and these rates can be compared with rates of 200–600 μmol mg^{-1} h^{-1} for a number of other C_3 species (e.g. Björkman and Gauhl, 1969; Ku, Gutierrez and Edwards, 1974).

There are considerable problems in the isolation and assay of enzyme systems from coniferous needles. The somewhat low measured activity of the photosystems and carboxylase and the high values of the K_M may result from inactivation or inhibition during isolation or assay and cannot properly represent the *in vivo* activity. These observations, therefore, cannot be taken as providing firm evidence of any real differences in photosynthetic capacity between conifers and other C_3 species, although they do show a tendency for a lower intrinsic capacity.

OVERALL LIMITATION WITHIN THE CELL

Of more direct relevance are comparisons of maximum measured mesophyll conductance, or minimum mesophyll (residual or intracellular) resistance. Mesophyll conductance is derived from the slope of the curve that relates CO_2 influx, F_c, to the mean intercellular space CO_2 concentration, C_i, as $g_m = dF_c/dC_i$ (Jarvis, 1971). The mesophyll resistance ($r_m = 1/g_m$) may be regarded as being in series with the stomatal resistance (r_s) so that the relative importance of r_m depends on its size in relation to r_s. As the stomatal conductance approaches infinity, i.e. $r_s = 0$, the rate of CO_2 assimilation becomes proportional to the mesophyll conductance. The rate of photosynthesis with stomatal conductance infinite can readily be found from the curve that relates F_c to C_i by equating C_i with the ambient CO_2 concentration, C_a. *Table 5.5* shows that the maximal mesophyll conductance of conifers is $\frac{1}{3}$ to $\frac{2}{3}$ that of other C_3 plants. These measurements made *in vivo* do, therefore, support the *in vitro* estimates of somewhat lower intrinsic rates of electron transport and of carboxylase activity.

PHOTORESPIRATION

From the evidence of the CO_2 compensation concentration (40–70 cm^3 m^{-3}) for most conifers, the efflux of CO_2 into CO_2-free air (e.g. 0.4 mg m^{-2} s^{-1}) the CO_2 efflux minimum (at 2.3 W m^{-2} in Sitka spruce), the Koch effect (at 2.3 W m^{-2} in Sitka spruce) and the inhibition of CO_2 influx by oxygen, it is clear that Sitka spruce and *Picea glauca* (Moench) Voss are normal C_3 plants with photorespiration (Poskuta, Nelson and Krotkov, 1967; Poskuta, 1968; Poskuta and Ostrowski, 1969; Ludlow and Jarvis, 1971; Cornic and Jarvis, 1972) and that this is probably also true of *Abies grandis* Lindl. (Maleszewski and Lewanty, 1972), Scots pine (Zelawski and Goral, 1966; Zelawski, 1967; Gordon and Gatherum, 1969) and *Pseudotsuga menziesii* (Mirb.) Franco (Brix,1968). So far there have been no reports of a C_4 conifer which lacks apparent photorespiration.

Rates of photorespiration estimated from CO_2 efflux or from measurements of photosynthesis in the presence and absence of oxygen appear to be similar to rates in other C_3 plants, e.g. about 30 per cent of current net photosynthesis. By comparing steady-state $^{12}CO_2$ and $^{14}CO_2$ uptake with $^{14}CO_2$ uptake measured over short periods of time, Neilson (1977) showed that photorespiration reduced photosynthesis of Sitka spruce by about 24 per cent.

Table 5.5   A COMPARISON BETWEEN MAXIMAL VALUES OF MESOPHYLL AND STOMATAL CONDUCTANCE FOR CO_2 UPTAKE IN CONIFERS AND OTHER C_3 SPECIES (PLAN SURFACE AREA BASIS)

Crop	Mesophyll conductance $(mm\ s^{-1})$	Stomatal conductance $(mm\ s^{-1})$
Coniferous		
Pinus halepensis Mill.	0.6	1.5
Pinus resinosa Ait.		2.0
Pinus contorta Dougl.	1.3	2.5
Pinus ponderosa Dougl.		0.7
Pinus sylvestris L.		3.1
Pinus radiata D. Don	1.0	4.8
Picea brewerana Wats.		4.5
Picea engelmanni (Parry) Engelm.		1.6
Picea sitchensis (Bong.) Carr.	2.0	5.5
Tsuga heterophylla (Raf.) Sarg.		1.4
Abies grandis Lindl.		2.6
Abies balsamea Mill.	1.1	3.0
Pseudotsuga menziesii (Mirb.) Franco.	1.5	5.5
Other C_3 species	3–6	8–14

Data from Dykstra (1974), Gifford (1974), Leverenz (1974), Jarvis, James and Landsberg (1976), Bennett and Rook (1978) and Körner, Scheel and Bauer (1979).

CARBON DIOXIDE DIFFUSION

The normal atmospheric CO_2 concentration is usually well below the concentration at which CO_2 saturation of photosynthesis occurs in conifers. Hence the rate of CO_2 fixation depends in part on the rate of diffusion of CO_2 into the leaf. The resistances to diffusion in the gas phase through the boundary layer, stomatal antechamber, stomatal pore and substomatal cavity can be regarded as being in series with the mesophyll resistance, so that their relative importance depends on their size in relation to r_m.

Because of the small diameter of the leaves, the boundary layer resistance from the bulk air to the leaf surface is much smaller in conifers than in most other plants and is unlikely to have a significant effect on CO_2 uptake. At a wind speed of $1\ m\ s^{-1}$ boundary layer conductances for CO_2 of needles on shoots are in the range of 50–$300\ mm\ s^{-1}$ (Jarvis, James and Landsberg, 1976) whereas the boundary layer conductance of most broad leaves is an order of magnitude smaller (e.g. Holmgren, Jarvis and Jarvis, 1965).

Estimates of stomatal conductance (g_s) in conifers vary widely (*Table 5.5*). The lower maximal conductances ($2\ mm\ s^{-1}$) are small compared with values for deciduous trees and herbaceous plants (Körner, Scheel and Baur, 1979)

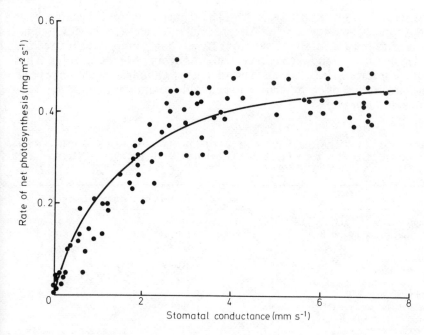

Figure 5.2   *The relationship between rate of net photosynthesis and stomatal conductance for water vapour transfer in Sitka spruce. The solid line is the hyperbolic curve of best fit and corresponds to a mesophyll conductance of 0.92 mm s⁻¹. (From Watts and Neilson, 1978)*

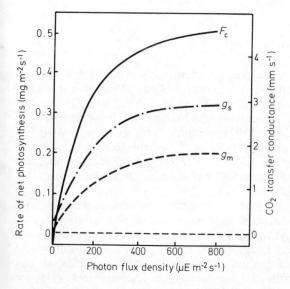

Figure 5.3   *The relationship between rate of net photosynthesis (F_c), stomatal conductance (g_s) and mesophyll conductance (g_m) for CO_2 and photon flux density in Sitka spruce. (From Ludlow and Jarvis, 1971)*

and must strongly limit CO_2 influx. The higher stomatal conductances of Sitka spruce and *Pinus resinosa* Ait. are similar to values for other C_3 plants (*Table 5.5*) and are not strongly limiting (*Figure 5.2*). Comparison between g_m and g_s in *Table 5.5* shows that in general g_m is about $\frac{1}{3}$ to $\frac{2}{3}$ the size of g_s. Thus in these species, at least, the main limitation to CO_2 influx lies within the cell. The smaller size of g_m in relation to g_s at light saturation in Sitka spruce is shown in *Figure 5.3*.

Table 5.6   A COMPARISON BETWEEN THE MAXIMAL VALUES OF PHOTOSYN-THESIS OF CONIFEROUS NEEDLES (EXCLUDING COTYLEDONS) AND OTHER C_3 SPECIES (PLAN LEAF AREA BASIS)

Crop	Rate of net photosynthesis ($mg\ m^{-2}\ s^{-1}$)
Coniferous	
11 species listed by Larcher	0.1–0.5
Pinus sylvestris L.	0.72
Pinus radiata D. Don	0.69
Larix decidua Mill.	0.62*
Picea sitchensis (Bong.) Carr.	0.55*
Abies balsamea Mill.	0.50
Pseudotsuga menziesii (Mirb.) Franco	0.50*
Tsuga heterophylla (Raf.) Sarg.	0.31*
Abies alba Mill.	0.31*
Pinus contorta Dougl.	0.28
Agricultural C_3 species	1.0–2.0

*With diffuse, bilateral or multidirectional illumination.
Data from Larcher (1969), Šesták, Jarvis and Čatský (1971), Dykstra(1974), Gifford (1974), Jarvis, James and Landsberg (1976), Szaniawski and Wierzbicki (1978), Troeng and Linder (1978) and Benecke (1980).

MAXIMUM LEAF PHOTOSYNTHESIS

The CO_2 exchange characteristics of coniferous shoots have been compre-hensively reviewed by Larcher (1969), Ferrell (1970) and Walker *et al.* (1972). Maximum rates of photosynthesis are listed in *Table 5.6*. They are low compared with CO_2 fluxes in deciduous and herbaceous C_3 plants (Larcher, 1969; Šesták, Jarvis and Čatský, 1971; Gifford, 1974), being only $\frac{1}{10}$ to $\frac{1}{2}$ of the rates found in many other C_3 species. From the foregoing, it would appear that the low rates of CO_2 influx result from a combination of moderate stomatal conductance, moderate rates of photorespiration and appreciable limitation to CO_2 fixation within the cell.

Limitation within the cell may result from low intrinsic carboxylase and photosystem activity, as discussed earlier, but other possibilities such as assimilate accumulation and light distribution within the needles also need to be considered. Little and Loach (1973) investigated the possibility that photosynthesis in conifers is limited by assimilate accumulation and concluded that this was not the case. However, there is considerable evidence to suggest that the rate of photosynthesis may often be limited by low quantum flux at the chloroplasts. Kramer and Clarke (1947) first showed that the grouping together of needles into fasicles or on shoots results in mutua[l] shading and prevents light saturation of all the chloroplasts in unidirectional irradiation, and Caldwell (1970) found that photosynthesis in *Pinus cembra* L. was reduced at high wind speeds which caused clumping of the needles and shading of one by the other.

The needles of conifers are much thicker than the leaves of many broad-leaved plants and are optically denser. Leaf thickness may be 600–1200 μm, and the transmittance of visible radiation 1–5 per cent (Jarvis, James and Landsberg, 1976) whereas leaves of broad-leaved plants are typically about 200–600 μm thick with a transmittance of visible radiation of 5–20 per cent (Gates *et al.*, 1965). Since chlorophyll contents per unit area are similar to those of other C_3 species, as noted above, the low transmittance must result from the particular arrangement of chlorophyll molecules within the leaf and may be the result of the greater leaf thickness and the larger Chl/P700 ratio than in leaves of other C_3 species. It is therefore unlikely that low absorptivity or low pigment content limits photosynthesis in conifers. However, the low rates of maximum photosynthesis may well result, at least partly, from inadequate quantum saturation of the pigments present because of the way in which the pigments are arranged within the leaf and the leaves are grouped together into shoots. Ludlow and Jarvis (1971) found that Sitka spruce had a higher rate of photosynthesis at the same irradiance when the source of radiant energy was diffuse rather than direct, and Zelawski *et al.* (1973) obtained very much higher rates of photosynthesis in young seedlings of

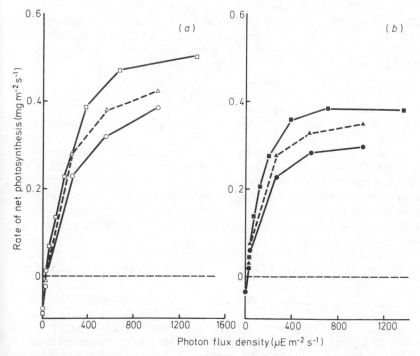

Figure 5.4 *The relationship between rate of net photosynthesis of (a) sun and (b) shade needles and total photon flux density. Illumination was bilateral (□, ■), on the top surface (△, ▲) or on the bottom surface (○, ●) of the needles. (From Leverenz and Jarvis, 1979)*

Scots pine with uniform, multilateral irradiation than with a unilateral source. The highest photosynthetic rate measured for a conifer so far is 0.65 mg m^{-2} s^{-1} by *Larix decidua* Mill. in multilateral illumination (Szaniawski and Wierzbicki, 1978). This is, however, still less than half the rate that many C_3 crop species are capable of (*Table 5.6*).

In Sitka spruce bilateral illumination of needles or shoots results in higher quantum efficiencies and light-saturated rates of photosynthesis, and lower quantum flux densities to reach saturation, than are found in unilateral illumination of the same total flux density (*Figure 5.4*; Leverenz and Jarvis, 1979, 1980). At the same total incident photon flux density, the rate of photosynthesis by shoots is substantially less than by unshaded needles and the differences in rate between bilateral and unilateral illumination are much larger for shoots than for unshaded needles. Thus both the needle anatomy and the grouping of needles into shoots limit photosynthesis by reducing the quantum flux density at the chloroplasts. However, shading of needles by one another and by the shoot axis has a larger limiting effect on the rate of photosynthesis than shading within the needle.

Photosynthetic Characteristics of Canopies

CANOPY PHOTOSYNTHESIS

In a crop stand with a closed canopy, several variables in addition to the maximum possible rate of net photosynthesis by single leaves influence the net influx of CO_2 to the canopy during the day. These variables include leaf area index, leaf area density, the grouping and inclination of leaves, photosynthetic acclimation to shade during development, and respiration by woody tissues and the soil.

While a very large number of studies has been made of photosynthesis of single shoots in forest canopies, using intensive techniques such as assimilation chambers, there have been very few attempts to measure CO_2 exchange by the canopy as a whole. There are considerable sampling problems, as well as problems of expense and maintenance, in attempting to derive the CO_2 influx to a canopy using assimilation chambers. On the other hand, less accurate, extensive methods, such as $^{14}CO_2$ feeding, can provide enough information from which to compute canopy CO_2 uptake. However, direct measurement of CO_2 uptake by the whole canopy can be obtained from micrometeorological measurements of CO_2 concentration gradients in the air above the canopy.

The uptake of CO_2 by the canopy, per unit area of ground, is estimated as the net vertical exchange of CO_2 between the canopy and the bulk air above as:

$$F_c = -K_E\rho(\delta\phi/\delta z) \tag{5.1}$$

where K_E is the eddy diffusivity derived from the energy budget, ρ is the density of CO_2 and $\delta\phi/\delta z$ is the gradient of CO_2 concentration. Because of the aerodynamic roughness of the canopy, K_E above forest is generally an order of magnitude larger than over field crops and lies in the range of 0.4–2.0 m^2 s^{-1}. The CO_2 concentration gradient, on the other hand, is much smaller because of the enhanced eddy diffusion and it rarely exceeds $0.5 \text{ cm}^3 \text{ m}^{-3}$. Hence a measurement resolution of $0.1 \text{ cm}^{-3} \text{ m}^3$ is necessary. The derivation of K_E depends on the accurate measurement of gradients of humidity and temperature, which are also small over coniferous forest and which require careful measurement (Jarvis, James and Landsberg, 1976).

Not many studies of CO_2 influx to coniferous forest canopies have been made in this way—possibly no more than half a dozen—so that it is not possible to draw many conclusions regarding the influence of species or management on CO_2 assimilation. However, the general magnitude of the CO_2 flux is in itself highly relevant to comparisons between the production of forest and agricultural crops.

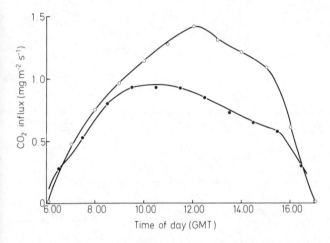

Figure 5.5   *The diurnal course of CO_2 influx to stands of Sitka spruce (○) and Scots pine (●) at Fetteresso Forest on 11 September 1973 and Thetford Forest on 10 September 1975, respectively. Midday solar irradiances were 600 W m^{-2} and 500 W m^{-2}, respectively*

Figure 5.5 shows the diurnal course of CO_2 influx to two forest canopies in the UK on sunny days. CO_2 influx increases with quantum flux density and decreases with increasing vapour pressure deficit in Sitka spruce (*Figure 5.6*) and probably also in Scots pine. The median value of CO_2 influx into a Sitka spruce forest canopy in north-east Scotland (Fetteresso Forest) at low vapour pressure deficits and corresponding moderate levels of irradiation was $1.3 \text{ mg m}^{-2} \text{ s}^{-1}$. Some higher values of up to $2 \text{ mg m}^{-2} \text{ s}^{-1}$ have been measured, but may not be reliable because of the several sources of error in the method (Jarvis and James, 1981). Although the data are not yet all available, a corresponding median figure for Scots pine in south-east England (Thetford Forest) is probably about $0.9 \text{ mg m}^{-2} \text{ s}^{-1}$. These figures compare well with similarly derived estimates of CO_2 influx for temperate agricultural crops (*Table 5.7*). Indeed, they suggest that mean hourly rates of CO_2 influx to coniferous forest canopies are at least as high as CO_2 influx to many crops and almost certainly much higher than the CO_2 influx into the vegetation which the forest has replaced.

In Sitka spruce at Fetteresso, we found little difference in CO_2 influx

96

Table 5.7 MAXIMAL RATES OF CO_2 INFLUX TO CANOPIES OF SOME FOREST AND C_3 AGRICULTURAL CROPS IN TEMPERATE CLIMATES

Crop	CO_2 influx $(mg\ m^{-2}\ s^{-1})$	Leaf area index (projected area)	Extinction coefficient (K)
Coniferous			
Picea sitchensis (Bong.) Carr.	2.0 (1.3)*	9.8	0.43
Picea abies (L.) Karst.	1.2	8.4	0.28
Pinus sylvestris L.	1.5 (0.9)*	3.5	0.46
Pinus taeda L.	1.4	2.6	—
Pinus radiata D. Don	3.5	—	—
Agricultural			
Wheat	1.6	6.1	0.40
Barley	1.1	5.9	0.48
Sugar beet	1.3	5.0	0.63
Potatoes	1.1	—	—
French bean	1.1	6.0	—
Cabbage	1.1	5.0	—
Red clover	1.7	3.5	0.90
Rye grass	2.2	8.3	0.35

*Median maxima. The values in the columns do not necessarily correspond to the identical stand.
Data from Lemon (1967), Denmead (1969), Monteith (1969), Rhodes (1971), Leafe (1972), Szeicz (1974), Sale (1974, 1975), Biscoe, Scott and Monteith (1975), Jarvis, James and Landsberg (1976) and Jarvis and James (1981).

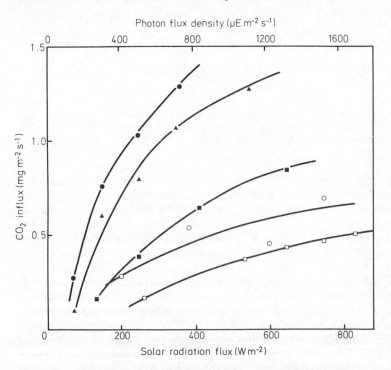

Figure 5.6 *The relationship between CO_2 influx to the Sitka spruce canopy at Fetteresso forest and solar irradiance for different classes of vapour pressure deficit: 0–2 mbar (•), 2–6 mbar (▲), 6–8 mbar (■), 8–10 mbar (○), 10–20 mbar (□). Points are medians of 6–20 points. (From Jarvis and James, 1981)*

between 1973 when there were approximately 4000 trees per hectare and 1975, one year after thinning in 1974, when the number of trees was reduced by half and the basal area and leaf area by about one third (Jarvis and James, 1981). The somewhat lower CO_2 influx to Scots pine at Thetford than to Sitka spruce at Fetteresso is probably because the leaf area index of the Scots pine was less than half that of the Sitka spruce and there was also proportionately more respiring wood present. In addition, as far as the Scots pine canopy is concerned, the figures are an overestimate because a part of the CO_2 influx must have been assimilated by the ground flora of *Pteridium aquilinum* (L.) Kuhn.

A CO_2 influx of 3.5 mg m^{-2} s^{-1}, three times the Sitka spruce maximum median value, has been reported for *Pinus radiata* Don when well supplied with water in a Mediterranean type of climate near Canberra, Australia (Denmead, 1969). This is consistent with the extremely high growth rates which *P. radiata* can achieve in that environment. Forrest and Ovington (1970) obtained net above-ground production rates of 19.4–25.9 Mg ha^{-1} yr^{-1} in 5–12-year-old plantations in New South Wales, while Miller (1971) estimated a current annual rate of production of 38 Mg ha^{-1} yr^{-1} in 26–29-year-old plantations in New Zealand. These figures demonstrate the extraordinary growth potential which some conifers have.

Despite comparatively low rates of photosynthesis by needles and shoots, the rate of CO_2 assimilation by the canopy of a Sitka spruce plantation is as high as that of agricultural crops which possess much higher rates of leaf photosynthesis. While the CO_2 influx to the Scots pine canopy is somewhat less, it too is within the range of other crops with much higher rates of leaf photosynthesis. What, then, are the features of spruce and pine canopies which result in such high rates of CO_2 assimilation per unit ground area?

In the first place, spruce canopies have a high leaf area index. With a leaf area index of 8, an average rate of photosynthesis of only 0.16 mg m^{-2} s^{-1} per unit leaf area is necessary to achieve a rate of 1.3 mg m^{-2} s^{-1} per unit ground area. Thus, very high rates of photosynthesis on a single-leaf basis are not necessary if the population of leaves contains a substantial proportion of physiologically active leaves, which are so placed that they obtain an adequate CO_2 supply and quantum flux density. The extent to which these considerations are met will be discussed in the following paragraphs.

NEEDLE POPULATION AND PHOTOSYNTHESIS

The total leaf area index is made up of partial indices of leaves of different ages, but the majority of leaves are of the current and previous year (*Figure 5.7*) and these leaves are largely in the upper canopy. In 1952, Freeland showed clearly in six species (*Pinus ponderosa*, Dougl., *Pinus nigra*, Arnold, Scots pine, *Pinus monticola*, Dougl. ex D. Don, *Picea pungens* Engelm. and *Abies concolor* (Gord.) Hildebrand) that the rate of photosynthesis was maximum in needles of the current year and declined in successive years. The greatest reduction occurred in needles of *Pinus ponderosa* followed by *Pinus nigra*. Subsequently, similar results have been obtained in other species (e.g. *Pinus resinosa*, Gordon and Larson, 1968), but Helms (1970) found that two-year-old needles had the highest photosynthetic rate in a *Pinus ponderosa* forest canopy.

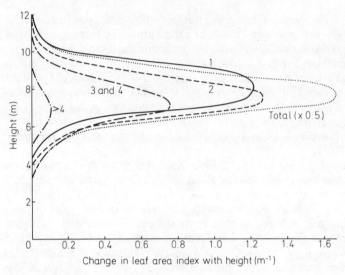

Figure 5.7   *The distribution of leaf area with height in the Sitka spruce forest canopy at Fetteresso Forest in 1971. The numbers on the curves are the ages of the needles in years. (From Norman and Jarvis, 1974)*

In the Sitka spruce canopy, the rate of photosynthesis is low immediately after bud break and increases as the shoots elongate to a maximum about eight weeks later. This increase in photosynthesis is correlated with a rise in g_m from below 0.13 mm s^{-1} to about 1.7 mm s^{-1}, and a rise in g_s from below 0.7 mm s^{-1} to about 2.5 mm s^{-1}. The decline in photosynthesis in subsequent years is associated with decreases in both g_s and g_m so that three years after emergence photosynthesis has fallen from 0.44 mg m^{-2} s^{-1} to 0.1 mg m^{-2} s^{-1}, and g_s and g_m have fallen to 0.4 mm s^{-1} and 0.5 mm s^{-1}, respectively (Ludlow and Jarvis, 1971). The decrease in g_s in older needles may in part be caused by a reduction in permeability of the matrix of wax tubes in the stomatal antechamber, as a result of the accumulation of pollutants, spores and microorganisms (Jeffree, Johnson and Jarvis, 1971).

Needles in different parts of the canopy are constantly exposed to large differences in the quantum flux density. Needles at the top of a tree are physiologically and anatomically acclimated to the high quantum flux densities which they experience ('sun' needles), whereas the needles lower down are structurally and physiologically shade-adapted ('shade' needles). Many experiments with conifer seedlings grown in different light regimes have established that shade-adapted needles have lower light-saturated photosynthetic rates on a unit needle area or chlorophyll basis than needles developed in high irradiances (e.g. Zelawski and Kinelska, 1967; Zelawski, Kinelska and Lotocki, 1968, with Scots pine).

In a forest canopy of large leaf area index, similar acclimation of photosynthesis to irradiance occurs. The maximum rate of photosynthesis is higher in sun needles than in shade needles and both maximum g_s and g_m decrease with increasing depth into the canopy (Jarvis, James and Landsberg, 1976; Schulze, Fuchs and Fuchs, 1977; Leverenz and Jarvis, 1979). In Scots pine there are fewer stomata and a smaller surface area of the mesophyll cells in shade needles than in sun needles, and these features may contribute to the lower maximal g_s and g_m of the shade needles (Zelawski, Kinelska and Lotocki, 1968). In shade needles of Sitka spruce, the dark respiration rates and compensation irradiances are lower than in sun needles: at the lowest level in the canopy they are about one quarter of the values at the uppermost level (Leverenz and Jarvis, 1979). In shade needles the amount of chlorophyll per unit area is less than in sun foliage (Wood, 1974; Boardman, 1977; Lewandowska, Hart and Jarvis, 1977) and, as a result, photosynthesis in shade foliage becomes light-saturated at lower photon flux densities than sun foliage (Leverenz and Jarvis, 1979). However, because of the low irradiances within the canopy, photosynthesis at depth is not often restricted by limitation to the light-saturated rate but is largely limited by the photochemical efficiency of the needles in the lower levels. This seems to be similar at all levels in the canopy (Jarvis, James and Landsberg, 1976; Leverenz and Jarvis, 1979; Neilson, personal communication), and, as a result, the rate of photosynthesis by shade needles is higher than that of sun needles at low and intermediate photon flux densities because of the low dark respiration rate (Leverenz and Jarvis, 1979).

CARBON DIOXIDE CONCENTRATION

The aerodynamic roughness of the canopy, which causes extremely small CO_2 gradients in the air above, also results in very little local CO_2 depletion within the canopy (Jarvis, James and Landsberg, 1976). Essentially, all the foliage is in air which contains average CO_2 concentrations within 1 per cent of that well above the canopy. While photosynthesis in conifers, as in other C_3 crops, is almost linearly dependent on CO_2 concentration at normal ambient concentrations, such small depletions have a negligible effect on the rate of photosynthesis. In contrast, significant CO_2 depletion occurs in field crops, which have much larger canopy boundary layer resistances, so that canopy photosynthesis may be reduced by 10–20 per cent.

LIGHT DISTRIBUTION

The albedo (solar radiation reflectance) of coniferous forest canopies is generally in the range of 5–15 per cent whereas the albedo of most agricultural crops and grassland is about 25 per cent (Jarvis, James and Landsberg, 1976). With a leaf area index of about 8, almost all the solar radiation which penetrates the canopy is absorbed. Thus about 90 per cent of the solar radiation is absorbed by the foliage of a spruce canopy, compared with not more than 75 per cent in a field crop (Landsberg, Jarvis and Slater, 1973).

However, probably the most important feature of the coniferous forest canopy is the way in which this radiation is distributed within the canopy as a result of properties of the foliage. In the first place, the foliage is not randomly distributed but is grouped. Three levels of grouping of foliage can readily be identified: grouping into conical crowns at the top of the canopy; grouping into spaced-out whorls of branches; and grouping into densely foliated shoots. This grouping results in a highly effective trapping of the incident radiation and very efficient transmission of radiation to the lower levels of the canopy.

Secondly, because of the small diameter of the individual needles and the spaced-out whorl distribution, most of the radiation which reaches the lower levels of the canopy is penumbral and consequently is at intermediate flux densities (Norman and Jarvis, 1974).

Thirdly, needles at the top of the canopy are almost uniformly distributed around the shoot axis so that the symmetry is radial and the geometry cylindrical (Norman and Jarvis, 1974). In addition, the shoots are held up so that the majority of the needles are at inclinations close to the vertical. In contrast, towards the base of the canopy the shoots have a pronounced dorsiventral structure with the needles mostly lying close to the horizontal plane. In addition, the needles are narrower than at the top of the canopy (Leverenz and Jarvis, 1979, 1980).

As a result of these features, the Beer's law extinction coefficient $K(= \ln(Q_z/Q_0)/L_z)$, where Q_z is the average radiation flux at a height z above which there is a leaf area index L_z, and Q_0 is incident irradiance) is substantially lower than would be expected from a random distribution of leaves. In coniferous forest canopies K is 0.3–0.6 (Kira, 1975; Jarvis, James and Landsberg, 1976) and, as a result, the radiation flux density at the base

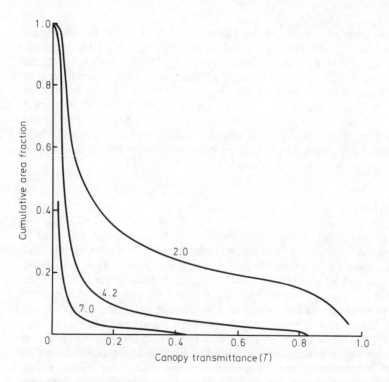

Figure 5.8   *The fraction of horizontal area occupied by transmittances greater than T as a function of transmittance, T. The numbers on the curves are the leaf area index above the measurement level. All curves are for a diffuse radiation fraction of 0.12–0.2 and a solar inclination angle of 32–45 degrees. (From Norman and Jarvis, 1974)*

of the canopy is approximately ten times what would be expected from random distribution of the needles (Norman and Jarvis, 1975).

In contrast, in agricultural crops K is generally in the range of 0.5–0.8 (Monteith, 1969) because of a more random distribution of foliage, smaller depth of canopy and negligible penumbral effects because of much larger leaves.

Thus a large proportion of the needles in the coniferous canopy are exposed to intermediate quantum flux densities of diffuse radiation (*Figure 5.8*). As with erectophile broad-leaved canopies, this would be expected to lead to high rates of canopy photosynthesis at high leaf area indices (Monsi, Uchijima and Oikowa, 1973). Experiments by Leverenz have shown that radiation incident on either surface of a shoot is utilized with almost equal efficiency, but that maximum photosynthesis results when both surfaces are illuminated (Leverenz and Jarvis, 1980). As *Figure 5.4* shows, a quantum flux of Q on each surface of a shoot is somewhat more effective than a quantum flux of $2Q$ on one surface. Consequently, the intermediate levels of diffuse radiation, which are found in the densest part of the canopy, result in substantial rates of photosynthesis, as demonstrated by $^{14}CO_2$ assimilation (Jarvis, James and Landsberg, 1976; Watts, Neilson and Jarvis, 1976).

Annual Growth Rates

The annual growth rates reviewed at the outset suggest that the average maximum current annual dry matter increment of conifers in the UK (*Table 5.2*, column 7) exceeds the annual yield of most agricultural crops (*Table 5.3*) by $1\frac{1}{2}$–2 times. Exceptions to this, for which the annual yield equals approximately the average maximum current annual increment, are fodder crops and high-yielding cash crops like sugar beet, which are generally grown in the most fertile regions of the country.

Maintenance and growth respiration would be expected to be substantially larger in forest crops than in agricultural crops because of the large amount of structural tissue which, although largely non-living, has associated with it a living peripheral layer that contains an active cambium. Reliable information on the respiration of the stems and branches of a stand is difficult to obtain.

The respiration rate of soil, branches and stem wood from the Sitka spruce canopy at Fetteresso during the daytime is probably about 0.4–0.8 mg $m^{-2} s^{-1}$ at typical summer temperatures of 10–20°C (Jarvis and James, 1981). This figure can be compared with daytime rates for cereals (e.g. Biscoe, Scott and Monteith, 1975), French beans, cabbages and potatoes (Sale, 1974, 1975) of 0.1–0.3 mg m^{-2} s^{-1} for the same temperature range.

Despite this rate of respiration, the average maximum current annual dry matter increment of Sitka spruce, yield class 11, exceeds the annual growth rate of most of the crops in *Table 5.3*. This is largely because coniferous tree crops carry their leaves for 365 days of the year, as compared with say 140 days for field crops, and in this climate have a positive net carbon balance in every month on average (Hagem, 1947, 1962; Bradbury and Malcolm, 1978).

The harvest index (proportion of total dry matter which is the economic yield) of agricultural crops varies widely. In forage crops and some tuber and root crops it is very large whereas in cereals and pulses it is much smaller (*Table 5.3*). Good estimates of the harvest index of forest trees are difficult to come by because the roots are not included in most production studies. In addition, harvest index is not a concept which has received much attention in forestry or tree breeding, although very impressive increases in agricultural production have resulted from improvements in harvest index. A comprehensive study by Albrektsson (1978) on 15 stands of Scots pine gave values for the harvest index of 61–69 per cent for trees over 15 years old, if the roots, butt and stump are included in the total dry matter and if the stem bark is excluded from the yield. Inclusion of the stem bark with the stem wood raises the harvest index by about 6 per cent, and exclusion of the roots from the total dry matter raises the harvest index by about 5 per cent. Other published data for *Pinus radiata*, *Picea abies* and *Pseudotsuga menziesii* lie within these ranges (Will 1966; Kestemont, Duvigneaud and Paulet, 1977; Madgwick, Jackson and Knight, 1977; Heth and Donald, 1978; van Laar and van Lill, 1978) and our own unpublished data for Scots pine (Whitehead) and Sitka spruce (Jarvis and Miller) are similar.

Conclusion

The conclusions are summarized by *Table 5.8*. The rate of photosynthesis by single needles of conifers is low compared with the rate in leaves of other C_3 species, probably because of a low intrinsic capacity of the electron transport and carboxylation processes. However, we know insufficient details about limitations to photosynthesis in conifers at the chloroplast level to be more specific.

Despite low rates of leaf photosynthesis, daily rates of CO_2 assimilation by conifer canopies are as large as by canopies of C_3 agricultural crops. The reason for this is probably the effective distribution of intermediate flux densities of radiation throughout the foliage, as a result of grouping of the needles, the small size of the needles, and the ability of the needles to utilize effectively low flux densities received on each surface.

Unfortunately, we have little factual information about the carbon balance of coniferous canopies. Despite possibly large rates of maintenance and growth respiration by the woody parts of the trees, maximum current annual increment of dry matter substantially exceeds that of most field crops even though coniferous tree crops are generally grown on very much poorer soils. The main reason for this is the retention throughout the year of a considerable leaf area, which results in a net carbon gain in almost every month of the year in temperate, oceanic climates. Only grassland, which also assimilates CO_2 throughout the year whenever the temperature allows, has comparable rates of current annual increment.

The *potential* current annual increment of forest crops probably greatly exceeds that of most agricultural crops, with the possible exception of grassland, in the UK. However, we know extremely little about the real growth potential of coniferous tree crops in our climate. To test the assertion that forest crops have enormous, unrealized growth potential we should set up some maximum production plots with optimization of fertilizer applica-

Table 5.8 RELATIVE EFFICIENCIES OF VARIOUS PARTIAL PHOTOSYNTHETIC PROCESSES IN TEMPERATE CONIFEROUS TREE CROPS AND TEMPERATE AGRICULTURAL CROPS. ALL FIGURES VERY APPROXIMATE

	Ratios of crop efficiencies, conifer/C_3 agricultural crop	
	Field crop	Forage crop
Photosynthetic unit size	1.6–3.6	
Electron transport	0.6	
Carboxylation process	0.6	
Photorespiration	1.0	
Boundary layer conductance	10.0	
Stomatal conductance	0.5	
Leaf photosynthesis		
unilateral illumination	0.4	
multilateral illumination	0.6	
Crop photosynthesis	1.0	
Retention after respiration	0.5	
Current total annual production		
actual	1–2	1.0
potential	2–3	1.0
Harvest index	1.3	0.8
Maximum mean annual yield		
actual	1.0	0.5
potential	0.8	0.4

tions and soil water in different parts of the country. Recent experience in the Swedish Coniferous Forest Project indicates that unexpectedly large increases in growth rate can be achieved if this is done (Ingestad and Linder, personal communication). This potential can only be fully realized in the field if forest crops are grown on much better soils than at present or are managed more intensively.

The harvest index for conifers seems to be high compared with crops, other than fodder crops, but we have totally inadequate information on how harvest index varies or what controls it in tree crops. The maximum mean annual increment of harvestable yield is approximately the same as that of a large number of widely grown agricultural crops, even though the latter are grown with intensive management and substantial inputs whereas the forest crops are grown with minimal inputs. It would seem, therefore, that there are enormous possibilities for improvement in the rate of yield production as a result of both more intensive management and manipulation of the harvest index.

Finally, we must conclude that we are remarkably ignorant about many of the factors which determine yield in coniferous crops. We know far less about yield determination in coniferous forest crops than in many far less important agricultural crops. What we need is a comparable crop approach to the production of wood in the UK.

Acknowledgements

I should like to thank Dr C. E. Jeffree for providing me with figures on the production of agricultural crops.

References

ALBERTE, R. S., MCCLURE, P. R. and THORNBER, J. P. (1976). *Pl. Physiol.*, **58**, 341–344

ALBREKTSSON, A. (1978). Swedish Coniferous Forest Project Internal Report No. 73

ANDREWS, T. J., BADGER, M. R. and LORIMER, G. H. (1975). *Arch. Biochem. Biophys.*, **171**, 93–103

BAHR, J. T. and JENSEN, R. G. (1974). *Pl. Physiol.*, **53**, 39–44

BEADLE, C. L. (1977). PhD Thesis, University of Aberdeen

BEADLE, C. L. and JARVIS, P. G. (1977). *Physiol. Plant.*, **41**, 7–13

BENECKE, U. (1980). *Oecologia (Berl.)*, **44**, 197–198

BENNETT, K. J. and ROOK, D. A. (1978). *Austr. J. Pl. Physiol.*, **5**, 201–230

BISCOE, P. V., SCOTT, R. K. and MONTEITH, J. L. (1975). *J. appl. Ecol.*, **12**, 269–293

BJÖRKMAN, O. (1973). *Photophysiol.*, **8**, 1–63

BJÖRKMAN, O. and GAUHL, E. (1969). *Planta*, **88**, 197–203

BOARDMAN, N. K. (1977). *Ann. Rev. Pl. Physiol.*, **23**, 355–377

BOGDANOVIĆ, M. (1973). *Physiol. Plant.*, **29**, 17–18

BRADBURY, I. K. and MALCOLM, D. C. (1978). *Can. J. For. Res.*, **8**, 207–213

BRIX, H. (1968). *Plant Physiol.*, **43**, 389–393

CALDWELL, M. M. (1970). *Cbl. ges. Forstwesen.*, **87**, 193–201

CHRISTIE, J. M. and LINES, R. (1979). *For. Ecol. Man.*, **2**, 75–102

CLARK, J. B. and LISTER, G. R. (1975). *Pl. Physiol.*, **55**, 401–406

COOPER, J. T. (1975). In *Photosynthesis and Productivity in Different Environments*, pp. 593–621. Ed. by J. B. Cooper. Cambridge University Press, Cambridge

CORNIC, G. and JARVIS, P. G. (1972). *Photosynthetica*, **6**, 225–239

DENMEAD, O. T. (1969). *Agric. Meteorol.*, **6**, 357–372

DE WIT, C. T. (1965). *Agric. Res.*, Report No. 663, pp. 1–57. Wageningen Centre of Agriculture Publications and Documentation

DE YOE, D. R. and BROWN, G. N. (1977). *Can. J. Bot.*, **18**, 2399–2407

DONALD, C. M. and HAMBLIN, J. (1976). *Adv. Agron.*, **28**, 361–405

DYKSTRA, G. F. (1974). *Can. J. For. Res.*, **4**, 201–206

EADIE, J. and CUNNINGHAM, I. M. M. (1971). In *Potential Crop Production*, pp. 239–249. Ed. by P. F. Wareing and J. P. Cooper. Heinemann, London

FARINEAU, J. and POPOVIC, R. (1975). *C. R. Acad. Sci. Paris D*, **281**, 1317–1320

FERRELL, W. K. (1970). Paper presented at the *First North American Forest Biology Workshop*, East Lansing, Michigan, 5–7 August

FORESTRY COMMISSION (1977). *The Wood Production Outlook in Britain*

FORESTRY COMMISSION (1980). *Forestry Facts and Figures 1979–80*

FORREST, W. G. and OVINGTON, J. D. (1970). *J. appl. Ecol.*, **7**, 177–186

FREELAND, R. O. (1952). *Pl. Physiol.*, **27**, 685–690

FRY, D. J. and PHILLIPS, I. D. J. (1976). *Physiol. Plant.*, **37**, 185–190

GABRIELSEN, E. K. (1948). *Physiol. Plant.*, **1**, 5–37

GATES, D. M., KEEGAN, H. J., SCHLETER, J. C. and WEIDNER, V. R. (1965). *Appl. Optics.*, **4**, 11–20

GEZÉLIUS, K. (1975). *Photosynth.*, **9**, 192–200

GIFFORD, R. M. (1974). *Austr. J. Pl. Physiol.*, **1**, 107–117

GORDON, J. C. and GATHERUM, P. R. (1969). *Bot. Gaz.*, **130**, 5–9

GORDON, J. C. and LARSON, P. R. (1968). *Pl. Physiol.*, **43**, 1617–1625

HAGEM, O. (1947). *Medd. Vest. Forst. Forskst.*, **8**, 1–315

HAGEM, O. (1962). *Medd. Vest. Forst. Forskst.*, **37**, 249–347

HAMILTON, G. J. (1975). Forestry Commission Booklet No. 39 HMSO, London

HAMILTON, G. J. and CHRISTIE, J. M. (1971). Forestry Commission Booklet No. 34. HMSO, London

HANLEY, D. P. (1976). *U/Idaho, Coll. of For. Wildl. and Range Sci. Bull.*, **14**, 1–15

HELMS, J. A. (1970). *Photosynth.*, **4**, 243–253

HETH, D. and DONALD, D. G. M. (1978). *S. Afr. For. J.*, **107**, 60–70

HOLMGREN, P., JARVIS, P. G. and JARVIS, M. S. (1965). *Physiol. Plant.*, **18**, 557–573

INOUE, Y., FURATA, S., OKU, T. and SHIBATA, K. (1976). *Biochim. Biophys. Acta.*, **449**, 357–367

JARVIS, P. G. (1971). In *Plant Photosynthetic Production—Manual of Methods*, pp. 566–631. Ed. by Z. Šesták, J. Čatský and P. G. Jarvis. Dr W. Junk, The Hague

JARVIS, P. G. and JAMES, G. B. (1981). *Pl., Cell Environ.*, **4**, in press

JARVIS, P. G., JAMES, G. B. and LANDSBERG, J. J. (1976). In *Vegetation and the Atmosphere*, Vol. 2, pp. 171–240. Ed. by J. L. Monteith. Academic Press, London

JEFFREE, C. E., JOHNSON, R. P. C. and JARVIS, P. G. (1971). *Planta*, **98**, 1–10

KESTEMONT, P. (1977). In *Productivité Biologique en Belgique*, pp. 177–189. Ed. by P. Duvigneaud and P. Kestemont. Duculot, Paris

KESTEMONT, P., DUVIGNEAUD, P. and PAULET, E. (1977). In *Productivité Biologique en Belgique*, pp. 161–176. Ed. by P. Duvigneaud and P. Kestemont. Duculot, Paris

KIRA, T. (1975). In *Photosynthesis and Productivity in Different Environments*, pp. 1–40. Ed. by J. P. Cooper. Cambridge University Press, Cambridge

KÖRNER, CH., SCHEEL, J. and BAUER, H. (1979). *Photosynth.*, **13**, 45–82

KRAMER, P. J. and CLARKE, W. S. (1947). *Pl. Physiol.*, **22**, 51–57

KRÓL, M. (1978). *Z. Pflanzenphysiol.*, **86**, 379–387

KU, S. B., GUTIERREZ, M. and EDWARDS, G. E. (1974). *Planta*, **119**, 267–278

KU, S. B., GUTIERREZ, M., KANAI, R. and EDWARDS, G. E. (1974). *Z. Pflanzenphysiol.*, **72**, 320–337

LANDSBERG, J. J., JARVIS, P. G. and SLATER, M. B. (1973). *Proceedings of the Uppsala Symposium, 1970, Ecology and Conservation*, Vol. 5, pp. 411–418. Ed. by R. O. Slatyer. UNESCO, Paris

LARCHER, W. (1969). *Photosynth.*, **3**, 167–198

LEAFE, E. L. (1972). In *Crop Processes and Controlled Environments*, pp. 175–184. Ed. by A. R. Rees, K. E. Cockshull, P. W. Hand and R. G. Hurd. Academic Press, London

LEMON, E. (1967). In *Harvesting the Sun*, pp. 263–290. Ed. by A. San Pietro, F. A. Greer and T. J. Army. Academic Press, New York

LEVERENZ, J. W. (1974). MSc Thesis, University of Washington

LEVERENZ, J. W. and JARVIS, P. G. (1979). *J. appl. Ecol.*, **16**, 919–932

LEVERENZ, J. W. and JARVIS, P. G. (1980). *J. appl. Ecol.*, **17**, 59–68

LEWANDOWSKA, M., HART, J. W. and JARVIS, P. G. (1976). *Physiol. Plant.*, **37**, 269–274

LEWANDOWSKA, M., HART, J. W. and JARVIS, P. G. (1977). *Physiol. Plant.*, **41**, 124–128

LEWANDOWSKA, M. and JARVIS, P. G. (1977). *New Phytol.*, **79**, 247–256
LEWANDOWSKA, M. and JARVIS, P. G. (1978). *Physiol. Plant.*, **42**, 277–282
LEWANDOWSKA, M. and ÖQUIST, G. (1980). *Physiol. Plant.*, **48**, 134–138
LINDER, S. (1972). *Stud. For. Suecica*, **100**, 1–37
LITTLE, C. H. A. and LOACH, K. (1973). *Can. J. Bot.*, **51**, 751–758
LOCKE, G. M. L. (1978). In *The Case for Regional Silviculture*, pp. 3–19. Ed. by J. Dewar. *Forestry*, **51**, 3–19
LOOMIS, R. S. and GERAKIS, P. A. (1975). In *Photosynthesis and Productivity in Different Environments*, pp. 145–172. Ed. by J. P. Cooper. Cambridge University Press, Cambridge
LUDLOW, M. M. and JARVIS, P. G. (1971). *J. appl. Ecol.*, **8**, 925–953
MADGWICK, H. A. I., JACKSON, D. S. and KNIGHT, P. J. (1977). *N. Z. J. For.*, **7**, 445–468
MALESZEWSKI, S. and LEWANTY, Z. (1972). *Z. Pflanzenphysiol.*, **67**, 305–310
MAFF (1974). *Agricultural Statistics 1972*. HMSO, London
MAFF (1976). *Output and Utilization of Farm Produce in the UK, 1968/69 to 1974/75*. HMSO, London
MARTIN, B., MÅRTENSSON, O. and ÖQUIST, G. (1978a). *Physiol. Plant.*, **43**, 297–305
MARTIN, B., MÅRTENSSON, O. and ÖQUIST, G. (1978b). *Physiol. Plant.*, **44**, 102–109
MARTIN, B. and ÖQUIST, G. (1979). *Physiol. Plant.*, **46**, 42–49
MICHEL-WOLWERTZ, M. R. and BRONCHART, B. (1974). *Pl. Sci. Lett.*, **2**, 45–54
MILLER, R. B. (1971). In *Productivity of Forest Ecosystems*, pp. 299–305. Ed. by P. Duvigneaud. UNESCO, Paris
MINDERMAN, G. (1967). *Pedobiol.*, **7**, 11–22
MONSI, M., UCHIJIMA, Z. and OIKAWA, T. (1973). *Ann. Rev. Ecol. System.*, **4**, 301–327
MONTEITH, J. L. (1969). In *Physiological Aspects of Crop Yield*, pp. 89–111. Ed. by J. D. Eastin, F. A. Haskins, C. Y. Sullivan and C. H. M. Van Bavel, American Society of Agronomy and Crop Science Society of America, Washington
MONTEITH, J. L. (1977). *Phil. Trans. R. Soc. Lond.*, **281**, 277–294
NEILSON, R. E. (1977). *Photosynth.*, **11**, 241–250
NIKOLIC, D. and BOGDANOVIĆ, M. (1972). *Protoplasma*, **75**, 205–213
NORMAN, J. M. and JARVIS, P. G. (1974). *J. appl. Ecol.*, **11**, 375–398
NORMAN, J. M. and JARVIS, P. G. (1975). *J. appl. Ecol.*, **12**, 834–879
OKU, T., INOUE, Y., SANADA, M., MATSUSHITA, K. and TOMITA, G. (1978). *Pl. Cell Physiol.*, **19**, 1–6
OKU, T., SUGAHARA, K. and TOMITA, G. (1974). *Pl. Cell Physiol.*, **13**, 175–178
OKU, T. and TOMITA, G. (1976). *Physiol. Plant.*, **38**, 181–185
ÖQUIST, G. and HELLGREN, N. O. (1976). *Pl. Sci. Lett.*, **7**, 359–369
ÖQUIST, G., MÅRTENSSON, O., MARTIN, B. and MALMBERG, G. (1978a). *Physiol. Plant.*, **44**, 187–192
ÖQUIST, G., MARTIN, B., MÅRTENSSON, O., CHRISTERSSON, L. and MALMBERG, G. (1978b). *Physiol. Plant.*, **44**, 300–306
OVINGTON, J. D. (1962). In *Advances in Ecological Research*, pp. 103–192. Ed. by J. B. Cragg. Academic Press, London
POSKUTA, J. (1968). *Physiol. Plant.*, **21**, 1129–1136

POSKUTA, J., NELSON, C. D. and KROTKOV, G. (1967). *Pl. Physiol.,* **42**, 1187–1190

POSKUTA, J. and OSTROWSKI, S. (1969). *Z. Pflanzenphysiol.,* **61**, 81

RHODES, I. (1971). *J. agric. Sci., Camb.,* **77**, 283–292

SALE, P. J. M. (1974). *Austr. J. Pl. Physiol.,* **1**, 283–296

SALE, P. J. M. (1975). *Austr. J. Pl. Physiol.,* **2**, 461–470

SCHULZE, E.-D., FUCHS, M. I. and FUCHS, M. (1977). *Oecologia,* **29**, 43–61

SENSER, M. and BECK, E. (1978). *Photosynth.,* **12**, 323–327

SENSER, M., SCHÖTZ, F. and BECK, E. (1975). *Planta,* **126**, 1–10

ŠESTÁK, Z., JARVIS, P. G. and ČATSKÝ, J. (1971). In *Plant Photosynthetic Production—Manual of Methods,* pp. 1–48, Ed. by Z. Šesták, J., Čatský and P. G. Jarvis, Dr W. Junk, N.V., The Hague

SOIKKELI, S. (1978). *Can. J. Bot.,* **56**, 1932–1940

SZANIAWSKI, R. K. and WIERZBICKI, B. (1978). *Photosynth.,* **12**, 412–417

SZEICZ, G. (1974). *J. appl. Ecol.,* **11**, 1117–1156

THOMPSON, R. and TAYLOR, H. (1979). *Sci. Hortic.,* **10**, 309–316

TROENG, E. and LINDER, S. (1978). Swedish Coniferous Forest Project, Internal Report, **83**, 1–20

VAN LAAR, A. and VAN LILL, W. S. (1978). *S. Afr. For. J.,* **107**, 71–76

WALKER, R. B., SCOTT, D. R. M., SALO, D. J. and REED, K. L. (1972). In *Proceedings of Research on Coniferous Forest Ecosystems,* pp. 211–226. Ed. by J. F. Franklin, L. J. Dempster and R. H. Waring. USDA, Portland, Oregon

WALLES, B. and HUDAK, J. (1975). *Stud. For. Suecica,* **127**, 5–22

WATTS, R. W. and NEILSON, R. E. (1978). *J. appl. Ecol.,* **15**, 245–255

WATTS, R. W., NEILSON, R. E. and JARVIS, P. G. (1976). *J. appl. Ecol.,* **13**, 623–638

WILL, G. M. (1966). *N.Z. For. Res. Notes,* **44**, 1–15

WOOD, G. B. (1974). *Austr. For. Res.,* **6**, 5–14

ZELAWSKI, W. (1967). *Bull. Acad. Sci. Pol.,* **15**, 565–569

ZELAWSKI, W. and GORAL, I. (1966). *Acta Soc. Bot. Pol.,* **35**, 587–598

ZELAWSKI, W. and KINELSKA, J. (1967). *Acta. Soc. Bot. Pol.,* **36**, 713–725

ZELAWSKI, W., KINELSKA, J. and LOTOCKI, A. (1968). *Acta. Soc. Bot. Pol.,* **38**, 505–518

ZELAWSKI, W., SZANIAWSKI, R., DYBCZYNSKI, W. and PIECHUROWSKI, A. (1973). *Photosynth.,* **7**, 351–357

6

PROCESSES LIMITING PHOTOSYNTHETIC CONDUCTANCE

JOHN A. RAVEN
SHEILA M. GLIDEWELL
Department of Biological Sciences, University of Dundee, UK

Introduction

The concept of photosynthetic conductance is a useful one in comparing rates of net photosynthesis in crop plants. It represents the increment in photosynthesis per unit increase in substrate (CO_2) concentration. When photosynthesis is measured in nmol (cm^2 area of one side of the leaf)$^{-1}$ s^{-1}, and the CO_2 concentration is measured in nmol cm^{-3}, the photosynthetic conductance has the units of cm s^{-1}. It is thus formally a conductance in the 'electrical analogue' of photosynthesis (Maskell, 1928; Penman and Schofeld, 1951; Heath, 1969; Jones, 1973). Here, the CO_2 concentration represents the driving force, and multiplication by the conductance yields the flux. For a certain CO_2 concentration (or CO_2 concentration change), a larger photosynthetic conductance means a higher rate of photosynthesis (or a larger increment in photosynthesis). It is our purpose in this paper to use the photosynthetic conductance as a way of analysing the differences in the rates of net photosynthesis which can be attained by crop plants under optimal conditions, and then to analyse the effects of changes from these optimal conditions.

Pursuing the concept of photosynthetic conductance in more detail, the relationship mentioned above is:

net flux = conductance × driving force
nmol CO_2 (cm^2 external leaf area)$^{-1}$ s^{-1} = cm s^{-1} × nmol cm^{-3}

$$(6.1)$$

A more readily manipulated form of the equation uses the reciprocal of the conductance (s cm^{-1}, which is a resistance):

nmol CO_2 (cm^2 external leaf area)$^{-1}$ s^{-1} = (s cm^{-1})$^{-1}$ × nmol cm^{-3}

$$(6.2)$$

The driving force in equation (6.2), when we are considering overall photosynthesis, is the CO_2 concentration in bulk air *minus* the CO_2 concentration after all the transport and chemical reactions of photosynthesis have taken place. In C_3 plants this is generally taken to be the CO_2 concentration at the CO_2 compensation point. Thus for a typical C_3 crop plant, fixing CO_2 under optimal conditions (growth under optimal temperature and irradiance, adequate nutrients and water; photosynthesis measured in a sun leaf under optimal light and temperature conditions in normal air) at 2 nmol (cm^2 area of one side of a leaf)$^{-1}$s^{-1} (Gifford, 1974) and with a bulk CO_2 concentration of 10 nmol cm^{-3} and a CO_2 compensation point of 1 nmol cm^{-3}, the photosynthetic conductance is 0.222 cm s^{-1}, i.e. the resistance is 4.5 s cm^{-1}. In this paper we shall be chiefly concerned with the limitations on photosynthesis which are associated with processes within the leaf cell rather than those which occur in the gas phase transport path from the bulk air to the mesophyll cell wall. The resistance which is associated with gas phase transport can be estimated from the rate of transpiration (Maskell, 1928; Penman and Schofield, 1951; Heath, 1969; Mansfield, this volume). For our C_3 crop plant under optimal conditions for net photosynthesis, the value of this gas phase transport resistance is some 1 s cm^{-1} (Gifford, 1974). The total resistance is the gas phase resistance plus the resistance associated with events in the mesophyll cells: thus the 'mesophyll resistance' in this case is 4.5−1.0 or 3.5 s cm^{-1}. The other name for mesophyll resistance is residual resistance, because it is obtained as the difference between two measured quantities. It is important to emphasize that the gas phase transport resistance is variable: 1 s cm^{-1} is its minimal value, and it can be increased by a decrease in stomatal aperture and by decreased convective mixing in the air outside the leaf. Stomatal aperture is under direct control by the plant, which can also control the value of convective mixing by means of its architecture. Granted the minimal value of 1 s cm^{-1}, the effect of a certain fractional increase in the gas phase resistance decreases the rate of photosynthesis by 1/3.5 of the amount that the same *fractional* increase in residual resistance would produce. Thus, under our standard conditions, the residual resistance must be regarded as the major limitation on the rate of photosynthesis.

Using a version of equation (6.2) we can estimate from the net CO_2 flux and the gas phase transport resistance, the CO_2 concentration difference between the bulk air and the mesophyll cell wall. This value is 8 nmol cm^{-3} if the flux is 2 nmol cm^{-2} s^{-1}, the external concentration is 10 nmol cm^{-3}, and the resistance is 1 s cm^{-1}. The next three sections will attempt to account for genetic differences in the value of the residual resistance (3.5 s cm^{-1} in the case discussed above, corresponding to a mesophyll (residual) conductance of 0.29 cm s^{-1}). The two components of the residual resistance are the transport of inorganic carbon to the carboxylation enzyme (ribulose-1,5-bisphosphate carboxylase-oxygenase, RuBPc-o, EC41139), and the activity of this enzyme and subsequent chemical processes. Analysis of these resistances is best conducted in terms of the *internal* leaf area exposed to the intercellular gas spaces, which can itself account for some of the variation in the residual resistance, the mesophyll conductance of the mesophyll increases in parallel with the internal leaf area, when both are expressed in terms of the external leaf area.

Transport of Carbon Dioxide and Oxygen in the Liquid Phase

We are dealing here with the movement of CO_2 and oxygen between the outside of the mesophyll cell wall and their site of consumption (CO_2) or production (O_2) in the chloroplast. The path traversed for CO_2 is, sequentially, the aqueous phase of the cell wall, the plasmalemma, some ground cytoplasm, the outer chloroplast envelope membrane, the inter-membrane space, the inner chloroplast envelope membrane, and the stroma up to the RuBPc molecule which assimilates the CO_2. For oxygen the same pathway is traversed in reverse, except that the oxygen is produced within the thylakoid so it must traverse a thylakoid membrane before reaching the stroma (*Table 6.1*).

As was pointed out in the introduction, the analysis is best carried out in terms of the area of mesophyll cells: the liquid phase flux takes place at right angles to this gas-water interface at the mesophyll cell surface. The common value for the ratio of intercellular area (A_{mes}) to external area (A) in our C_3 crop plants is perhaps 20 (Turrell, 1936, 1944; El-Sharkaway and Hesketh, 1965; Nobel, 1974). The net CO_2 flux of 2 nmol $(cm^2\ A)^{-1}\ s^{-1}$ thus becomes 100 pmol $(cm^2\ A_{mes})^{-1}\ s^{-1}$; the total resistance on this basis is 90 s cm^{-1} and the residual resistance is 70 s cm^{-1}, while the photosynthetic conduct-ances are 0.011 cm s^{-1} (total) and 0.014 cm s^{-1} (residual).

In order to analyse the fraction of the 70 s cm^{-1} of the residual resistance which can be attributed to the transport process in the liquid phase, we can examine the path length (Rackham, 1966; Woolhouse, 1968; Raven, 1970, 1977a; Nobel, 1974; Ticha and Catsky, 1977) and the composition of the various parts of the pathway enumerated above. At a given temperature, the composition allows us to estimate the solubility of the gas in these various parts of the pathway and the diffusivity of the gas in that part of the pathway. For 25°C, the concentration of CO_2 in the aqueous portions of the pathway in equilibrium with air is very similar to that in air (10 nmol cm^{-3}), although the concentration in the lipid phase of the membranes is different; we can correct for this by 'adjusting' the diffusivity in the lipid material to take into account the partition coefficient between water and lipid. This gives us effective diffusion coefficients for the aqueous and the lipid portions of the pathway in terms of the concentration in the aqueous phase. The diffusivities, D, obtained in this way are related to the conductance or permeability coefficient (P) for that part of the pathway by the relationship:

$$P\ (cm\ s^{-1}) = D\ (cm^2\ s^{-1})/l\ (cm), \tag{6.3}$$

where l is the length of that portion of the pathway. Values of the resistance (s cm^{-1}: l/P) for CO_2 in these various portions of the pathway are given in *Table 6.1*, on the assumption that only CO_2 is involved in the transport processes throughout this pathway. A great complication here is the interconversion of CO_2 and $H_2CO_3/HCO_3^-/CO_3^{2-}$. The discussion in the Appendix shows that equilibration of CO_2 and HCO_3^- cannot enhance inorganic carbon transport (by increasing the effective concentration of diffusing species) except in the stroma. Even here, the participation of HCO_3^-, the carbonic anhydrase-inorganic carbon complexes, and inorganic carbon bound to the relatively slow-acting RuBPc-o probably only increase the partial pressure of carbon dioxide ($P\ co_2$) by some 2–3-fold (*Table 6.1*). The total transport resistance in the mesophyll is thus some 10–15 s cm^{-1}.

The other important photosynthetic gas is oxygen. For diffusion through cell membranes and aqueous phases, in the absence of facilitation by chemical combination, it is likely that the resistance for oxygen transport is similar to that for CO_2 (Samish, 1975; Raven, 1977a; Berry and Farquhar, 1978). While work on isolated chloroplasts has been interpreted by Steiger,

Table 6.1 RESISTANCES OF PORTIONS OF THE LIQUID PHASE TRANSPORT PATHWAY FOR CO_2

Segment of pathway	Length (μm)	Composition	Diffusing species	Resistance ($s\ cm^{-1}$) on A_{mes} basis	References
Cell wall	0.1	Polysaccharide, water	CO_2	~2	Rackham (1966); Raven (1970); Nobel (1974); Appendix
Plasmalemma	0.01	Phospholipid, protein, cholesterol	CO_2	~1	Raven (1970, 1977a); Nobel (1974); Gutknecht, Bisson and Tosteson (1977); Kaethner and Bangham (1977)
Ground cytoplasm	0.1	Water, protein	CO_2	<1	Raven (1970, 1977a); Nobel (1974)
Outer membrane of chloroplast envelope	0.01	Galactolipid, protein, phospholipid	CO_2	<1	Raven (1970, 1977a); Heldt and Sauer (1971); MacKendor and Leech (1971); Werdan and Heldt (1972); Douce, Moltz and Benson (1973); Nobel (1974); Gutknecht, Bisson and Tosteson (1977); Shiraiwa and Miyachi (1978)
Intermembrane space	0.01	Water, protein	CO_2	<1	
Inner membrane of chloroplast envelope	0.01	Galactolipid, protein, phospholipid	CO_2	<1	
Stroma	1.0	Water, protein	CO_2 (CO_2, HCO_3^-, carbonic anhydrase and RuBPc-o inorganic carbon complexes)	~10–15 } ~(3–5)	Appendix
Total pathway (assuming some facilitated transport in the stroma)				~10–15	

The analysis presented here assumes CO_2 transport from a planar source (the outer surface of the mesophyll cell walls) to a parallel array of sinks (the RuBPc-o molecules), whereas the real transport pathway is closer to a radial path from the surface towards the centre of mesophyll cells, which can be approximated to spheres or cylinders with hemispherical ends. However, the error introduced by the planar assumption is small, since in the highly vacuolate mesophyll cells of C_3 plants the path length from the outer surface of the cell wall to the RuBPc-o molecules is a small fraction of the cell radius (Jones and Slatyer, 1972; Raven, 1977b).

Beck and Beck (1977) as indicating a very low oxygen permeability, especially at low ambient oxygen, it seems that the data can be better explained in terms of the kinetics of catalase activity (Grodzinski, 1980). Accepting an effective permeability to oxygen in the mesophyll, equivalent to that for 'uncatalysed' CO_2, the total resistance in the mesophyll for oxygen is probably some 20 s cm^{-1} (compared with the estimate for facilitated CO_2 diffusion of $10\text{–}15 \text{ s cm}^{-1}$).

From the point of view of the influence of these diffusion resistances for CO_2 and oxygen for the functioning of RuBPc-o (*see* next section) we need to know the concentration of the CO_2 and oxygen at the site of RuBPc-o. In the introduction we suggested that the CO_2 concentration in the gas phase at the mesophyll cell wall is 8 nmol cm^{-3}; at $25°C$, where equilibrium CO_2 concentrations in the gas and aqueous phases are equal, the resistance (on an A_{mes} basis) for inorganic carbon transport means a CO_2 concentration at the site of RuBPc-o of some 6.5 nmol cm^{-3}. For oxygen, the very much higher concentration of oxygen in the gas phase (some $10.2 \text{ μmol cm}^{-3}$) means that the oxygen concentration in the gas phase at the mesophyll cell wall is not significantly different from the bulk gas phase concentration (granted a flux of $2 \text{ nmol } (cm^2 \ A)^{-1} \text{ s}^{-1}$, i.e. a photosynthetic quotient of 1, and a gas phase resistance of 1 s cm^{-1}). At $25°C$, the equilibrium oxygen concentration in the mesophyll cell walls is some 250 nmol cm^{-3}; with a liquid-phase diffusion resistance (for fluxes on an A_{mes} basis) of 20 s cm^{-1} and a flux of 0.1 nmol $(cm^2 \ A_{mes})^{-1} \text{ s}^{-1}$, the oxygen concentration in the centre of the chloroplast is some 252 nmol cm^{-3}. As we shall see, the decrement in CO_2 concentration has much more significance for RuBPc-o activity than does the increment in oxygen concentration.

Biochemistry of Net Carbon Dioxide Fixation: Ribulose-1,5-bisphosphate Carboxylase Oxygenase (RuBPc-o)

This enzyme (EC41139) catalyses both carboxylase (equation (6.4)) and oxygenase (equation (6.5)) activities:

$$RuBP + CO_2 + H_2O \xrightarrow{\text{RuBPc}} 2PGA \qquad (6.4)$$

$$RuBP + O_2 \xrightarrow{\text{RuBPo}} PGA + phosphoglycolate \qquad (6.5)$$

where PGA is phosphoglyceric acid. It is almost certain that reactions (6.4) and (6.5) are alternative and competitive activities of a single enzyme (Jensen and Bahr, 1977; McCurry *et al.*, 1978; cf. Bränden, 1978). The scheme in *Figure 6.1* depicts RuBPc-o as being responsible for all the light-dependent CO_2 fixation (via the reaction of equation (6.4) as part of the PCRC) and light-dependent CO_2 production (via the reactions of the PCOC whose substrate, glycolate, is derived from the phosphoglycolate generated in reaction (6.5)). Before considering the evidence that the *in vivo* activity of RuBPc-o is a major limitation on net photosynthesis in air under our optimal conditions, and indeed is responsible for much of the mesophyll resistance, we must consider the *in vivo* evidence for the adequacy of the scheme in *Figure 6.1* to describe the metabolism of C_3 crop plants.

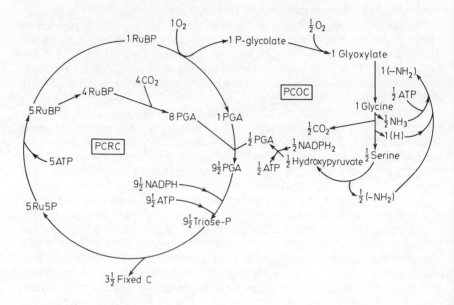

Figure 6.1　*The photosynthetic carbon reduction cycle (PCRC) and the photorespiratory carbon dioxide cycle (PCOC) of C_3 plants. The stoichiometry indicated is for a RuBPc/RuBPo ratio of 4. The energetics of the PCOC involves the glutamine synthetase-glutamate synthetase pathway for the scavenging of NH_3; the stoichiometry of reductant transfer in this reaction does not imply that the reductant generated in the glycine-to-serine conversion is used in the glutamate synthetase reaction. (From Keys et al., 1978)*

The tracer evidence for RuBPc-o as the major pathway for CO_2 fixation in C_3 leaves under the conditions specified in the introduction (Black, 1973; Benedict, 1978) is very strong. However, there is also a significant fixation of inorganic carbon (with HCO_3^- as the immediate substrate) by phosphoenolpyruvate carboxylase (PEPc, EC41131) in illuminated C_3 plant leaves (Raven, 1970). While the anaplerotic role of this fixation in the biosynthetic operation of the tricarboxylic acid cycle (TCAC) has been stressed (Raven, 1970), it is clear that considerations of pH regulation limit the *net* fixation by this enzyme in shoots under certain conditions of nitrogen nutrition (Raven and Smith, 1976; Raven, 1977b, 1977c). Carbamyl phosphate synthetase probably has a small role in terms of total CO_2 fixation (Raven, 1970), while the status of a possible direct conversion of CO_2 to formate remains to be elucidated under optimal photosynthetic conditions (Kent, 1972, 1977; Kent, Pinkerton and Strobel, 1974).

For glycolate synthesis there is good ^{14}C evidence for the origin of glycolate from intermediates of the PCRC (Agrawal and Fock, 1978; Andrews and Lorimer, 1978), and for the operation of the PCOC as shown in *Figure 6.1* (cf. Gerster, Dimon and Peyberres, 1974). The mechanism of reaction (6.5) incorporates one atom of oxygen from O_2 into the carboxyl group of each phosphoglycolate molecule, so the use of [18]O_2 permits a further test of the quantitative importance of RuBPo in glycolate synthesis. Studies with [18]O_2, in which ^{18}O incorporation into glycolate in short-term experiments equals the rate of glycolate turnover, and in which the steady-state enrichment of ^{18}O in the carboxyl group of glycolate (and of the acids of the PCOC) equals that in the supplied $^{18}O_2$, would rule out significant contributions to glycolate synthesis from reactions which did not incorporate ^{18}O from $^{18}O_2$. No process other than reaction (6.5) has so far been described which gives ^{18}O incorporation.

Such ^{18}O-labelling experiments have been carried out (Andrews, Lorimer and Tolbert, 1971; Gerster and Tournier, 1977; Lorimer, Krause and Berry, 1977; Andrews and Lorimer, 1978; Lorimer *et al.*, 1978) on leaves and chloroplasts of C_3 plants (*see* page 136, Note 1). The initial rate and final extent of incorporation in these experiments was consistent with at least 0.6–0.8 of the glycolate arising from reaction (6.5). The lack of complete labelling was generally attributed to the presence of storage pools of the PCOC intermediates, and to dilution of $^{18}O_2$ by photosynthetic $^{16}O_2$. The experiments of Gerster and coworkers and Lorimer and coworkers were carried out either close to the CO_2 compensation point or at CO_2 saturation. We have recently carried out similar experiments on *Spinacia oleracea* leaves (Glidewell and Raven, 1979) using an apparatus (Jones and Milburn, 1974) which keeps the CO_2 concentration constant at close to the air level during the course of $^{18}O_2$ (or $^{13}CO_2$) feeding. *Table 6.2* shows some of the results we obtained. The ^{18}O enrichment (atom O incorporated from atmospheric O_2 per molecule of PCOC intermediate) is expressed in two ways. One involves the *maximum* ^{18}O enrichment seen per molecule of intermediate based on the total extracted intermediates; the other involves the *mean* ^{18}O enrichment per molecule of intermediate in the 'metabolic' pool. The metabolic pool is defined on the basis of the fractional enrichment of ^{13}C per atom of C in the intermediates in experiments in which $^{13}CO_2$ was supplied in experiments otherwise similar to those in which $^{18}O_2$ was provided.

The data in *Table 6.2* show that the directly measured oxygen enrichment

Table 6.2 INCORPORATION OF ^{18}O FROM $^{18}O_2$ INTO INTER-MEDIATES OF THE PHOTORESPIRATORY CARBON DIOXIDE CYCLE (PCOC) IN *SPINACIA OLERACEA* LEAVES*

PCOC intermediate	^{18}O enrichment in metabolic pool as a fraction of supplied O_2 (mean)**	Maximum ^{18}O enrichment of total leaf pool as a fraction of supplied O_2	Maximum fraction of total leaf pool which is in metabolic pool as defined by ^{13}C labelling
Glycolate	0.35	0.48	0.38
Glycine	0.33	0.46	0.95
Serine†	(0.27)	0.36	(0.95)
Glycerate	0.18	0.22	0.74

*Experiments were carried out in saturating light on detached leaves in a gas phase of 21–24 per cent oxygen and 345–430 ppm CO_2; $^{18}O_2$ (^{18}O enrichment 40–70 per cent) *or* $^{13}CO_2$ (^{13}C enrichment 15–18 per cent) was supplied for 20–40 minutes. Techniques followed Andrews, Lorimer and Tolbert (1971) for glycine, serine and glycerate; and Mahon, Egle and Fock (1975) for glycolate.
**The 'metabolic pool' as a fraction of the total leaf content of intermediates was estimated from the ^{13}C enrichment (per atom) of the intermediates relative to the ^{13}C enrichment in the supplied CO_2. It was assumed that the length of the $^{13}CO_2$ ($^{18}O_2$) feeding period was sufficient to give steady-state labelling of the PCRC and the PCOC intermediates.
†^{13}C incorporation into serine could not be accurately measured; the measured ^{18}O incorporation was used, together with the ^{13}C enrichment in glycine, for the estimation of the ^{18}O enrichment in the metabolic pool.

in glycolate and other PCOC intermediates does not exceed one half. The lack of the fractional labelling of one, which would be expected if all of the glycolate were produced by RuBPo, could be partly due to dilution of $^{18}O_2$ by respiratory $^{16}O_2$ to a greater extent than that to which the $^{13}CO_2$ is diluted by (photo-)respiratory $^{12}CO_2$. This is likely to be the case (Agrawal and Fock, 1978; cf. Jackson and Volk, 1970), although the arguments used in the previous section suggests that such dilution of $^{18}O_2$ is unlikely to account for such a large discrepancy. An additional possibility is tracer carbon (but not tracer oxygen) flux into the PCOC from the PCRC via the phosphoserine pathway (Daley and Bidwell, 1977; cf. Chapman and Leech, 1976). The results of the experiments with isotopes of carbon and oxygen suggest that RuBPo (equation (6.5)) is the major pathway for glycolate synthesis in C_3 plants, while the PCOC (*Figure 6.1*) is the major source of photorespiratory CO_2 (cf. Glidewell and Raven, 1976; Grodzinski, 1978). Accepting that *Figure 6.1* does represent the pathways for CO_2 fixation and (with 'dark' respiratory pathways) CO_2 production in the light, we attempt to answer another quantitative question: is the activity of RuBPc-o, on the basis of A, a major determinant of the net photosynthesis (P_{net}) measured on the same basis? If the activity measured *in vitro* is substantially below that required to account for the *in vivo* rates of gas exchange we must look for additional reactions to catalyse the pathways of *Figure 6.1*. If, on the other hand, the activity *in vitro* considerably exceeds that required to account for the *in vivo* rates, then clearly the *capacity* of RuBPc-o *in vivo* is unlikely to be the major determinant of P_{net} under our standard conditions.

Semiquantitative data, which suggest that the catalytic capacity of RuBPc-o may indeed be a limiting factor in determining P_{net}, come from experiments in which different genotypes of plants and the same genotype under different cultivation conditions, have *in vivo* P_{net} and *in vitro* RuBPc activities in the same rank order (for reviews *see* Treharne, 1972; Björkman, 1973; Taylor and Pearcy, 1976). However, in much of this work it is not possible to make a quantitative comparison of P_{net} *in vivo* and RuBPc-o *in vitro* because RuBPo activity was not measured, and the RuBPc activity measured may have been an underestimate due to the incomplete activation of the enzyme before assay (Lorimer, Badger and Andrews, 1977).

The basis of such comparisons is explained in a number of papers (Badger and Andrews, 1974; Laing, Ogren and Hagemann, 1974; Peisker, 1974; Berry and Farquhar, 1978; Lorimer *et al.*, 1978). Net CO_2 fixation (P_{net}) is given by:

$$P_{net} = v_c - \tfrac{1}{2}v_o \tag{6.6}$$

where v_c and v_o are respectively the *in vivo* activities of RuBPc and RuBPo, measured in the same units as P_{net} (i.e. nmol $(cm^2\ A)^{-1}s^{-1}$); the factor of $\tfrac{1}{2}$ is a product of the stoichiometry of oxygen uptake and CO_2 release in the PCOC (*Figure 6.1*). The *in vivo* values of v_c and v_o are computed from equations (6.7) and (6.8), respectively:

$$v_c = \frac{V_c \cdot [CO_2]}{[CO_2] + K_{1/2}co_2 \cdot (1 + [O_2]/K_i o_2)} \tag{6.7}$$

$$v_o = \frac{V_o \cdot O_2}{[O_2] + K_{1/2}o_2 \cdot (1 + [CO_2]/K_i co_2)} \tag{6.8}$$

where: V_c and V_o are respectively the maximum *in vitro* catalytic activities of the RuBPc and RuBPo activities of RuBPc-o (nmol $(cm^2\ A)^{-1}\ s^{-1}$); $[CO_2]$ and $[O_2]$ are the concentrations of the respective gases at the *in vivo* site of RuBPc-o activity (nmol cm^{-3}); and $K_{1/2}$ and K_i are respectively the half-saturation (Michaelis–Menten) and inhibitor constants for the specified gases (nmol cm^{-3}).

There is, alas, only a very limited number of cases for which all the data required for this comparison are available; even then the CO_2 and the oxygen concentrations used in equations (6.7) and (6.8) are those in the bulk gas phase or, at best, in the intercellular air spaces (Lorimer *et al.*, 1978) or in the medium which bathes isolated chloroplasts (Lilley and Walker, 1975). *Table 6.3* shows the comparisons which are possible, based on measurements of *in vivo* P_{net} and values of V_o, V_c, $K_{1/2}CO_2$, and K_iO_2 for RuBPc-o *in vitro* on the same organism. It will be seen that the ratio of the rate predicted from RuBPc-o characteristics to those measured is between 0.47 and 1.5. These ratios would be decreased if the CO_2 and oxygen concentrations at the site of RuBPc-o activity were used rather than the values outside the chloroplasts, leaf cells or leaves, which were actually used.

The limited data available suggest that the activity of RuBPc-o *in vitro* is barely sufficient to account for the *in vivo* rate of P_{net}. In view of the very strong isotopic evidence for RuBPc activity, which accounts for at least 90 per cent of gross photosynthetic CO_2 fixation, and RuBPo activity, which accounts for at least 60 per cent of glycolate synthesis and hence CO_2 evolution by the PCOC, it appears that we must conclude that the RuBPc-o activity in C_3 plant leaves is barely sufficient to account for P_{net} under our standard conditions, and is very likely to be the rate-limiting biochemical process for P_{net} under these conditions. This is supported by measurements of the activities of other enzymes of the PCOC (Treharne, 1972) and of the reactions of adenosine triphosphate (ATP) and reduced nicotinamide adenine dinucleotide phosphate ($NADPH_2$) generation in the 'light reactions' (Lilley and Walker, 1975; Baker, Heath and Leech, 1978). Under CO_2 saturated conditions it may well be that the light reactions are limiting (Grahl and Wild, 1972; Lilley and Walker, 1975; Collatz, 1978). The apparent overcapacity of RuBPc-o activity at CO_2 saturation is related to the provision of adequate catalytic capacity at normal CO_2 levels.

Further data consistent with the view that the limited activity of RuBPc-o is a major component of the mesophyll resistance come from measurements (with labelled CO_2) of P_{gross}, P_{net} and CO_2 efflux in *Helianthus annuus* (Ludwig and Canvin, 1971a, 1971b) and *Triticum aestivum* (Thomas, Hall and Merrett, 1978). In the case of the latter, there are also data on V_c and on RuBPo activity in 21 per cent oxygen in extracts. With 'promiscuous' use of the $K_{1/2}$ and K_i measurements for CO_2 and O_2 from *Spinacia* (Badger and Andrews, 1974) and *Atriplex* (Badger and Collatz, 1977) we can compare predicted with measured (in parentheses) values for P_{gross} of 1.15–1.54 (1.6), P_{net} 0.89–1.44 (1.2) and photorespiratory CO_2 efflux 0.1–0.17 (0.4), with all rates in nmol $(cm^2\ A)^{-1}\ s^{-1}$. The agreement is poorest for photorespiration: the RuBPo/RuBPc activity which can be deduced from the data of Thomas, Hall and Merrett (1978) is lower than that found by most other recent investigators.

Table 6.3 SUMMARY OF RATES OF PHOTOSYNTHESIS PREDICTED FROM A CONSIDERATION OF THE KINETICS OF RuBPc-o WITH THE MEASURED RATE

Organism	Parameter, units	Rate		Ratio Predicted/measured	References
		Predicted	Measured		
Glycine max leaves	P_{net}, nmol cm^{-2} s^{-1}	1.0	1.92	0.52	Laing, Ogren and Hagemann (1974); Keck and Ogren (1976); Servaites and Ogren (1977)
Atriplex patula ssp. hastata leaves	P_{net}, µmol (mg chl.h)$^{-1}$ (based on ambient CO$_2$)	89	73	1.22	Lorimer et al. (1978)
	(based on internal gas space CO$_2$)	56	73	0.77	
Atriplex patula ssp. glabriuscula leaves	P_{net}, nmol cm^{-2} s^{-1}	0.94	2	0.47	Björkman et al. (1972); Björkman, Mooney and Ehleringer (1975); Björkman, Boynton and Berry (1976); Badger and Collatz (1977)
Spinacia oleracea leaves	P_{net}, µmol (mg chl.h)$^{-1}$	66	60–80	0.83–1.1	Badger and Andrews (1974); Lilley and Walker (1975)
Spinacia oleracea chloroplasts	v_c, µmol (mg chl.h)$^{-1}$	60	40	1.50	Lilley and Walker (1975)

All measurements at 25°C, except Spinacia oleracea chloroplasts at 20°C; all leaf areas are A rather than A_{mes}.

Further 'promiscuous' use of RuBPc-o data and tracer CO_2 fluxes permits the conclusion that the measured P_{gross}, P_{net} and photorespiration rates can be accounted for in terms of predicted relative v_c and v_o values (from equations (6.6)–(6.8)): this is well exemplified by calculations based on the *in vivo* measurements on *Helianthus annuus* (Ludwig and Canvin, 1971a, 1971b) in conjunction with *in vitro* RuBPc-o data on *Spinacia oleracea* (*Table 6.4*; Badger and Andrews, 1974). However, it must be borne in mind that this agreement does not necessarily mean that the total catalytic activity of RuBPc-o is a major limitation on the rate of CO_2 metabolism, since any restriction of RuBPc-o activity to lower values tends to maintain the ratio of RuBPc activity to RuBPo activity constant (Andrews and Lorimer, 1978). The same considerations apply to the agreement between prediction and measurement when RuBPc-o activity is used to predict the CO_2 compensation point (Laing, Ogren and Hagemann, 1974).

Table 6.4 MEASURED CO_2 FLUXES IN GROSS AND NET PHOTOSYNTHESIS, AND IN PHOTORESPIRATION, COMPARED WITH PREDICTED RATES FROM RuBPc-o KINETICS

Gas phase		Gross photosynthesis		Net photosynthesis		Photorespiratory CO_2 efflux	
$\%O_2$	ppm CO_2	(measured)	(v_c)	(measured)	$(v_c - {}_{1/2}v_o)$	(measured)	(v_o)
21	53	0.447	0.422	0.069	0.006	0.378	0.416
21	150	1.096	1.052	0.750	0.674	0.346	0.378
21	293	1.966	1.783	1.619	1.443	0.347	0.340
1	53	0.542	0.460	0.540	0.426	0.038	0.034
1	150	1.418	1.134	1.537	1.107	(−0.119)	0.027
1	285	2.577	2.577	2.747	2.553	(−0.170)	0.024

Measured rates for *Helianthus annuus* (Ludwig and Canvin, 1971a, 1971b); RuBPc-o kinetics for *Spinacia oleracea* (Badger and Andrews, 1974). The values of v_c and v_o for *Spinacia* were calculated for RuBPc-o using the kinetics from Badger and Andrews on the basis of rates per mg protein, and 'normalized' to the value of gross photosynthesis in *Helianthus* in 1 per cent oxygen/0.0285 per CO_2 in the gas phase. The computed values of v_c and v_o involve the conversion of gas phase concentrations to aqueous phase equilibrium concentrations at 25°C.

A final line of evidence which is compatible with a major contribution of RuBPc-o activity to mesophyll resistance, comes from work on isotopic discrimination. The ${}^{13}C/{}^{12}C$ ratio in C_3 land plants is rather higher than that found for RuBPc-o *in vitro*, but is much lower than the ratio in atmospheric CO_2 (Estep *et al.*, 1978). This suggests that RuBPc-o *in vivo* can exhibit much of its innate preference for ${}^{12}CO_2$ over ${}^{13}CO_2$, and is not restricted to the near-atmospheric ratio of ${}^{13}C$ to ${}^{12}C$ which it would encounter if net fixation were strictly diffusion limited (*see* page 136, Note 2).

The evidence we have reviewed in the foregoing is consistent with a major role for the biochemical resistance (particularly restrictions on RuBPc-o activity) in the overall mesophyll resistance. This contrasts with conclusions from other analyses (Chartier, Chartier and Catsky, 1970; Jones and Slatyer, 1972; Prioul, Reyss and Chartier, 1975) where a major contribution for the transport resistance to the overall mesophyll resistance has been suggested. While we would claim that the sort of model we propose here is based on more direct data on transport resistances and enzyme activities than the models which predict a high transport resistance component of the mesophyll resistance, we must admit that these models are better for explaining the rapid approach to saturation in the relationship between net photosynthesis and external (or intercellular space) CO_2 concentration. However, we feel that the evidence we have reviewed is generally in favour of the major role for the limited activity of enzymes in determining the mesophyll resistance, and further discussion will be based on this view.

Leaf Anatomy and Mesophyll Conductance: The Balancing of Diffusive and Chemical Limitations

The experimental observations discussed in the two preceding sections suggest that the major determinant of the rate of net CO_2 fixation in C_3 crop plants under our standard conditions is the enzymic capacity of RuBPc-o (intrinsic and as regulated by related biochemical processes) rather than by transport reactions. It is our contention that this state of affairs is intimately related to leaf anatomy, and particularly to the ratio of A_{mes} (the area of mesophyll exposed to intercellular gas spaces) to A (the external area of one side of the leaf). For a given capacity of RuBPc-o per unit A we can qualitatively predict the effect of extremes of A_{mes}/A ratios. A very high value means that there are very few RuBPc-o molecules per unit of A_{mes}, so that they are disposed in a very thin layer around the cell periphery. The short aqueous diffusion distance for CO_2 from the gas spaces to the RuBPc-o molecules means that each RuBPc-o molecule is exposed to CO_2 and oxygen concentrations close to those in the cell wall in equilibrium with the gas phase in the leaf. However, this very efficient use of individual RuBPc-o molecules is only purchased at the expense of a leaf with a very large load of cell walls to support the large A_{mes}. Conversely, a very low ratio of A_{mes}/A means that there are many RuBPc-o molecules per unit of A_{mes}, many of which must be far from the cell periphery and thus separated from gas phase CO_2 by a long liquid phase diffusion path. These innermost RuBPc-o molecules are thus doubly disadvantaged with respect to net carbon fixation, being exposed to both a low CO_2 concentration and a low CO_2/O_2 ratio (Raven, 1977a, 1977b; Sinclair, Goudriaan and de Wit, 1977).

In quantitative terms, the limitation on the number of RuBPc-o molecules which can be usefully associated with unit area of A_{mes} stems from the high molecular weight of the enzyme (530 000) and its relatively low specific activity: its V_c of perhaps 10 s^{-1} (i.e. mol CO_2 (mol enzyme)$^{-1}$ s^{-1}) at 25°C (Badger and Collatz, 1977; Jensen and Bahr, 1977), which corresponds to a v_c in air-equilibrated solution of perhaps some $1–2 \text{ s}^{-1}$. If the RuBPc-o concentration in the chloroplast stroma is 1 µmol cm^{-3} (Raven, 1977a, 1977b), our net CO_2 fixation of 100 pmol cm^{-2} s^{-1} on the basis of A_{mes} requires at least 1 µm thickness of stroma if v_c is 1 s cm^{-1}. Allowing for an incomplete cover of chloroplasts over A_{mes} and the presence of starch (Nafziger and Coller, 1976), this estimate is in reasonable agreement with the value used in the calculation on page 112. Under these conditions, the 'mean' RuBPc-o molecule that is 1 µm into the stroma is exposed to a CO_2 concentration of 6.5 nmol cm^{-3}.

If we now notionally add another 2 µm layer of RuBPc-o molecules (and other photosynthetic components) inside the first layer, we double the

RuBPc-o *capacity* per unit of A_{mes}. However, an iterative analysis similar to that quoted by Raven (1977b) but making quantitative use of equations (6.7)–(6.8) suggests that the net CO_2 flux per unit of A_{mes} is only increased from 100 pmol cm^{-2} s^{-1} to some 140 pmol cm^{-2} s^{-1}. Thus, each RuBPc-o molecule is only turning over in P_{net} at 0.7 of its rate with the original RuBPc-o/A_{mes}, while the *additional* RuBPc-o molecules are only turning over at 0.4 of the rate of the *original* RuBPc-o molecules in the earlier model with just 2 μm chloroplast thickness.

This sort of analysis, while admittedly involving a number of assumptions, suggests that the ratio of RuBPc-o to A_{mes} can have a very important influence on the efficiency with which the average RuBPc-o molecule is used in terms of its expressed turnover *in vivo*. It is difficult to find quantitative evidence from the literature on the RuBPc-o/A_{mes} ratio in C_3 plant leaves: there are data on A_{mes}/A (Turrell, 1936, 1944; El-Sharkaway and Hesketh, 1965; Rackham, 1966; Nobel, 1974, 1977; Nobel, Zanagoza and Smith, 1975; Bunce *et al.*, 1977) and on RuBPc-o/A (Treharne, 1972; Björkman, 1973; Taylor and Pearcey, 1976), although there is very little overlap in genotypes between the two sets of data. What conclusions can be drawn suggests that there *is* a correlation of RuBPc-o and A_{mes} (both expressed on the basis of A). The lack of correlations between total leaf N, and A_{mes} in *Brassica* cultivars (Sasahara, 1971) is not unexpected, since there is little correlation between RuBPc-o activity and total N in *Solidago* leaves (Björkman and Holmgren, 1963; Björkman, 1968).

Further evidence consistent with the critical importance of A_{mes}/A in accounting for carboxylation efficiency differences (on an A basis) between genotypes, comes from work which correlates high P_{net} (on an A basis) with small mesophyll cells in *Lolium* (Wilson and Cooper, 1969) and in the wild diploid ancestors of modern wheat compared with the modern hexaploid *Triticum aestivum* (Evans and Dunstone, 1970; Dunstone and Evans, 1974; Evans and Wardlaw, 1976). It appears likely that the smaller cells mean a higher A_{mes}/A, although cell number /A should be checked. Further evidence from *Triticum* is consistent with the hypothesis that additional RuBPc-o molecules need not necessarily allow increased P_{net}, even if the original number of RuBPc-o molecules is close to limiting P_{net}; this evidence comes from the work of Thomas and Thorne (1975) on the effect of N fertilizer levels on P_{net} and RuBPc-o levels in *Triticum* leaves.

The Effects of Variations from the Optimal Conditions on Mesophyll Conductance in C_3 Plants

Our definition of standard or optimal conditions for photosynthesis in the introduction implies that changes in conditions (either acute or chronic) reduce the rate of photosynthesis and the mesophyll conductance. The exception (with respect to photosynthesis but *not* mesophyll conductance) is increased CO_2 or decreased oxygen concentrations: such changes are not common in commercial practice (with the exception of CO_2 fertilization). We shall deal individually with the effects of various perturbations which are met commonly in commercial practice, and will consider both acute and chronic effects of the treatments.

INSUFFICIENT WATER SUPPLY

Hsaio (1973) and Boyer (1976) have produced lucid reviews of plant responses to large evaporative demand relative to root supply. In homoiohydric vascular plants they attribute the major acute effects on photosynthesis of slight or moderate water stress to stomatal closure. This decreases both the CO_2 concentration and the CO_2/O_2 ratio at the site of RuBPc-o activity and hence decreases P_{net} (*see* page 116). This effect is exacerbated if stomatal closure increases leaf temperature and thus decreases the solubility of CO_2 more than that of oxygen (*see* page 123). More severe stresses increase the 'residual resistance' (Hsaio, 1973; Boyer, 1976), which involves inhibition of both RuBPc-o and of ATP synthesis and primary photochemical reactions. Long-term effects of water stress involve changes in leaf anatomy (Cutler, Rains and Loomis, 1977).

LOW IRRADIANCE

Immediate effects of suboptimal irradiance occur on the rate of generation of ATP and $NADPH_2$ in the photochemical reactions of photosynthesis. The rates of the PCRC reactions, including RuBPc-o, are inhibited in parallel

Table 6.5 KINETIC PARAMETERS OF RuBPc-o FROM *ATRIPLEX GLABRIUSCULA*, AND A COMPARISON OF P_{net} PREDICTED FROM THESE PARAMETERS WITH THAT OBSERVED *IN VIVO*, AT THREE TEMPERATURES

Parameter	Value at temperature of:		
	15°C	25°C	35°C
1. $[CO_2]$, nmol cm^{-3}, in water in equilibrium with 0.03 per cent in gas phase	14	10.5	8
2. $[O_2]$ nmol cm^{-3}, in water in equilibrium with 21 per cent in gas phase	320	260	230
3. V_c, mol CO_2 (mol RuBPc-o·s)$^{-1}$	9.9	13.7	32.6
4. V_o, mol O_2 (mol RuBPc-o·s)$^{-1}$	1.01	2.51	5.65
5. $K_{1/2}CO_2$ RuBPc, nmol cm^{-3}	21	27	48
6. $K_{1/2}CO_2$ RuBPo, nmol cm^{-3}	21	27	499
7. $K_{1/2}O_2$ RuBPo, nmol cm^{-3}	181	328	499
8. $K_{1/2}O_2$ RuBPc, nmol cm^{-3}	181	328	499
9. V_{cs} mol CO_2 (mol-RuBPc-o·s)$^{-1}$ (in an air-equilibrated solution, computed from equation (6.7))	1.93	2.45	3.34
10. v_o, mol O_2 (mol RuBPc-o·s)$^{-1}$ (in air-equilibrated solution, computed from equation (6.8))	0.52	0.91	1.60
11. P_{net}, mol CO_2 (mol RuBPc-o·s)$^{-1}$ (computed from equation (6.6))	1.67	1.99	2.54
12. P_{net}, nmol cm^{-2} s^{-1} (computed from line (11) and RuBPc-o V_c on an area basis in *Atriplex hastata*	0.79	0.94	1.19
13. P_{net}, nmol cm^{-2} s^{-1} (measured *in vivo*)	1.8	2.0	1.7

Data in lines (1) and (2) from Ku and Edwards (1977a, 1977b).
Data in lines (3)–(8) from Badger and Collatz (1977), assuming that $K_{1/2}CO_2$, $K_{1/2}O_2$ for RuBPc activity is equal to K_iO_2 for RuBPc activity.
Data in line (9) from lines (1), (2), (3), (5) and (8), using equation (6.7).
Data in line (10) from lines (1), (2), (4), (6) and (7), using equation (6.8).
Data in line (11) from lines (9) and (10), using equation (6.6).
Data in line (12) from line (11), using as a conversion factor from mol RuBPc-o to cm^2 of external leaf surface (one side) the ratio of RuBPc-o to soluble leaf protein in *Atriplex glabriuscula* given in Björkman, Boynton and Berry (1976) and the soluble protein/external leaf area in *Atriplex hastata* (Björkman *et al.*, 1972).
Data in line (13) from Björkman, Mooney and Ehleringer (1975).

with this reduced rate of cofactor synthesis, but by 'messengers' other than ATP and $NADPH_2$ (Werden, Heldt and Milovancev, 1975; Anderson and Avron, 1976). Thus low irradiance increases the mesophyll resistance: it can also increase stomatal resistance via a direct light effect on stomata, and the indirect effect of the increased intercellular CO_2 concentration caused by the increased mesophyll resistance. In general, the effects on mesophyll resistance are larger, so the CO_2 supply rate is decreased less than v_c and v_o. The end result is an unchanged V_c/V_o (Andrews and Lorimer, 1978; Salaja and McFadden, 1978) despite a decreased turnover of RuBPc-o and an increased CO_2 concentration at the site of RuBPc-o. This should increase v_c relative to v_o (equations (6.7) and (6.8)); however, the carbon balance of the leaf is also influenced by 'dark' respiration (Raven, 1972a, 1972b), which probably becomes a larger fraction of P_{net} as defined in equation (6.6) at lower irradiances (McCree, 1974).

Long-term (growth) effects of low irradiance on leaf photosynthesis in C_3 crop plants lead to a well-characterized range of phenotypic changes in their photosynthetic apparatus; the 'shade-adapted' leaves resemble the leaves of genetically shade-adapted plants (Björkman, 1968, 1973; Boardman, 1977; Nobel, 1977; Clough, Terri and Alberte, 1979). On an external area basis, the shade-adapted plants have a considerably decreased capacity for net photosynthesis under optimal light: the reduced carboxylation efficiency on the basis of A is paralleled by a decreased RuBPc-o activity, and rather smaller decreases in A_{mes} and in the capacity for ATP and $NADPH_2$ generation. The possible adaptive significance of this combination of characteristics, which have implications for the light-use efficiency and dark respiratory properties, are discussed by Raven and Glidewell (1975).

TEMPERATURE

We have defined our standard conditions such that any increase or decrease in temperature involves a short-term decrease in P_{net}. The effects of temperature on the various component resistances which determine mesophyll conductance are widespread. Starting with the biochemical resistance and RuBPc-o activity, the values of V_c and V_o increase with temperature over the entire growth range for plants in the UK, while P_{net} peaks at some intermediate value (*Table 6.5*). Further data for the temperate American C_3 *Atriplex glabriuscula* in *Table 6.5* show that $K_{1/2}$ and K_i for both CO_2 and oxygen increase with temperature. This means that the values of v_c and v_o (equations (6.7) and (6.8)) fall relative to V_c and V_o at higher temperatures if the dissolved CO_2 and oxygen concentrations are held constant. If account is taken of the decreased solubility of oxygen and, more particularly, of CO_2 at higher temperatures (*Table 6.4*), we can see that the v_c and v_o values achieved in air-equilibrated solution show an even smaller fall in v_o and (particularly) v_c with temperature.

Ignoring temperature effects on diffusive resistances (*see* later), we can see from *Table 6.4* that the computed P_{net} (equations (6.6)–(6.8)) and the measured P_{net} change in parallel between 15 and 25°C, which suggests that the model of limitation by RuBPc-o fits the data in this range. The model does *not* predict the decline in P_{net} between 25 and 35°C, which suggests that some other limitation must impose a 'ceiling' lower than the 'roof' imposed by RuBPc-o (cf. Björkman, 1973; Tieszen and Sigurdsen, 1973; Armond, Schreiber and Björkman, 1978; Mooney, Björkman and Collatz, 1978). Data presented by Mooney, Björkman and Collatz (1978) on *Larrea divaricata* photosynthesis at light and CO_2 saturation, as a function of temperature, are consistent with limitation by RuBPc-o activity, operating in this case via V_c, if the effect of temperature on V_o is similar to that shown in *Table 6.5*.

Thus far we have ignored the effect of temperature on diffusive resistance. Relevant temperature effects here are those on the diffusion coefficient and the solubility of the gases (Ku and Edwards, 1977a, 1977b). For *Atriplex glabriuscula* (*Table 6.4*) the decrease in flux (P_{net}) parallels the decrease in diffusion coefficient, so the diffusive driving force (equation (6.1)) remains constant. However, the increase in gas solubility at low temperatures means that the diffusive driving force for CO_2 entry (CO_2 *concentration difference*) is imposed on a larger total CO_2 *concentration*, so the CO_2 concentration at the site of RuBPc-o activity at the lower temperature should be *higher* than at the higher temperature. This applies *a fortiori* if the Q_{10} of photosynthesis is higher than 1.2 (as it commonly is; Björkman, 1973), unless a phase transition of the lipid components of the transport pathway occurs, with a corresponding decline in CO_2 permeability below the transition temperature (Champigny and Moyse, 1975; Bagnall and Wolfe, 1978).

On the high temperature side of the optimum, a similar analysis suggests that the CO_2 concentration at the RuBPc-o site probably does not decrease from that found at the optimal temperature for photosynthesis.

To conclude our discussion of the acute effects of temperature on net photosynthesis, on the high temperature side of the optimum for net photosynthesis some mesophyll limitation other than RuBPc-o or gas solubility and transport accounts for the decrease (Tiezen and Sigurdson, 1973). At temperatures below the optimum, photosynthesis commonly declines more rapidly with decreasing temperature than a consideration of gas solubility and diffusivity, and of the kinetics of RuBPc-o, would suggest (cf. Selwyn, 1966).

Turning to chronic exposure to non-optimal temperatures (but those which still permit reasonable rates of growth), a general response by a given genotype is a shift in the temperature optimum towards the growth temperature (Björkman, 1973). For a decreased growth temperature, the rates of net photosynthesis at all temperatures below the new optimum are greater than those found for plants grown at the old optimum temperature, and the reverse is true for plants grown at superoptimal temperatures. In a number of cases, the change in photosynthetic rate at temperatures below the 'adapted' optimum has been correlated with increased RuBPc-o activity (both P_{net} and enzyme activity being expressed per unit A) (Björkman and Gauhl, 1969; Björkman, Nobs and Heisey, 1969; Peet, Ozbun and Wallace, 1977; cf. Downton and Slatyer, 1972). This 'capacity adaptation' (Hochachka and Somero, 1973; Hazel and Prosser, 1974; Raven and Smith, 1978b) does not appear to involve changes in the kinetic properties of the enzyme, but rather the content of RuBPc-o protein. In some cases this increase in RuBPc-o at low growth temperatures is related to an increased leaf thickness (Peet, Ozbun and Wallace, 1977), which probably means an increased A_{mes}/A (Nobel, 1974, 1977). Further work is needed to see if capacity adaptation involves an increased RuBPc-o/A_{mes} at low temperatures: our earlier computations on diffusion resistances suggest that such an increase would not be required to permit similar CO_2 concentrations at RuBPc-o at the same distance from the intercellular gas space.

As with the *acute* effects of temperature, the *adaptive* effects are more easily explained in as much as they change rates of net photosynthesis below the optimal temperature: it is not immediately obvious why the increased RuBPc-o activity in plants adapted to lower temperatures should not permit more rapid photosynthesis at high temperatures (Björkman, 1973).

Finally, the evolutionary significance of capacity adaptation may be related to such phenomena as the timing of reproduction in relation to the seasonal availability of pollinating and dispersing agents, i.e. ensuring that the plant arrives at a certain stage of its life cycle at a certain time of year despite environmental differences from year to year (Calow, 1977).

NUTRIENT SUPPLY

Inhibition of photosynthesis is a consequence of deficiency of many nutrients (Possingham, 1971). Our discussion here relates solely to the large fraction of the total leaf protein, which is associated with the single protein RuBPc-o (fraction 1 protein), e.g. 60 mg g^{-1} dry weight out of a total protein concentration of 260 mg g^{-1} dry weight in *Atriplex glabriuscula* (Björkman, Boynton and Berry, 1976). The 22 per cent or so of total leaf protein which is found in this single enzyme means that it acts as a protein reserve: some indication of this possibility is found in the work of Thomas and Thorne (1975), which was mentioned in the previous section; they showed that *Triticum aestivum* with 'luxury' nitrogen supply accumulated additional RuBPc-o in their leaves without increasing P_{net}. Conversely, the production of the high-protein fruits of *Glycine max* can involve removal of amino acids from the leaves, with corresponding decreases in RuBPc-o activity and leaf photosynthesis (Sinclair and de Wit, 1975, 1976). This self-destructive nature of reproduction in crop plants is not unexpected in view of their 'ruderal' ancestry (Cohen, 1971; Grime, 1977, 1979). The fraction of plant nitrogen that is involved in the production of RuBPc-o is discussed in the next section in the context of possible adaptive advantages of the C_4 pathway (Brown, 1978; cf. Doliner and Joliffe, 1979).

LEAF AGE

Young, growing leaves often have high RuBPc-o activity relative to their capacity for photosynthesis (Taylor and Pearcey, 1976); some ceiling below the roof imposed by RuBPc-o activity must be operative here. Postmature leaves show a decline in P_{net} on the basis of A. An early and prescient investigation of this phenomenon (Woolhouse, 1967, 1968) showed a close parallel of decreases in P_{net}, RuBPc-o activity and RuBPc-o protein, while stomatal conductance, photochemical activity and liquid phase permeability (as judged from leaf structure) do *not* decrease as much as P_{net}. In the flag leaf of *Triticum aestivum*, however, senescence involves a parallel decrease in

P_{net}, RuBPc and RuBPo activity, but with a much smaller decrease in RuBPc-o enzyme protein on the basis of A. The decline in RuBPo is not, however, reflected in a decreased photorespiratory CO_2 efflux (Hall, Keys and Merrett, 1978; Thomas, Hall and Merrett, 1978). While it is possible that a decreased stomatal conductance might so reduce the CO_2 and oxygen concentrations at the site of RuBPc-o activity as to allow this increase in the ratio of measured CO_2 efflux to RuBPo activity, the possibility of additional sources of glycolate cannot be ruled out. Yet another pattern of RuBPc-o changes in senescence is found in *Capsicum annuum* (Hall and Brady, 1977). Here the RuBPc-o activity, which is barely adequate to account for the P_{net} in the mature leaf, declines much more rapidly than does P_{net} during senescence. The decrease in RuBPc-o protein is rather less than that of RuBPc-o activity. These three cases suggest that there is no currently useful generalization which can be made about the relationship between P_{net} and the activity (or turnover) of RuBPc-o.

CONCLUSIONS

Divergence from standard conditions, in the short term, decreases the efficiency with which RuBPc-o is used. This can come about from changes in the intrinsic catalytic activity of RuBPc-o (non-optimal temperature), or from changes in the supply of CO_2 in absolute terms or relative to oxygen (high temperatures, partial stomatal closure) or from partial repression of RuBPc-o activity by some feedback effects (non-optimal temperature, or low irradiance).

The importance of reduced enzyme capacity relative to diffusion capacity in the responses to non-optimal temperature, or low irradiance, is supported by an increased isotopic discrimination between ^{12}C and ^{13}C (Christeller, Laing and Troughton, 1976; Smith, Oliver and McMillan, 1976; Estep *et al.*, 1978), although such effects are not found by Troughton and Card (1976).

Even under optimal conditions, net photosynthesis can only use some 0.2 of the already low specific activity of the large, and thus nitrogen- and energy-expensive, RuBPc-o molecule. The C_4 pathway represents a way of partially circumventing the limitations imposed by RuBPc-o when plants are growing with a limited supply of water and of nitrogen under high-light and high-temperature conditions.

The C_4 Pathway

A generalized version of the chemical and structural characteristics of the C_4 pathway is shown in *Figure 6.2*. The essence of the pathway seems to be a 'CO_2 pump', which increases the CO_2 concentration and the CO_2/O_2 ratio at the site of RuBPc-o activity. The details of C_4 metabolism are discussed by Black (1973), Hatch and Osmond (1976) and Osmond and Smith (1976).

An essential feature of C_4 metabolism is spatial differentiation: RuBPc-o and the complete suite of PCRC and PCOC enzymes are located in bundle sheath cells. These are chlorophyllous cells that surround vascular bundles with a limited access to intercellular gas spaces, which are separated from the chloroplasts by a 10–30 µm liquid-phase diffusion path. Thus, CO_2 supplied

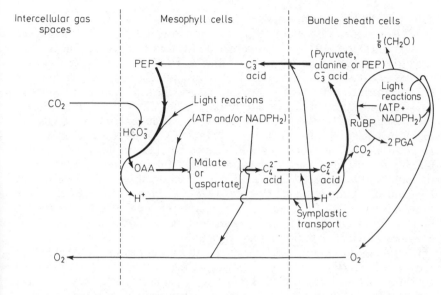

Figure 6.2   *The C_4 pathway of photosynthetic carbon metabolism. Thick arrows: C_3–C_4 cycle. Thin arrows: C path for $CO_2 \rightarrow$ sucrose*

to these chloroplasts by the CO_2 pump is prevented from leaking to the intercellular gas spaces, and the high CO_2 concentration which can be maintained at the site of RuBPc-o activity permits biochemical resistance to CO_2 fixation to be minimized. Quantitative aspects of these resistances are discussed by Raven (1972b, 1977a), Osmond and Smith (1976) and by Berry and Farquhar (1978). These computations show that the fluxes are in balance:

$$CO_2 \text{ influx via } CO_2 \text{ pump} = \text{fixation by RuBPc-o} + \text{efflux by leakage}$$

with 20–50 nmol cm^{-3} of CO_2 at the site of RuBPc-o. An important aspect of the activity of RuBPc-o is the extent of oxygen build-up in the bundle sheath cells, since the barrier to CO_2 efflux also restricts oxygen efflux. The extent of this build-up depends on the fraction of total photosystem II activity which occurs in the bundle sheath cells, and the length of the aqueous diffusion path (Samish, 1975; Raven, 1977a; Berry and Farquhar, 1978). The oxygen concentration in the bundle sheath chloroplasts in different C_4 plants varies from near-air equilibrium to twice that value (250–500 nmol cm^{-3} at 25°C). It appears that despite this increased oxygen content, the CO_2 concentration of 20–50 nmol cm^{-3} is sufficient to account for the efficient use of RuBPc activity and the low rate of glycolate synthesis and thus PCOC activity in bundle sheath cells.

The high CO_2 concentration and the high ratio of CO_2/O_2 in the bundle sheath cells is dependent on the CO_2 pump. The basic C_4 dicarboxylic acid/C_3 monocarboxylic acid shuttle, which constitutes this CO_2 pump, is shown in *Figure 6.2*. From the carboxylation viewpoint we must look at the initial CO_2 fixation reaction catalysed by PEPc (*see* page 113). PEPc has a rather lower molecular weight than RuBPc-o (1.7–3.5×10^5; cf. 5.5×10^5: Enama, 1976: Jensen and Bahr, 1977), and a higher intrinsic catalytic activity (6000, cf. 10 nmol C fixed (nmol enzyme)$^{-1}$ s^{-1}; Jensen and Bahr, 1977; Reibach and Benedict, 1977). Furthermore, it is insensitive to oxygen (Bowes and Ogren, 1972). The apparent affinity for CO_2 is very pH dependent since the true substrate for carboxylation is HCO_3^- with a $K_{1/2}(HCO_3^-)$ of 110 µM (Reibach and Benedict, 1977). Thus, cytoplasmic pH regulates not only the intrinsic activity of the enzyme but also the availability (at a constant CO_2 concentration) of its true substrate. The net CO_2 influx and fixation in the mesophyll cells of C_4 plants is probably a little higher (on the basis of mesophyll cell area) than that in comparable C_3 plants (El-Sharkaway and Hesketh, 1965). The properties of PEPc mean that less protein is involved in making the PEPc in these C_4 mesophyll cells than for the RuBPc-o in C_3 mesophyll cells. Indeed, leaf N per cm^2 of external leaf area is generally lower in C_4 than C_3 plants (Brown, 1978), which suggests that the extra protein required for PEPc and the CO_2 pump is more than offset by savings on RuBPc-o working under more nearly optimal conditions, and on PCOC enzymes which have to process less glycolate.

Overall, the C_4 pathway is associated with a lower residual resistance than the C_3 pathway (Gifford, 1974), which is probably a function of the improved conditions of work for RuBPc-o. Thus, a C_4 plant can *either* have a higher rate of net CO_2 fixation than a C_3 plant (a lower residual resistance in the C_4 plant, together with a slightly higher stomatal resistance even under plentiful water supply; Gifford, 1974), *or* it can have a similar net CO_2 fixation rate despite a considerably higher stomatal resistance. In the second case, when the same stomata are more nearly closed than in a C_3 plant that photosynthesizes at the same rate, the quantity of water transpired per unit of carbon fixed can be almost halved (Black, 1973). This decreased water loss frequently involves a larger increment of leaf temperature above ambient, which is often high in any case in the habitats where C_4 plants are found. Again, the CO_2 pump can offset the adverse effects of high temperature on gas solubility, which may affect the efficiency of RuBPc-o use in C_3 plants. Finally, we have already mentioned the economy of enzyme protein on a leaf area basis in C_4 plants.

These considerations suggest that the C_4 plant can fix more carbon per unit of water lost or per unit of carbon invested in leaf protein, and it may be better able to photosynthesize at high leaf temperatures. With the visual acuity so characteristic of hindsight, we can thus predict quite precisely the environmental conditions where wild C_4 plants are, in fact, found (Doliner and Joliffe, 1979). What, then, are the prospects for the large-scale growth of commercial C_4 crops in cool temperate climates? There are relatively few native C_4 plants in north-west Europe. In North America there are a number of C_4 plants in temperate habitats: it appears that these 'cool-adapted' C_4 plants do not necessarily 'use' their C_4 advantages under these conditions. Caldwell *et al.* (1977) investigated seasonal changes in carbon exchange and water loss in sympatric C_3 and C_4 shrubs. They found that the net carbon fixation and net water transpired per unit ground area are the same for the C_3 as for the C_4 species. Thus, while there seems to be no intrinsic difficulty in making C_4 metabolism work in cool temperate environments, its advantages in terms of productivity or water use are debatable under these conditions (Björkman, Boynton and Berry, 1976; Caldwell, Osmond and Nott, 1977; Long and Woolhouse, 1978; Thomas and Long, 1978).

Appendix

The possible participation of inorganic carbon species other than CO_2 in the transport of CO_2 from intercellular gas spaces to the site of carboxylation

The other inorganic carbon species at the oxidation level of CO_2, which are found in solution, include H_2CO_3 and HCO_3^-; the proportions of these depend on pH. The initial reaction upon solution of CO_2 in water is the formation of H_2CO_3:

$$CO_2 + H_2O \underset{k_2}{\overset{k_1}{\rightleftharpoons}} H_2CO_3 \qquad (6.9)$$

with the equilibrium constant $K(k_1/k_2) = 2.59 \times 10^{-3}$ at 25°C (Lucas, 1975). The hydration and dehydration reactions are relatively slow: the rate constant for hydration (k_1) is $0.052 \ s^{-1}$. H_2CO_3 is a fairly strong acid, which dissociates very rapidly relative to the rate of the hydration and dehydration reactions:

$$H_2CO_3 \rightleftharpoons H^+ + HCO_3^- \qquad (6.10)$$

The dissociation is generally expressed in terms of the concentration of free $(CO_2 + H_2CO_3)$: the pK_a for the combined hydration-dissociation is some 6.43 at 25°C and low ionic strength.

It is readily shown that uncatalysed hydration of CO_2 from the atmosphere in the liquid-phase diffusion pathway is too slow to be a significant reaction for photosynthesis. This is best illustrated in the mesophyll cell wall, where there is no carbonic anhydrase (Raven, 1970; Jones and Slatyer, 1972; Ullrich-Eberius, Lüttge and Neher, 1976) to catalyse reaction (6.9). For a wall which is half water (*see* page 111 and *Table 6.1*) the CO_2 concentration in the wall at 25°C is 5 nmol cm^{-3} in equilibrium with air that contains 10 nmol cm^{-3}. For a wall that is 0.1 μm thick (*Table 6.1*), the CO_2 content is 5×10^{-2} pmol cm^{-2}. With a flux of 100 pmol cm^{-2} s^{-1} (page 120) the turnover time of CO_2 in the wall is 0.5 ms, compared with the uncatalysed $t_{1/2}$ for CO_2 hydration at 25°C of 19 000 ms. Thus (even if the cell wall pH were high enough) HCO_3^- cannot be significant in inorganic carbon transport through the cell wall. HCO_3^- is not involved in transport through any of the membranes in the liquid-phase diffusion pathway (*Table 6.1*), and the cytoplasm has a very low resistance even without the participation of HCO_3^- because of its very short path length (*Table 6.1*). Carbonic anhydrase is present at high activity in the stroma of C_3 plants (Jacobsen, Kong and Heath, 1975), so forms of inorganic carbon other than free CO_2 could, on kinetic grounds, be involved in inorganic carbon transport through the stroma. A simple model is presented in *Figure 6.3*. Essentially, the scheme involves instantaneous equilibration of incoming CO_2 at the 'source' end of the pathway with HCO_3^- via the enzyme CO_2 and enzyme-HCO_3^- intermediate forms of carbonic anhydrase. We can translate 'instantaneous' into spatial terms by calculating the distance from the inner envelope membrane at which the CO_2 that enters at 100 pmol cm^{-2} s^{-1} is fully equilibrated with HCO_3^-, granted the observed concentration and properties of carbonic anhydrase. In *Spinacia* chloroplasts there is some 2 μmol cm^{-3} of carbonic anhydrase (Jacobsen, Kong and Heath, 1975), with a turnover at

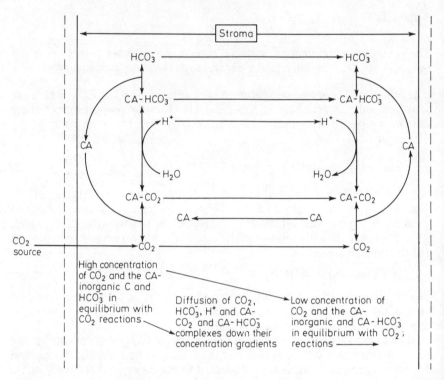

Figure 6.3 *Inorganic carbon transport in the stroma by* CO_2, HCO_3^- *and their complexes with carbonic anhydrase*

CO_2 saturation for the hydration reaction of some 3×10^5 mol CO_2 (mol enzyme·s)$^{-1}$ and a $K_{1/2}$ for CO_2 of 2 µmol cm^{-3} at 25°C (Pocker and Ng, 1974, 1975). The distance for complete equilibration is less than 0.1 µm compared with a 'mean' inorganic carbon diffusion distance of 1 µm (*Table 6.1*). A similarly rapid dehydration can occur at the sink end of the pathway (i.e. close to a RuBPc-o molecule) to regenerate the true substrate of RuBPc-o, i.e. CO_2 (Cooper *et al.*, 1969). We can thus regard all the components of the equilibrium catalysed by carbonic anhydrase as contributing to inorganic carbon diffusion. Since the pH of the stroma is at least 1 pH unit above the pK_{a1} of CO_2 (Werdan, Heldt and Milovancev, 1975), the equilibrium HCO_3^- concentration will be at least ten times that of CO_2; the results of Raven and Smith (1978a) suggest that the ratio is temperature-invariant (cf. page 124). Thus, with 8 nmol cm^{-3} of CO_2 at the sink end, there will be some 188 nmol cm^{-3} of HCO_3^-, while the arguments of Jacobsen, Kong and Heath (1975) and the kinetic data of Pocker and Ng (1974, 1975) suggest that the concentration of the carbonic anhydrase-inorganic carbon complexes in the steady state are also about 100 nmol cm^{-3} (CO_2 complex + HCO_3^- complex).

Granted the presence of these inorganic carbon species at the concentrations we have calculated, in rapid equilibrium with CO_2 throughout the chloroplast, we can compute their contribution to inorganic carbon transport from their diffusivities. HCO_3^- probably has a similar diffusivity to CO_2, so a concentration of ten times higher means that HCO_3^- participation could reduce the apparent resistance to CO_2 transport by an order of magnitude. While the carbonic anhydrase complexes are present at similar concentrations to free HCO_3^-, the diffusion constant for a protein of molecular weight 212 000 is little more than 10^{-7} cm^2 s^{-1} (Nobel, 1974; Kandel *et al.*, 1978), i.e. about two orders of magnitude less than CO_2 or HCO_3^-; thus, these complexes could only reduce the transport resistance by 10 per cent or so.

Two constraints on the scheme in *Figure 6.3* must be noted before we quantify the resistance to inorganic carbon movement. A minor one is that of back-diffusion of 'free' carbonic anhydrase, which must occur if the (quantitatively minor) contribution from the carbonic anhydrase complexes is to be expressed. Since the enzyme-inorganic carbon complexes only involve some 100 nmol cm^{-3} out of a total enzyme concentration of 2000 nmol cm^{-3} even at the source side where the CO_2 concentration is highest, back-diffusion of free carbonic anhydrase is not a problem.

More serious for transport of inorganic carbon as HCO_3^- (either free or enzyme-bound) is the need for an equal transport of H^+ if gross pH imbalances (a pH fall at the source end and an increase at the sink end) are not to occur. The arguments employed by Raven (1977c) suggest that buffer-mediated H^+ fluxes of the required 100 pmol H^+ (cm^2 s)$^{-1}$ could occur with pH differences between the two ends of the pathway of less than 0.1 pH units.

A possible additional diffusive inorganic carbon species is the RuBPc-o-CO_2 complex. Its behaviour would be analogous to that of the carbonic anhydrase complexes, but with the complication that the CO_2 would be consumed by the enzyme *en route*. An upper limit on the contribution of this complex to the transport of CO_2 can be obtained by ignoring net CO_2 fixation, and using an argument similar to that used above for carbonic anhydrase. With a RuBPc-o concentration of 1 µmol cm^{-3}, a $K_{1/2}$ for CO_2 of 20 nmol cm^{-3}, and a CO_2 concentration of 8 nmol cm^{-3}, the concentration of the RuBPc-o-CO_2 complex is less than 250 nmol cm^{-3}. With a diffusion constant of 10^{-7} cm^2 s^{-1} for RuBPc-o-CO_2 (Nobel, 1974), this complex cannot add to inorganic carbon diffusivity more than 0.3 of that due to free CO_2.

Thus the major enhancement is due to HCO_3^-, and the calculations performed above suggest that HCO_3^- production in the stroma, catalysed by the very active carbonic anhydrase, might reduce the transport resistance for inorganic carbon in the stroma to 0.1 of that for free CO_2 as the sole diffusing species. However, direct measurements in animal tissues with high carbonic anhydrase activity suggest that the maximum enhancement is nearer three-fold rather than ten-fold (Raven, 1977a). While admitting that there are differences between the stroma and a metazoan cell (notably, for our analysis, the high free CO_2 concentration in land vertebrate cells), it would perhaps be unwise to assume that the full computed capacity of carbonic anhydrase-mediated transport is available to the stroma. We conclude that the resistance of the 'mean' 1 µm diffusion path for inorganic carbon in the stroma may be reduced from the 10 s cm^{-1} for free CO_2 alone to 3–5 s cm^{-1} in the presence of high activities of carbonic anhydrase (both values being expressed on the basis of A_{mes}).

Acknowledgements

The work on ^{18}O and ^{13}C incorporation into metabolites in *Spinacia* leaves was supported by a Research Grant from the Science Research Council (UK). We are most grateful to Dr J. D. Baty, Department of Biochemical Medicine, University of Dundee for the use of gas chromatography-mass spectrometry facilities.

References

AGRAWAL, P. K. and FOCK, H. (1978). *Planta*, **138**, 257–261
ANDERSON, L. E. and AVRON, M. (1976). *Pl. Physiol.*, **57**, 209–213
ANDREWS, T. J. and LORIMER, G. H. (1978). *FEBS Lett.*, **90**, 1–9
ANDREWS, T. J., LORIMER, G. H. and TOLBERT, N. E. (1971). *Biochem.*, **10**, 4777–4782
ARMOND, P. A., SCHREIBER, U. and BJÖRKMAN, O. (1978). *Pl. Physiol.*, **61**, 411–415
BADGER, M. R. and ANDREWS, T. J. (1974). *Biochem. biophys. Res. Commun.*, **60**, 204–210
BADGER, M. R. and COLLATZ, G. J. (1977). *Yearbook of the Carnegie Institution of Washington*, Vol. 76, pp. 355–361
BAGNALL, D. J. and WOLFE, J. A. (1978). *J. exp. Bot.*, **29**, 1231–1242
BAKER, N. R., HEATH, R. L. and LEECH, R. M. (1978). *Pl. Sci. Lett.*, **12**, 317–322
BENEDICT, C. R. (1978). *Ann. Rev. Pl. Physiol.*, **29**, 67–93
BERRY, J. A. and FARQUHAR, G. (1978). In *Photosynthesis 77*, pp. 119–132. Ed. by D. O. Hall, J. Coombs and T. W. Goodwin. Biochemical Society, London
BJÖRKMAN, O. (1968). *Physiol. Plant.*, **21**, 84–99
BJÖRKMAN, O. (1973). *Photophysiol.*, **8**, 1–63
BJÖRKMAN, O., BOARDMAN, N. K., ANDERSON, J. M., THORNE, S. W., GOODCHILD, D. J. and PYLIOTIS, N. A. (1972). *Yearbook of the Carnegie Institution of Washington*, Vol. 71, pp. 115–135
BJÖRKMAN, O., BOYNTON, J. and BERRY, J. (1976). *Yearbook of the Carnegie Institution of Washington*, Vol. 75, pp. 400–407
BJÖRKMAN, O. and GAUHL, E. (1969). *Planta*, **88**, 197–203
BJÖRKMAN, O. and HOLMGREN, P. (1963). *Physiol. Plant.*, **16**, 899–914
BJÖRKMAN, O., MOONEY, H. A. and EHLERINGER, J. (1975). *Yearbook of the Carnegie Institution of Washington*, Vol. 74, pp. 743–748
BJÖRKMAN, O., NOBS, M. A. and HEISEY, W. M. (1969). *Yearbook of the Carnegie Institution of Washington*, Vol. 68, pp. 614–620
BLACK, C. C., JR (1973). *Ann. Rev. Pl. Physiol.*, **24**, 253–283
BOARDMAN, N. K. (1977). *Ann. Rev. Pl. Physiol.*, **28**, 355–377
BOWES, G. and OGREN, W. L. (1972). *J. biol. Chem.*, **247**, 2171–2176
BOYER, J. S. (1976). In *Water Deficits and Plant Growth*, Vol. IV, pp. 153–190. Ed. by T. T. Kozlowski. Academic Press, New York
BRÄNDEN, R. (1978). *Biochem. biophys. Res. Commun.*, **81**, 539–546
BROWN, R. H. (1978). *Crop Sci.*, **18**, 93–98
BUNCE, J. A., PATTERSON, D. T., PEET, M. M. and ALBERTE, R. S. (1977). *Pl. Physiol.*, **60**, 255–258

CALDWELL, M. M., OSMOND, C. B. and NOTT, D. L. (1977). *Pl. Physiol.*, **60**, 157–164

CALDWELL, M. M., WHITE, R. S., MOORE, R. T. and CAMP, L. B. (1977). *Oecologia*, **39**, 275–300

CALOW, P. (1977). *Adv. ecol. Res.*, **10**, 1–62

CHAMPIGNY, M. L. and MOYSE, A. (1975). *Biochem. Physiol. Pl.*, **168**, 575–583

CHAPMAN, D. G. and LEECH, R. M. (1976). *FEBS Lett.*, **68**, 160–164

CHARTIER, P., CHARTIER, M. and CATSKY, J. (1970). *Photosynth.*, **4**, 48–57

CHRISTELLER, J. T., LAING, W. A. and TROUGHTON, J. M. (1976). *Pl. Physiol.*, **57**, 580–582

CLOUGH, J. M., TEERI, J. A. and ALBERTE, R. S. (1979). *Oecologia*, **38**, 13–22

COHEN, D. (1971). *J. theoret. Biol.*, **33**, 299–307

COLLATZ, G. J. (1978). *Pl. Physiol.*, **61**, 109s

COOPER, T. G., FILNER, D., WISHNIK, M. and LANE, D. M. (1969). *J. biol. Chem.*, **244**, 1081–1083

CUTLER, J. M., RAINS, D. W. and LOOMIS, R. E. (1977). *Physiol. Plant.*, **40**, 255–260

DALEY, L. S. and BIDWELL, R. G. S. (1977). *Pl. Physiol.*, **60**, 109–114

DOLINER, L. M. and JOLIFFE, P. A. (1979). *Oecologia*, **38**, 23–34

DOUCE, R., MOLTZ, R. B. and BENSON, A. A. (1973). *J. biol. Chem.*, **248**, 7215–7222

DOWNTON, J. and SLATYER, R. O. (1972). *Pl. Physiol.*, **50**, 518–522

DUNSTONE, R. L. and EVANS, L. T. (1974). *Austr. J. Pl. Physiol.*, **1**, 157–165

EL-SHARKAWAY, M. and HESKETH, J. (1965). *Crop Sci.*, **5**, 517–521

ENAMA, M. (1976). *Yearbook of the Carnegie Institution of Washington*, Vol. 75, pp. 409–410

ESTEP, M. F., TABITA, R. F., PARKER, P. L. and VAN BAALEN, C. (1978). *Pl. Physiol.*, **61**, 680–687

EVANS, L. T. and DUNSTONE, R. L. (1970). *Austr. J. biol. Sci.*, **23**, 725–741

EVANS, L. T. and WARDLAW, I. F. (1976). *Adv. Agron.*, **28**, 301–359

GERSTER, R., DIMON, B. and PEYBERRES, A. (1974). In *Proceedings of the Third International Congress on Photosynthesis*, pp. 1589–1600. Ed. by M. Avron. Elsevier, Amsterdam

GERSTER, R. and TOURNIER, P. (1977). In *Abstracts of the Fourth International Congress on Photosynthesis*, pp. 129–130. Compiled by J. Coombs, UK, ISESS, Albemarle St., London

GIFFORD, R. M. (1974). *Austr. J. Pl. Physiol.*, **1**, 107–117

GLIDEWELL, S. M. and RAVEN, J. A. (1976). *J. exp. Bot.*, **27**, 200–204

GLIDEWELL, S. M. and RAVEN, J. A. (1979). *Pl. Physiol.*, **63**, 111a

GRAHL, H. and WILD, A. (1972). *Z. Pfl. Physiol.*, **67**, 443–453

GRIME, J. P. (1977). *Am. Naturalist*, **111**, 1190–1194

GRIME, J. P. (1979). *Plant Strategies and Vegetation Processes*. John Wiley, Chichester

GRODZINSKI, B. (1978). *Planta*, **144**, 31–37

GRODZINSKI, B., O'CONNOR, M. D. L. and VUCINIC, Z. (1980). In *Plant Membrane Transport: Current Conceptual Issues*, pp. 451–452. Ed. by R. M. Spanswick, W. J. Lucas and J. Dainty. Elsevier-North Holland, Amsterdam

GUTKNECHT, J., BISSON, M. A. and TOSTESON, D. C. (1977). *J. gen. Physiol.*, **69**, 779–794

HALL, A. J. and BRADY, C. J. (1977). *Austr. J. Pl. Physiol.*, **5**, 771–784

HALL, N. P., KEYS, A. J. and MERRETT, M. J. (1978). *J. exp. Bot.*, **29**, 31–37

HATCH, M. D. and OSMOND, C. B. (1976). In *Encyclopedia of Plant Physiology, New Series*, Vol. III: *Transport in Plants*, pp. 144–184. Ed. by C. R. Stocking and U. Heber. Springer-Verlag, Berlin

HAZEL, J. R. and PROSSER, C. L. (1974). *Physiol. Rev.*, **54**, 620–677

HEATH, O. V. S. (1969). *The Physiological Aspects of Photosynthesis*. Heinemann, London

HELDT, H. W. and SAUER, F. (1971). *Biochim. Biophys. Acta*, **234**, 83–91

HOCHACHKA, P. W. and SOMERO, G. N. (1973). *Strategies of Biochemical Adaptation*. W. B. Saunders, Philadelphia

HSAIO, T. C. (1973). *Ann. Rev. Pl. Physiol.*, **24**, 519–570

JACKSON, W. A. and VOLK, R. J. (1970). *Ann. Rev. Pl. Physiol.*, **21**, 385–432

JACOBSEN, B. S., KONG, F. and HEATH, R. L. (1975). *Pl. Physiol.*, **55**, 468–474

JENSEN, R. G. and BAHR, J. T. (1977). *Ann. Rev. Pl. Physiol.*, **28**, 379–400

JONES, H. G. (1973). *New Phytol.*, **72**, 1089–1094

JONES, H. G. and OSMOND, C. E. (1973). *Austr. J. biol. Sci.*, **26**, 15–25

JONES, H. G. and SLATYER, R. O. (1972). *Pl. Physiol.*, **50**, 283–288

JONES, M. B. and MILBURN, T. R. (1974). *New Phytol.*, **25**, 595–597

KAETHNER, T. H. and BANGHAM, A. D. (1977). *Biochim. Biophys. Acta*, **468**, 157–161

KANDEL, M., GORNALL, A. G., CYBULSKY, D. L. and KANDEL, S. I. (1978). *J. biol. Chem.*, **253**, 679–685

KECK, R. W. and OGREN, W. L. (1976). *Pl. Physiol.*, **58**, 552–555

KENT, S. S. (1972). *J. biol. Chem.*, **247**, 7293–7302

KENT, S. S. (1977). *Pl. Physiol.*, **60**, 274–276

KENT, S. S., PINKERTON, F. D. and STROBEL, G. A. (1974). *Pl. Physiol.*, **55**, 491–495

KEYS, A. J., BIRD, I. F., CORNELIUS, M. J., LEA, P. J., WALLSGROVE, R. M. and MIFLIN, B. J. (1978). *Nature*, **275**, 741–743

KU, S. B. and EDWARDS, G. E. (1977a). *Pl. Physiol.*, **59**, 986–990

KU, S. B. and EDWARDS, G. E. (1977b). *Pl. Physiol.*, **59**, 991–999

LAING, W. A., OGREN, W. L. and HAGEMANN, R. H. (1974). *Pl. Physiol.*, **54**, 678–685

LILLEY, R. MCC. and WALKER, D. A. (1975). *Pl. Physiol.*, **55**, 1087–1092

LONG, S. P. and WOOLHOUSE, H. W. (1978). *J. exp. Bot.*, **29**, 803–814

LORIMER, G. H., BADGER, M. R. and ANDREWS, T. J. (1977). *Anal. Biochem.*, **78**, 66–75

LORIMER, G. H., KRAUSE, G. H. and BERRY, J. A. (1977). *FEBS Lett.*, **78**, 199–202

LORIMER, G. H., WOO, K. C., BERRY, J. A. and OSMOND, C. B. (1978). In *Photosynthesis 77*, pp. 311–322. Ed. by D. O. Hall, J. Coombs and T. W. Goodwin. Biochemical Society, London

LUCAS, W. J. (1975). *J. exp. Bot.*, **26**, 331–346

LUDWIG, L. J. and CANVIN, D. T. (1971a). *Can. J. Bot.*, **49**, 1299–1313

LUDWIG, L. J. and CANVIN, D. T. (1971b). *Pl. Physiol.*, **48**, 712–719

MacKENDOR, R. O. and LEECH, R. M. (1971). In *Proceedings of the Second International Congress on Photosynthesis*, pp. 1431–1440. Ed. by G. Forti, M. Avron and A. Melandri. Dr W. Junk. The Hague

MAHON, J. D., EGLE, K. and FOCK, H. (1975). *Can. J. Biochem.*, **53**, 609–614

MASKELL, E. J. (1928). *Proc. R. Soc. Lond. B*, **102**, 488–533

MCCREE, K. J. (1974). *Crop Sci.*, **14**, 509–514

MCCURRY, S. D., HALL, N. P., PIERCE, J., PAECH, C. and TOLBERT, N. E. (1978). *Biochem. biophys. Res. Commun.*, **84**, 895–900

MOONEY, H. A., BJÖRKMAN, O. and COLLATZ, G. J. (1978). *Pl. Physiol.*, **61**, 406–410

NAFZIGER, E. D. and KOLLER, H. R. (1976). *Pl. Physiol.*, **57**, 560–563

NOBEL, P. (1974). *An Introduction to Biophysical Plant Physiology*. W. H. Freeman, San Francisco

NOBEL, P. (1977). *Physiol. Plant.*, **40**, 137–144

NOBEL, P. S., ZANAGOZA, L. J. and SMITH, W. K. (1975). *Pl. Physiol.*, **55**, 1067–1070

OSMOND, C. B. and SMITH, F. A. (1976). In *Intracellular Communication in Plants: Studies on Plasmodesmata*, pp. 229–241. Ed. by B. E. S. Gunning and A. W. Robards. Springer-Verlag, Berlin

PEET, M. M., OZBUN, J. L. and WALLACE, D. H. (1977). *J. exp. Bot.*, **28**, 57–69

PEISKER, M. (1974). *Photosynth.*, **8**, 47–50

PENMAN, H. L. and SCHOFELD, R. K. (1951). *Symp. Soc. exp. Biol.*, **5**, 115–129

POCKER, Y. and NG, J. S. Y. (1974). *Biochem.*, **12**, 5127–5134

POCKER, Y. and NG, J. S. Y. (1975). *Biochem.*, **13**, 5116–5120

POSSINGHAM, J. V. (1971). in *Recent Advances in Plant Nutrition*, Vol. 1, pp. 155–165. Ed. by R. M. Sanish. Gordon and Breach, New York

PRIOUL, J. L., REYSS, A. and CHARTIER, P. (1975). In *Environmental and Biological Control of Photosynthesis*, pp. 17–28. Ed. by R. Marcelle. Dr W. Junk, The Hague

RACKHAM, O. (1966). In *Light as an Ecological Factor*, pp. 167–185. Ed. by R. Bainbridge, G. C. Evans and O. Rackham. Blackwell Scientific Publications, Oxford

RAVEN, J. A. (1970). *Biol. Rev.*, **45**, 167–221

RAVEN, J. A. (1972a). *New Phytol.*, **71**, 227–247

RAVEN, J. A. (1972b). *New Phytol.*, **71**, 995–1014

RAVEN, J. A. (1977a). *Curr. Adv. Pl. Sci.*, **9**, 579–590

RAVEN, J. A. (1977b). In *Advances in Botanical Research*, Vol. 5, pp. 153–219. Ed. by H. W. Woolhouse. Academic Press, London

RAVEN, J. A. (1977c). *New Phytol.*, **79**, 465–480

RAVEN, J. A. and GLIDEWELL, S. M. (1975). *Photosynth.*, **9**, 361–371

RAVEN, J. A. and SMITH, F. A. (1976). *New Phytol.*, **76**, 415–431

RAVEN, J. A. and SMITH, F. A. (1978a). *J. exp. Bot.*, **29**, 853–866

RAVEN, J. A. and SMITH, F. A. (1978b). *Pl., Cell Environ.*, **1**, 185–197

REIBACH, P. H. and BENEDICT, C. R. (1977). *Pl. Physiol.*, **59**, 564–568

SALAJA, A. K. and MCFADDEN, B. A. (1978). *FEBS Lett.*, **96**, 361–363

SAMISH, Y. B. (1975). *Photosynth.*, **9**, 372–375

SASAHARA, T. (1971). *Jap. J. Breeding*, **21**, 61–68

SELWYN, M. J. (1966). *Biochim. Biophys. Acta*, **126**, 214–224

SERVAITES, J. C. and OGREN, W. L. (1977). *Pl. Physiol.*, **60**, 693–696

SHIRAIWA, Y. and MIYACHI, S. (1978). *FEBS Lett.*, **95**, 207–210

SINCLAIR, T. R. and DE WIT, C. T. (1975). *Science*, **189**, 565–567

SINCLAIR, T. R. and DE WIT, C. T. (1976). *Agron. J.*, **68**, 319–324

SINCLAIR, T. R., GOUDRIAAN, J. and DE WIT, C. T. (1977). *Photosynth.*, **11**, 56–65

SMITH, B. N., OLIVER, J. and MCMILLAN, C. (1976). *Bot. Gaz.*, **137**, 99–104

STEIGER, H. M., BECK, E. and BECK, R. (1977). *Pl. Physiol.*, **60**, 903–906

TAYLOR, R. J. and PAERCY, R. W. (1976). *Can. J. Bot.*, **54**, 1094–1103
THOMAS, S. M., HALL, N. P. and MERRETT, M. J. (1978). *J. exp. Bot.*, **29**, 1161–1168
THOMAS, S. M. and LONG, S. P. (1978). *Planta*, **142**, 171–174
THOMAS, S. M. and THORNE, G. N. (1975). *J. exp. Bot.*, **26**, 43–51
TICHA, I. and CATSKY, J. (1977). *Photosynth.*, **11**, 361–366
TIEZEN, L. L. and SIGURDSON, D.C. (1973). *Arct. alp. Res.*, **5**, 59–66
TREHARNE, K. J. (1972). In *Crop Processes in Controlled Environments*, pp. 285–303. Ed. by A. R. Rees. Academic Press, London
TROUGHTON, J. H. and CARD, K. A. (1975). *Planta*, **123**, 185–190
TURRELL, F. M. (1936). *Am. J. Bot.*, **23**, 255–264
TURRELL, F. M. (1944). *Bot. Gaz.*, **105**, 413–425
ULLRICH-EBERIUS, C. I., LÜTTGE, U. and NEHER, L. (1976). *Z. Pfl. Physiol.*, **79**, 336–346
WERDAN, K. and HELDT, H. W. (1972). *Biochim. Biophys. Acta*, **283**, 430–441
WERDAN, K., HELDT, H. W. and MILOVANCEV, M. (1975). *Biochim. Biophys. Acta*, **396**, 276–292
WILSON, D. and COOPER, J. P. (1969). *New Phytol.*, **68**, 627–644
WOOLHOUSE, H. W. (1967). *Symp. Soc. exp. Biol.*, **21**, 179–214
WOOLHOUSE, H. W. (1968). *Hilger J.*, **9**, 7–12

Notes added in Proof

(1)   Reference to later work on $^{18}O_2$ incorporation into PCOC intermediates *in vivo* may be found in BERRY, J. and BADGER, M. R. (1979). *Yearbook of the Carnegie Institution of Washington*, Vol. 78, pp. 175–178; and in CANVIN, D. T., BERRY, J. A., BADGER, M. R., FOCK, H. and OSMOND, C. B. (1980). *Pl. Physiol.*, **66**, 302–307.

(2)   Results of an elegant analysis of the carbon isotope discrimination in C_3 plants are given in FARQUHAR, G. D. (in press). In *Carbon Dioxide and Climate: Australian Research*. Ed. by G. I. Pearman. Australian Academy of Science, Canberra. The rapid approach to saturation in the relationship between photosynthesis and CO_2 concentration in intact leaves can be explained in terms of rate-limitation at CO_2 saturation by redox reactions, and is accordingly consistent with a small liquid-phase transport resistance (FARQUHAR, G. D. and VON CAEMMERER, S. (in press). *Proceedings of the Fifth International Congress on Photosynthesis*).

7

PHOTORESPIRATORY CARBON DIOXIDE LOSS

A. J. KEYS
C. P. WHITTINGHAM
Rothamsted Experimental Station, Hertfordshire, UK

Introduction

In C_3 plants the rate of oxygen uptake in the light is much greater than in the dark due to the process called photorespiration. There is an accompanying release of CO_2 so that photorespiration has the opposite effect to photosynthesis; thus the rate of gross photosynthesis is reduced to a rate of net photosynthesis. In C_4 plants, a specialized leaf anatomy is associated with the assimilation of CO_2 in two stages. This results in decreased photorespiratory metabolism and reassimilation internally of any CO_2 that is produced. Hence for the C_4 leaf as a whole there is little or no loss to be set against gross photosynthesis. As an alternative to the development of a specialized leaf structure and additional metabolic steps, photorespiration might also be controlled at the cellular level by either chemical or genetic means. At present such possibilities seem to provide the best potential for increasing photosynthetic production in C_3 plants.

Photorespiratory Metabolism

When the concentration of CO_2 is reduced, particularly in the presence of higher concentrations of oxygen, glycolate or glycine accumulate as photosynthetic products (Warburg and Krippahl, 1960; Coombs and Whittingham, 1966). These compounds are derived from intermediates in the Calvin photosynthetic cycle, from which phosphoglycolate is produced within the chloroplast. This is dephosphorylated and the glycolate formed moves into the peroxisome (Tolbert, 1971). In the peroxisome, glycolate is oxidized to glyoxylate, which is subsequently aminated by glutamic glyoxylate and serine glyoxylate aminotransferases to make glycine. Glycine then moves to the mitochondria where it is converted to serine and CO_2 (Kisaki, Imai and Tolbert, 1971; Bird, *et al.*, 1972a, 1972b), in a reaction catalysed by an enzyme complex associated with the inner membrane (Woo and Osmond, 1976; Douce, Moore and Neuburger, 1977). The decarboxylation of glycine and the subsequent synthesis of serine in mitochondrial preparations isolated from wheat, tobacco and spinach is fast enough to account for the rate of photorespiration *in vivo*; especially since only intact mitochondria have activity and up to 75 per cent of mitochondria may be damaged during extraction (Woo and Osmond, 1976). Studies with radioactive phosphate showed that the oxidation of glycine to serine was accompanied by the formation of adenosine triphosphate (ATP) and considerations of the effects of inhibitors and of the response to added adenosine diphosphate (ADP), indicate three phosphorylations per atom of oxygen that is taken up. The reaction is not light-dependent and therefore can account for the release of CO_2 in the postillumination burst shown by C_3 plants.

The rate of release of $^{14}CO_2$ from wheat leaves, cultivar Kolibri, previously fed $^{14}CO_2$, was measured when the atmosphere surrounding the leaf was rapidly changed to contain only $^{12}CO_2$ (Kumarasinghe, Keys and Whittingham, 1977a). A significant increase in the rate of $^{14}CO_2$ released was observed when the concentration of $^{12}CO_2$ in the surrounding atmosphere was increased. There was a parallel increase in the rate of loss of radioactivity from glycine. This suggests that as the partial pressure of $^{12}CO_2$ surrounding the tissue is increased, the probability that any $^{14}CO_2$ released is refixed, is diminished. Such measurements probably provide the best method of determining the activity of the enzyme glycine decarboxylase *in vivo*. Our data show that in wheat leaves in normal air the maximum rate of CO_2 released is equal to $\frac{1}{2}$ to $\frac{1}{3}$ the net rate of photosynthesis. The interpretation of analytical data which follow carbon tracer through intermediates is complicated by the fact that there is probably more than one active metabolic pool of glycine in leaves. Our present view is that there are multiple active pools of glycine and serine but one very small, highly active pool of glycine which can lead directly to the formation of serine, thus explaining the observation that in leaves which photosynthesize in $^{14}CO_2$ there is no significant lag in the labelling of serine compared to glycine (Atkins, Canvin and Fock, 1971; Canvin *et al.*, 1976).

For every molecule of CO_2 released in the conversion of glycine to serine, one molecule of ammonia is released from organic combination (Kisaki, Yoshida and Imai, 1971). If this ammonia were to be reassimilated by the action of glutamate dehydrogenase in the mitochondria, it would utilize the reduced coenzyme produced from the conversion of glycine to serine and so prevent it from being used for ATP synthesis by oxidative phosphorylation. This is to assume that both systems are present in the mitochondrial matrix and can equally react with the endogenous pyridine nucleotide.

However, recent experiments suggest that the ammonia is reassimilated through glutamine synthetase, not glutamic dehydrogenase. Ammonia was produced from glycine and accumulated in wheat leaves if they were simultaneously supplied with a specific inhibitor of glutamine synthetase (Keys *et al.*, 1978). According to the mechanism proposed, oxoglutarate reacts with glutamine to form two molecules of glutamate in the chloroplast. Thus, the reductive reaction can take place utilizing the reduced ferredoxin which is directly produced in the chloroplast as a result of light absorption. The glutamate can then in part be converted to glutamine in the cytoplasm using 1 mole of ATP per mole, and in part used to transaminate glyoxylate to glycine. Thus two molecules of glutamate are converted to one of glutamine and one of oxoglutarate by reaction with the photorespiratory sequence. These two molecules then revert to the chloroplast where they are converted to glutamate. Glutamine synthetase was also shown to permit the assimilation of ammonia produced by isolated mitochondria. When glycine was supplied the mitochondria released CO_2 and ammonia but, in the presence of appropriate amounts of glutamate, ATP and glutamine synthetase, little ammonia accumulated and the production of CO_2 was accelerated.

Inhibition of the Photorespiratory Pathway

Alpha-hydroxypyridine methane sulphonate (α-HPMS) was shown by Zelitch (1957) to be an inhibitor of glycolate oxidase and to increase CO_2 fixation by 50 per cent when supplied to tobacco leaf discs (Zelitch, 1966). Under the same conditions, glycolate was shown to accumulate but this was greatly decreased if the leaf discs had been previously floated on a solution of L-glutamate (Oliver and Zelitch, 1977). The mechanism by which glutamate inhibits glycolate synthesis is still uncertain. The inhibition of glycolate synthesis was accompanied by a marked decrease in the rate of photorespiratory CO_2 release. However, this effect has not been observed with intact leaves.

Another chemical, sodium 2-hydroxy-3-butynoic acid, is an irreversible inhibitor of glycolate oxidase (Jewess, Kerr and Whitaker, 1975). Exposure of illuminated leaf discs to this compound causes an accumulation of glycolate, although in this case there was no observed increase in net CO_2 assimilation, possibly because in other experiments the compound has been shown to inhibit photosynthesis (Kumarasinghe, Keys and Whittingham, 1977b).

Glycidate (2,3-epoxy propionate), which is an analogue of glycolate, has been shown to inhibit synthesis of glycolic acid in tobacco leaf discs (Zelitch, 1974). There was a corresponding increase in CO_2 uptake. However, in isolated intact chloroplasts, glycidate stimulated glycolate synthesis and photosynthesis (Chollet, 1976). Glycidate may modify the charge on the membrane to result in the excretion of 3-phosphoglyceric acid (PGA) instead of dihydroxyacetone phosphate.

Isonicotinyl hydrazide (INH) inhibits the decarboxylation of glycine to form serine and CO_2 (Kisaki and Tolbert, 1970). But it has been shown that in bean leaves (Smith, Tolbert and Ku, 1976) the addition of INH has resulted in an increase in the CO_2 compensation point. However, INH can also inhibit photosynthesis in the intact leaf (Kumarasinghe, Keys and Whittingham, 1977b).

Servaites and Ogren (1977) treated isolated soya bean mesophyll cells with three of the inhibitors just mentioned. With preincubation of cells under acid conditions, HPMS increased $^{14}CO_2$ incorporation into glycolate but severely inhibited photosynthesis. Sodium 2-hydroxy-3-butynoic acid completely and irreversibly inhibited glycolate oxidase, which resulted in an accumulation of ^{14}C into glycolate. INH increased the incorporation of $^{14}CO_2$ into glycine. However, the rate of photosynthesis was diminished in the presence of these compounds when in the presence of oxygen, but unaffected in the absence of oxygen. This suggests that carbon from the photorespiratory pathway may not be able wholly to recycle back into the Calvin cycle. Except under special conditions with leaf discs of tobacco, in all experiments undertaken so far, inhibition of the glycolate oxidative pathway under atmospheric conditions of CO_2 and oxygen has resulted in inhibition of photosynthesis. Accumulation of carbon in any of the compounds of the photorespiratory sequence represents a diversion of carbon out of the main intermediates of the photosynthetic cycle. According to this view, in order to minimize photorespiration it is necessary to prevent the entry of carbon into the glycolate pathway from the photosynthetic cycle.

Ribulose 1,5-biphosphate Carboxylase/Oxygenase (RuBPc/o)

A considerable body of evidence suggests that phosphoglycolate is formed via the oxidative cleavage of RuBP (Bowes and Ogren, 1972). This reaction, in which oxygen is taken up, is catalysed by the same enzyme that catalyses the incorporation of CO_2 into RuBP to form two molecules of PGA. It is often held that the formation of phosphoglycolate, and hence photorespiration, is an unavoidable consequence that arises from the fact that the enzyme-bound intermediate formed from RuBP is capable of reaction either with oxygen or with CO_2. The ratio of carboxylation to oxygenation is determined simply by the relative concentrations of CO_2 and oxygen.

According to this mechanism, enrichment of ^{18}O in the oxygen present should result in an equal enrichment of ^{18}O in the carboxyl group of glycolate but not in the other reaction product, phosphoglycerate. Experiments using this method have shown that at least 90 per cent of the glycolate formed has come about by oxidation of RuBP (Berry, Osmond and Lorimer, 1978). The results are somewhat equivocal since the degree of enrichment of the isotope within an intact leaf is difficult to measure because of the dilution of the isotope by photosynthetically produced oxygen. Present evidence indicates that carboxylation and oxygenation probably occur at the same site. Detailed studies of the protein which catalyses the reaction show that it consists of a larger and a smaller unit; the larger unit is synthesized in the chloroplast; the smaller is formed in the cytoplasm and must then be transported across the chloroplast envelope to be joined to the larger within the chloroplast. The catalytic site for both carboxylation and oxidation resides in the larger unit; the smaller unit exerts only a regulatory effect. Moreover, within the chloroplast the enzyme is probably bound to internal membranes and this again may have significance for the mechanism of the reaction. Most recent studies show that during purification of the enzyme the carboxylase and oxygenase activities vary in parallel. Earlier evidence tended to suggest that certain treatments elicited a differential response in the two activities. Most recent work indicates that the activity ratio remains constant throughout purification and subsequent storage of the enzyme. The pH activity profile indicates a very close similarity for the two activities.

The carboxylase/oxygenase is inhibited by a number of chemicals. Thus, Tolbert and Ryan (1975) claimed that ribose 5-phosphate increased the carboxylase activity and inhibited the oxygenase but that since it also inhibited phosphoglycolate phosphatase it could not be used as an effective stimulator of the carboxylase relative to the oxygenase. They also considered that fructose diphosphate could inhibit the carboxylase and stimulate oxygenase. Xylitol 1,5-diphosphate is another powerful inhibitor of the enzyme. It, however, inhibits the carboxylase and oxygenase to the same extent. Subsequent work by Chollet and Anderson (1976), who worked with crystalline tobacco enzyme, has shown that all the compounds which have previously been claimed selectively to inhibit the enzyme, affect both carboxylase and oxygenase equally. The claims for differential inhibition all refer to experiments in which impure enzyme was used. Furthermore, they were complicated by the failure to activate the catalyst fully and to do so equally under the conditions used both for the assay of the carboxylase and for the assay of the oxygenase.

There are, however, some reports of genetic differences in the kinetic properties of RuBP carboxylase/oxygenase. Garrett (1977) has reported differences in ryegrass according to ploidy. For four diploid cultivars the K_mCO_2 was about 50 μM but for four tetraploid cultivars it was only 22 μM. The K_mO_2 was approximately the same for each group at 510 μM oxygen. Thus the diploid cultivars should have half the rate of photosynthesis relative to photorespiration compared to tetraploids, and this was confirmed by direct measurements of the CO_2 compensation point and the postillumination burst.

Keck and Ogren (1976) have shown that *Panicum milioides* apparently has an RuBP carboxylase with an increased affinity for CO_2 relative to oxygen. They explained the low CO_2 compensation point, and the decreased response of the compensation point to oxygen, partly in terms of this property of the carboxylase and partly by the presence of increased phosphoenolpyruvate carboxylase (PEPc). Quebedeaux and Chollet (1977) showed that *Panicum milioides* has reduced photorespiration, higher photosynthetic rates than most C_3 plants and a higher dry matter production. They suggested that this may be entirely due to an enhanced PEPc activity.

Photorespiration and Plant Productivity

Present data indicate that in C_3 plants under normal field conditions in the UK, approximately 25 per cent of the net carbon uptake is re-evolved in photorespiration (*Figure 7.1*). We have made some investigations to find how this percentage is modified under different growing conditions.

Wheat is normally grown with a considerable application of nitrogen fertilizer. Three varieties of winter wheat, Maris Huntsman, Maris Fundin and Cappelle Desprez, were grown in the field at Rothamsted with different levels of nitrogen from 0 to 200 kg ha^{-1} (Pearman, Thomas and Thorne, 1977). The rate of true photosynthesis per unit leaf area decreased with increase in nitrogen, partly due to increased mutual shading, but the photosynthetic activity per m^2 of land progressively rose with increased nitrogen. However, grain yield reached an optimum at about 150 kg ha^{-1} and then subsequently fell so that increased total photosynthetic activity of the crop was not reflected in an increased grain yield. There is a number of possible explanations for this phenomenon, which would extend discussion beyond our present purpose. In these experiments the rate of photorespiration was measured by comparing the true rate of photosynthesis measured during short-term exposure to $^{14}CO_2$ with the steady-state net rate of photosynthesis measured by exchange of $^{12}CO_2$. The experiments showed that the rate of photorespiration progressively rose with increased nitrogen application and when expressed as a percentage of photosynthesis, increased from 20 to 40 per cent. There was also an increase in dark respiration rate at higher levels

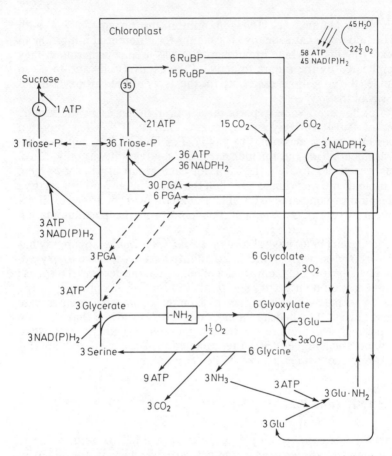

Figure 7.1 *The stoichiometry between the photosynthetic Calvin cycle and the photorespiratory cycle when the rate of photorespiration is 25 per cent of that of net photosynthesis. 58 mol ATP and 45 mol NAD(P)H are required to make 1 mol of sucrose. Without photorespiration the corresponding requirement would be 37 ATP and 24 NAD(P)H*

of nitrogen and the increases of photorespiration and dark respiration explained in large part the decrease in leaf efficiency. In order to determine whether additional nitrogen application had changed the relative activity of the two reactions catalysed by RuBPc, the enzyme was extracted from the plants and its activity determined. Both the catalysis of carboxylation by the RuBPc activity and the catalysis of oxidation catalysed by RuBPo activity decreased with time of development after anthesis. There was, however, no indication that extra nitrogen significantly changed the ratio of the two types of activity.

In other experiments, Lawlor and Fock (1977) observed the effect of water stress on photosynthesis and photorespiration in sunflower, a C_3 species. The plants were grown in nutrient solution in a controlled environment and their water potential decreased by increasing osmolarity of the culture solution. Leaf water potential was measured with a pressure chamber. The photosynthetic activity was measured at 25°C with an irradiance of

$400 \, \mu E \, m^{-2} \, s^{-1}$ with 300 v.p.m. CO_2 and either 1.5 or 21 per cent oxygen gas mixtures. The plants were chemically analysed to determine the distribution of radiocarbon within photosynthetic metabolism. 3-Phosphoglyceric acid was saturated with ^{14}C after ten minutes in unstressed plants at both oxygen concentrations, but with increasing water stress both the rate of accumulation and the specific activity were decreased. By contrast, with decreasing water potential a greater amount of radioactivity appeared in glycine and serine— the intermediates of the glycolate pathway. Other amino acids showed a much smaller effect, except at very great degrees of stress when the amino acid proline increased significantly. Because more radioactivity accumulated in amino acids, less accumulated in sugars, organic acids and sugar phosphates, and the specific activities of these compounds were also decreased. The experiments clearly indicated that water stress resulted in an increase of photorespiration relative to photosynthesis, and this conclusion was confirmed by comparing the results in the presence of 21 and 1.5 per cent oxygen. The principal effect of 1.5 per cent oxygen was to decrease the amount of ^{14}C in glycine and to result in some increase in the proportion of serine compared to glycine. There was also a greater accumulation in sugars and sugar phosphates. The results clearly indicate that under conditions of only moderate water stress the significance of photorespiration is increased and, correspondingly, the photosynthetic efficiency of C_3 plants is decreased.

The Energetics of Photorespiration and its Significance

Ehleringer and Björkman (1977) have measured the quantum yields of the C_3 plant *Atriplex gabriuscula* and the C_4 *Atriplex argentea*. The quantum yield for the C_3 plant was 1 mol CO_2 per 20 absorbed Einstein in the presence of normal air, which increased to 1 mol CO_2 per 13 Einstein in low oxygen. C_4 plants had a quantum yield that was the same as that for C_3 in air. Thus in these experiments the inhibition of photosynthesis by oxygen in C_3 plants imposed an energetic penalty equal to that of the additional phosphorylations required in the 4C acid cycle of C_4 photosynthesis.

It must be appreciated that there is no rigid stoichiometry between the photosynthetic Calvin cycle and the photorespiratory cycle. Only at the compensation point, where carbon fixed by the photosynthetic cycle is balanced by the loss of carbon from the photorespiratory cycle, is there a one-to-one stoichiometry. In order to assess the energy penalty imposed by photorespiration it is necessary to define under a given set of conditions the proportion of the carbon that enters the carbon cycle, which flows around the photorespiratory cycle. If we assume a value which is well substantiated for air, namely that one fifth of the carbon that enters the photosynthetic cycle is released as CO_2 in photorespiration, the overall energetic requirement per mole of sucrose synthesized is raised to 58 mol of ATP and 45 mol NAD(P)H as compared to 37 ATP and 24 NAD(P)H when photorespiration is zero. From this it becomes clear that although the conversion of glycine to serine produces a certain amount of ATP, there are no means by which the photorespiratory metabolism could produce sufficient reduced coenzyme in the cytosol to support conversion of serine to triosephosphate. It is generally assumed that glyceric acid produced as the end-point of glycolate metabolism must be transported back into the chloroplast where photochemistry can supply the necessary energy for its reduction to triosephosphate. The only alternative would be to formulate a shuttle mechanism whereby the reducing power generated in the chloroplast could be transferred to the cytosol by means of some secondary carbon compound. Thus the photorespiratory pathway is a photorespiratory carbon oxidation cycle (PCOC) contrasted with the photosynthetic carbon reduction cycle (PCRC).

Several plant physiologists regard photorespiration as a mechanism for dissipating excess light energy under conditions when CO_2 supply is limited and which might otherwise result in the photodestruction of photosynthetic pigments and reaction centres. For example, when bean leaflets were exposed to high light intensities in CO_2-free nitrogen with 1 per cent oxygen, there was a resultant decrease in maximal rates of photosynthesis and an increase in quantum requirement; the effects were reduced by the addition of 7–21 per cent oxygen, which probably induced photorespiration and allowed excess energy to be dissipated (Powles and Osmond, 1978). However, the presence of RuBPo activity in non-photosynthetic organisms may argue against photorespiration being a process specifically adapted for the dissipation of excess photochemical energy. An alternative view is that photorespiration is the inevitable consequence of using a carboxylase which is also an oxygenase. According to this view, photorespiration is a vestigial characteristic left over from earlier times when plants evolved in an atmosphere which contained a higher concentration of CO_2 and a lower concentration of oxygen.

Conclusions

We need to know a great deal more about the nature of the enzyme and the catalytic mechanism before it can be clearly established that it would be impossible to conceive of an oxygenase-less RuBP carboxylase. Present evidence suggests that such an enzyme is unlikely since the carboxylation and oxygenation are generally considered to involve catalysis at a common active centre and enzyme-RuBP complex. We should consider modifications of the active centre of existing RuBP carboxylases. This should be helped by comparisons with structure and mechanism of other carboxylases which involve other CO_2 acceptors, such as those employed in some photosynthetic bacteria.

In the meanwhile, we should not neglect other aspects of photosynthetic and photorespiratory carbon metabolism since, having achieved a plant that does not photorespire, we may find that some other functions are affected in such a way that irrigation and current agricultural practices cannot compensate.

References

ATKINS, C. A., CANVIN, D. T. and FOCK, H. (1971). In *Photosynthesis and Photorespiration*, pp. 497–505. Ed. by M. D. Hatch, C. B. Osmond and R. O. Slatyer. Wiley Interscience, New York
BERRY, J. A., OSMOND, C. B. and LORIMER, G. H. (1978). *Pl. Physiol.*, **62**, 954–967

BIRD, I. F., CORNELIUS, M. J., KEYS, A. J. and WHITTINGHAM, C. P. (1972a). *Phytochem.*, **11**, 1587–1594

BIRD, I. F., CORNELIUS, M. J., KEYS, A. J. and WHITTINGHAM, C. P. (1972b). *Biochem. J.*, **128**, 191–192

BOWES, G. and OGREN, W. L. (1972). *J. biol. Chem.*, **247**, 2171–2176

CANVIN, D. T., LLOYD, N. D. H., FOCK, H. and PRZYBYLLA, K. (1976). In *CO₂ Metabolism and Plant Productivity*, pp. 161–176. Ed. by R. H. Burris and C. C. Black. University Park Press, Baltimore

CHOLLET, R. (1976). *Pl. Physiol.*, **57**, 237–240

CHOLLET, R. and ANDERSON, L. L. (1976). *Arch. Biochem. Biophys.*, **176**, 344–351

COOMBS, J. and WHITTINGHAM, C. P. (1966). *Phytochem.*, **5**, 643–651

DOUCE, R., MOORE, A. L. and NEUBURGER, M. (1977). *Pl. Physiol.*, **60**, 625–628

EHLERINGER, J. and BJÖRKMAN, O. (1977). *Pl. Physiol.*, **59**, 86–90

GARRETT, M. K. (1977). *Nature*, **274**, 913–915

JEWESS, P. J., KERR, M. W. and WHITAKER, D. P. (1975). *FEBS Lett.*, **53**, 292–296

KECK, R. W. and OGREN, W. L. (1976). *Pl. Physiol.*, **58**, 552–555

KEYS, A. J., BIRD, I. F., CORNELIUS, M. J., LEA, P. J., WALLSGROVE, R. M. and MIFLIN, B. J. (1978). *Nature*, **275**, 741–743

KISASKI, T., IMAI, A. and TOLBERT, N.E. (1971). *Pl. Cell Physiol.*, **12**, 267–273

KISASKI, T. and TOLBERT, N. E. (1970). *Pl. Cell Physiol.*, **11**, 247–258

KISASKI, T., YOSHIDA, N. and IMAI, A. (1971). *Pl. Cell Physiol.*, **12**, 275–288

KUMARASINGHE, K. S., KEYS, A. J. and WHITTINGHAM, C. P. (1977a). *J. exp. Bot.*, **28**, 1247–1257

KUMARASINGHE, K. S., KEYS, A. J. and WHITTINGHAM, C. P. (1977b). *J. exp. Bot.*, **28**, 1163–1168

LAWLOR, D. W. and FOCK, H. (1977). *J. exp. Bot.*, **28**, 320–328

OLIVER, D. J. and ZELITCH, I. (1977). *Pl. Physiol.*, **59**, 688–694

PEARMAN, I., THOMAS, S. M. and THORNE, G. N. (1977). *Ann. Bot.*, **41**, 93–108

POWLES, S. B. and OSMOND, C. B. (1978). *Austr. J. Pl. Physiol.*, **5**, 619–629

QUEBEDEAUX, B. and CHOLLET, R. (1977). *Pl. Physiol.*, **59**, 42–44

SCHOU, L., BENSON, A. A., BASSHAM, J. A. and CALVIN, M. (1950). *Physiol. Plant.*, **3**, 487–495

SERVAITES, J. C. and OGREN, W. L. (1977). *Pl. Physiol.*, **60**, 461–466

SMITH, E. W., TOLBERT, N. E. and KU, H. S. (1976). *Pl. Physiol.*, **58**, 143–146

TOLBERT, N.E. (1971). *Ann. Rev. Pl. Physiol.*, **22**, 45–74

TOLBERT, N. E. and RYAN, K. (1975). *CO₂ Metabolism and Plant Productivity*, pp. 141–159. Ed. by R. H. Burris and C. C. Black. University Park Press, Baltimore

WARBURG, O. and KRIPPAHL, G. (1960). *Z. Naturforsch. B.*, **15**, 197–199

WOO, K. C. and OSMOND, C. B. (1976). *Austr. J. Pl. Physiol.*, **3**, 771–785

ZELITCH, I. (1957). *J. biol. Chem.*, **224**, 251–260

ZELITCH, I. (1966). *Pl. Physiol.*, **41**, 1623–1631

ZELITCH, I. (1974). *Arch. Biochem. Biophys.*, **163**, 367–377

8

LEAF ONTOGENY AND PHOTOSYNTHESIS

ZDENĚK ŠESTÁK
Institute of Experimental Botany, Czechoslovak Academy of Sciences,
Prague, Czechoslovakia

Introduction

The life of a leaf from unfolding to drop is marked by changes in its morphology and composition. The enlargement of area of leaf blade is usually more rapid and stops later than the increase in leaf thickness, while dry matter of one leaf and dry matter per unit leaf area sometimes continue to increase slowly almost throughout the whole leaf life span.

The increase in leaf area is usually paralleled by the increase in cell volume and chloroplast number (Kameya, 1972). The dimensions of chloroplasts, sizes of thylakoid surfaces and the number of thylakoids in the grana increase with the size of leaf blade and chlorophyll content (Holowinsky, Moore and Torrey, 1965; Miller and Nobel, 1972; Havelange, 1977). An increase in leaf size is also accompanied by changes in frequency of juvenile, mature and degenerating chloroplasts (Ehara and Misawa, 1975; for further information on chloroplast development, *see* Woolhouse and Batt, 1976; Kirk and Tilney-Bassett, 1978).

Ontogenetic changes in morphology and cell ultrastructure of a leaf are naturally accompanied by changes in its chemical composition (Woolhouse, 1967, 1974) and metabolic activities. The most important metabolic activity of leaves—photosynthesis—has often been studied with respect to leaf ontogeny. Ontogenetic changes in photosynthetic gas exchange, contents of photosynthetic pigments, activities of photosystems and composition of the electron transport chain, as well as activities of carboxylation enzymes and pathways of carbon fixation have been described in more than 1000 papers. Nevertheless, changes in photosynthetic characteristics during leaf development are only rarely described in textbooks of plant physiology and biochemistry. One possible reason for this is that the vast literature published during the 70 years since the first observations by Lyubimenko (1910) has only rarely been reviewed. In addition to the above-mentioned reviews by Woolhouse on plant senescence, detailed reviews of this field have been prepared over the last few years in our laboratory: review on chlorophyll content and photosynthetic rate (Šesták and Čatský, 1967), its continuation dealing with chlorophylls (Šesták, 1977a), reviews on photosystems, components of electron transport chain and photophosphorylations (Šesták, 1977b, 1978b, 1979), carotenoids (Šesták, 1978a), carbon fixation pathways, their enzymes and products (Zima and Šesták, 1979), and photosynthetic gas exchange and resistances which limit it (Tichá and Čatský, 1981a, 1981b).

147

The aim of this paper is to show, with some examples, the general shape of photosynthetic activities in the course of leaf ontogeny, disregarding the exceptions and extreme cases. For this reason only herbs and the normal leaf development in light and under conditions as optimal as possible are considered. We are omitting those papers in which the photosynthetic characteristics are given only for two or three intervals or for a brief phase of leaf ontogeny. Only some randomly selected, interesting articles are cited here; for further literature the reader should consult the list of references in the cited papers and in the above-mentioned reviews.

Carbon Dioxide Exchange and Conductances for Carbon Dioxide Transfer

The general course of photosynthetic rate per unit leaf area during leaf ontogeny is a rather steep incline to a maximum value, achieved often before completion of leaf expansion and reaching the highest amount of chlorophyll, the peak being usually followed by a slower decline in photosynthetic rate (*Figure 8.1*). In leaves formed in later phases of plant ontogeny, additional peaks in net photosynthesis may occur in the phases of flowering and fruit formation (Woodward and Rawson, 1976). The rates of dark respiration are high after leaf unfolding, at a minimum during the time of maximal photosynthesis, and they increase very slowly during leaf senescence (Čatský, Tichá and Solárová, 1976). A peak in respiration may be observed before the end of the leaf life span (Hardwick, Wood and Woolhouse, 1968). The photorespiration rate per unit leaf area peaks during the time of developing to maximum leaf blade area. In the period of maximum photorespiration the photosynthetic rate declines rather slowly, evidently due to reutilization of produced CO_2.

The CO_2 influx to the reaction sites inside the chloroplasts is controlled by conductances to CO_2 transfer. When the photochemical and carboxylating mechanisms in chloroplasts are active enough, the sum of individual conductances to CO_2 transfer limits the CO_2 influx and hence the photosynthetic rate. The boundary-layer conductance which peaks in the period before the maximum of photosynthetic rate has rather high values and thus cannot be the limiting factor of CO_2 influx (*Figure 8.1*). Stomatal conductance is rather low after leaf unfolding and in the second half of the leaf's life span, and it contributes to the low photosynthetic rates in these phases of leaf life. The main limiting factor is the intracellular (mesophyll) conductance. Čatský, Tichá and Solárová (1976) calculated its upper and lower limits (taking or not taking into account the reassimilation of photorespired CO_2): their shape is similar to the shape of a measure of gross photosynthesis computed as a sum of rates of net photosynthesis, photorespiration and one quarter of dark respiration (*Figure 8.1*). Internal leaf structure may limit CO_2 transfer in photosynthetically mature leaves which are exerting maximal photosynthetic activity (Tichá and Čatský, 1977). Nevertheless, no one of these individual factors is solely responsible for the whole ontogenetic course of net photosynthesis of the leaf. It is probable that

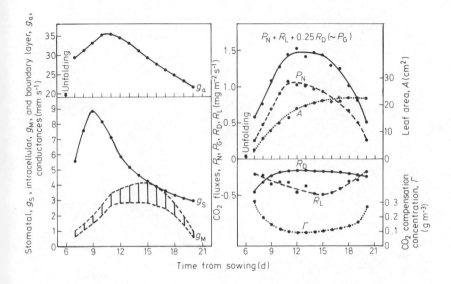

Figure 8.1 *Changes in leaf area (A), net photosynthetic rate (Pₙₑₜ), respiration (R_D), photorespiration (R_L), calculated value of gross photosynthetic rate (P_G ~ Pₙₑₜ + R_L + 0.25 R_D), CO₂ compensation concentration (Γ), and boundary layer (g_a), stomatal (g_s) and intracellular (g_M—two ways of calculation representing potential maximum and minimum values) conductances during ontogeny of a primary leaf of* Phaseolus vulgaris *L. cv. Jantar. (After Čatský, Tichá and Solárová, 1976)*

their contribution to the control of photosynthetic rate during ontogeny is different in various plant taxons and ecotypes (Ludlow and Wilson, 1971).

Each of the successively formed leaves on a plant passes through a course of ontogenetic changes in photosynthetic characteristics. The shapes of these courses are more or less similar, but they differ in the extent of values. These differences are induced by the respective phase of ontogeny of the plant and the microclimate in which the leaf develops (most important being the irradiance). In herbs, maximal net photosynthetic rates per unit area are often reached by leaves that are formed when the plant passes the first third of the vegetative phase of its life. The life span of these leaves is usually the longest (*Figure 8.2*) but they need not reach the largest possible leaf area (*Figure 8.3*). In some plants these maximum photosynthetic activities are reached in the largest leaves formed during the flowering period (Woodward, 1976). In the leaf insertion gradient, the most photosynthetically active are often the third to fifth leaves starting from the youngest one (Šesták and Čatský, 1962; Rawson and Hackett, 1974; Tanaka, Fujita and Kikuchi, 1974). Individual leaves differ also in the relative time of maximum photosynthetic activity (Hodáňová, 1981). The relative contribution of individual leaves to the sum of photosynthesis of all leaves on a plant is thus different in various phases of plant development (*Figure 8.4*).

Detailed analysis shows that the changes during leaf ontogeny are even more complicated than shown above. Václavík (1975) compared the net CO₂ uptake by the upper (adaxial) and lower (abaxial) tobacco leaf surfaces during the ontogeny of the youngest leaf with the net photosynthetic rate in

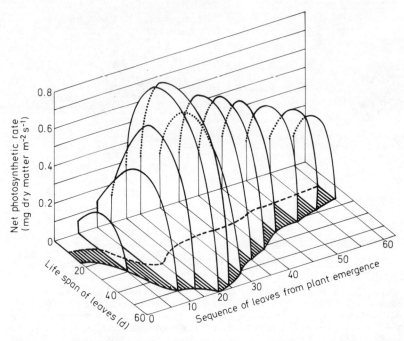

Figure 8.2   *Ontogenetic change in the net photosynthetic rate of successively formed leaves of sugar beet cv. Dobrovická A. Solid line on the base shows the life span of leaves, dashed line the attainment of the maximum leaf area. (After Hodáňová, 1981)*

leaves of various insertion levels (*Figure 8.5*). The differences found in the insertion level gradient are not equal to those found during the ontogeny of one leaf and hence cannot be used in its modelling. In addition to this, the peak values were observed sooner in the adaxial than in the abaxial surface, but the values for abaxial surface were much higher in both the ontogenetic course and the insertion level gradient.

Also, other photosynthetic characteristics of leaves, including their response to irradiance, CO_2 supply, etc. are different in leaves of various ages. Old leaves are saturated with radiant energy sooner and reach much lower photosynthetic rates than photosynthetically mature leaves, while the youngest leaves are less active, especially at low radiant flux densities (*Figure 8.6*). The CO_2 compensation concentration is much higher in young and senescent leaves than in photosynthetically mature leaves (*Figure 8.1*). In a rather wide range of CO_2 concentration (up to five times the normal air concentration) net photosynthetic rate per unit leaf area always increases up to a certain leaf age and then decreases; the dependence of CO_2 influx on CO_2 concentration in ambient air for a leaf of given age is linear (*Figure 8.7*).

Carboxylation Enzymes and Carbon Fixation Pathways

According to the prevailing pathway of photosynthetic carbon fixation, plant species are divided into the C_3, C_4 and crassulacean acid metabolism (CAM)

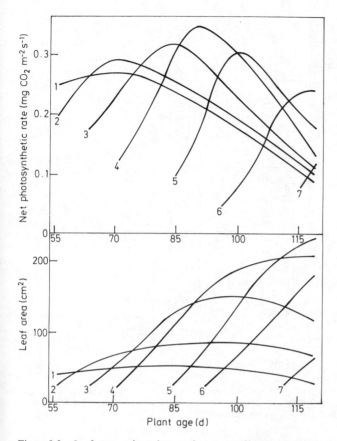

Figure 8.3 *Leaf areas and net photosynthetic rates of leaves successively formed (figures on curves)
on a* Plectranthus fruticosus *L Herit. plant cultivated in a growth chamber under dry and warm
conditions. (After Tichá, 1968)*

groups, which also differ in leaf anatomy and hence transport pathway for
CO_2, the presence of photorespiration and thus the possibility of CO_2
reutilization and CO_2 compensation concentration, responses to photon flux
density, temperature, CO_2 and oxygen concentrations, etc.

The activity of the main carboxylation enzyme in C_3 plants—ribulose 1,5-
bisphosphate carboxylase (RuBPc)— is more or less linearly correlated with
the amount of the so-called fraction 1-protein (F1P). The maximum in
RuBPc activity or F1P content has been found mostly prior to or at the end
of leaf blade expansion (Smillie, 1962; Lloyd, 1976); the rather rapid increase
to the maximum was followed by a slow decline. F1P and RuBPc formation
depend on radiant energy supply (Blenkinsop and Dale, 1974). Some authors
have found a correlation between the maximum RuBPc activity and
maximum photosynthetic rate, while other authors deny this. The source of
discrepancies may be that enzyme activity determined in extract and related
to various units is compared with the photosynthetic rate measured in whole
leaves. The activity of phosphoenolpyruvate carboxylase (PEPc) in C_3 plants

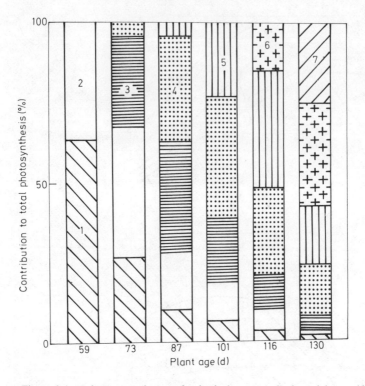

Figure 8.4 *Relative contribution of individual successively formed leaves (figures in areas of different patterns) to the sum of photosynthesis of all leaves on a plant of* Plectranthus fruticosus L. Herit. *grown as shown in Figure 8.3. (After Tichá 1976)*

is much lower than that of RuBPc, but their ontogenetic courses are usually similar.

In C_4 plants, an increase to a maximum followed by a decline has usually been found for both PEPc and RuBPc, and here the difference in activity between the main enzyme PEPc, which catalyses the primary fixation of CO_2 from the air, and RuBPc, which catalyses the secondary fixation of CO_2 in bundle sheath chloroplasts, is much smaller. In maize plants, used most often for these measurements, the evaluation is complicated by the heterogeneity of the long leaf blade. In the leaf insertion gradient the second-formed leaf reached the highest PEPc activities (Möller, Stamp and Geisler, 1977).

Exceptionally, a change in the main path of carboxylation has been found in the course of leaf ontogeny. Mokronosov (1973) observed in 2–3 d leaves of the C_3 tomato plant a high rate of PEP carboxylation (70–80 per cent of total CO_2 fixation), while in 25 d and older leaves 80–90 per cent of CO_2 was fixed by means of RuBPc. Similarly, Passera (1975) found an increase in the ratio of RuBPc to PEPc activities from 5 d (3.7) to 15 d (17.0) barley plants. Inverse changes in the carbon pathway were found in the leaf insertion gradients of wheat and oat plants (Wirth *et al.*, 1977).

153

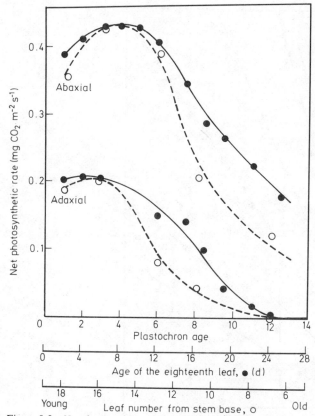

Figure 8.5   *Net photosynthetic rate at 300 W m⁻² (400–700 nm) as amount of CO₂ absorbed by the upper (adaxial) and lower (abaxial) leaf surfaces during ontogenesis of the eighteenth leaf of* Nicotiana tabacum *L. cv. Wisconsin 38 and in leaves in the descending insertion level. Plastochron age (LPI) is given for comparison. (After Václavík, 1975)*

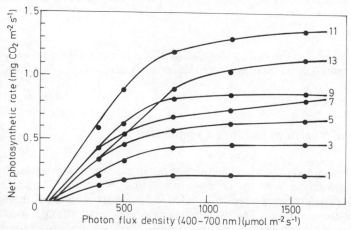

Figure 8.6   *Dependence of net photosynthetic rate on photon flux density in leaves of sugar beet cv. Dobrovická A of different ages (figures on curves indicate the leaf insertion level starting with the oldest leaf (Hodáňová, 1979)*

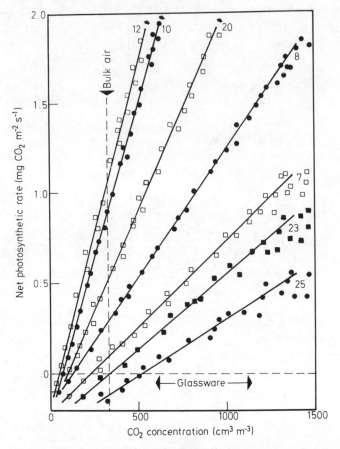

Figure 8.7   *Dependence of net photosynthetic rate on carbon dioxide concentration in primary leaves of* Phaseolus vulgaris *cv. Jantar of different ages (figures on curves = plant age, d). (Modified from Tichá, Čatský and Peisker, 1980)*

Leaf age also affects the composition of products of short-term and long-term $^{14}CO_2$ metabolism, mainly saccharides and amino acids (Zima and Šesták, 1979).

Photochemical Reactions and their Components

It is not easy to compare the ontogenetic changes in gas exchange, measured usually in whole leaves, with the changes in activities of photosystems and photophosphorylations, measured in isolated chloroplasts suspended in an artificial reaction medium. The isolation procedures may change some properties of the chloroplasts. As the preparations lose their activities rather soon, the data reported in the literature are far from uniform (Šesták, 1977b).

The activity of photosystem II (PSII) has been frequently measured in chloroplasts from leaves of varying age, the first reported being by Clendenning and Gorham (1950). In the last ten years the measurements of photosystem II activity (mostly as a Hill reaction) have sometimes been complemented with measurements of photosystem I and photophosphorylation activities. In the rather rare cases when the whole course of leaf ontogeny has been followed, up to three peaks in activities are observed. They are most expressed when calculated per unit amount of chlorophyll, but remain discernible after theoretical recalculation per area or dry matter of the source leaf. The peaks for both photosystems and photophosphorylations appear simultaneously. The first peak appears just after leaf unfolding, the second peak mostly after the end of main growth of leaf blade and chlorophyll accumulation, while the third peak is sometimes observed at the end of positive photosynthetic balance of the leaf (*Figure 8.8*). Artificial prolongation of the leaf life span (*Figure 8.8*, right) shifts the appearance of peaks by a few days. The ratio of activities of photosystem I/photosystem II slowly declines in the course of leaf ontogeny. The dependence of chloroplast activities on radiant flux density changes during leaf life, at the end of which they acquire

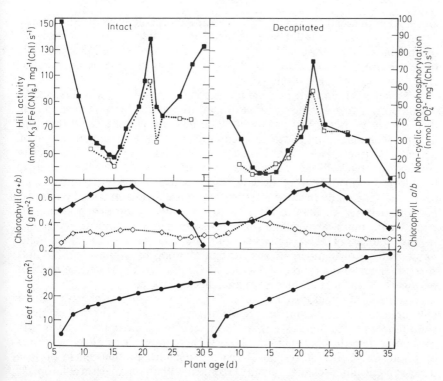

Figure 8.8 *Changes in the rates of Hill activity (■) and non-cyclic photophosphorylation (□) in chloroplasts isolated from primary leaves of Phaseolus vulgaris L. cv. Jantar in the course of their ontogeny. Leaves grew on intact plants or plants of which newly formed leaves were pinched off immediately after their appearance ('decapitated' plants). For comparison, changes in leaf area (●) and chlorophyll (a + b) content (♦) and a/b ratio (◇) are shown. (After Šesták, Zima and Strnadová, 1977)*

a shade character. The changes in photosystem activities are in agreement with the changes in relative contents of P700 and chlorophyll *in vivo* forms, amounts of components of the electron transport chain (cytochromes, quinones and ferredoxin), distribution and properties of separable chloroplast fragments, etc. Hence the changes in photosystem and photophosphorylation activities during leaf ontogeny are connected with the changes in size and composition of the photosynthetic unit. In young maize leaves the photosynthetic unit contains 290 chlorophyll molecules, while in old leaves there are 863 molecules (Keresztes and Faludi-Dániel, 1973).

Differences like those found during leaf ontogeny may be observed on the area of a developing leaf blade (Baker and Leech, 1977) or in the leaf insertion gradient.

The ontogenetic changes in photochemical activities need not occur in parallel with those of photosynthetic gas exchange, but the second and third photochemical peaks are reflected at least as shoulders in the declining part of the ontogenetic gas exchange curve (Šesták *et al.*, 1975).

Summary

During leaf ontogeny not only are its size and morphology changed, but also its metabolic activities. These changes have often been neglected by biochemists and by those studying plant physiology and agriculture, although there are over 1000 papers which describe changes in some photosynthetic characteristics during leaf ontogeny.

The photosynthetic rate of a leaf increases to a maximum, which is usually attained prior to reaching maxima of leaf area and chlorophyll content; it then decreases as the leaf grows old. The increase is more rapid than the decline. The maximum of stomatal conductance precedes the maximum of net photosynthetic rate. The shape of intracellular conductance changes is similar to that of the calculated sum of rates of net CO_2 influx, photorespiration and part of the dark respiration. The maximum rate of dark respiration is usually reached after leaf unfolding, while maximal photorespiratory activity appears in a leaf with a finished blade expansion. With regard to the main carboxylation enzymes, some authors found an agreement in the courses of their activities and the courses of CO_2 influx and photosynthate formation patterns during leaf ontogeny, while others did not. Also, changes in the ratio of activities of C_3, C_4 and CAM carbon fixation pathways during leaf development have been observed.

Activities of photosystems I and II and photophosphorylation in chloroplasts isolated from leaves of increasing age reach two or three simultaneous ontogenetic maxima, the effect of which may be reflected only as shoulders on the course of net photosynthetic rate. The ratio of activities of photosystem I/photosystem II usually declines in the course of leaf development. These changes are in agreement with the changes in chloroplast ultrastructure and chlorophyll *a* forms *in vivo*.

The ontogenetic course of each leaf is affected also by the respective phase of ontogeny of the plant and the microclimate in which it develops. In herbs, maximal net photosynthetic rates are usually reached by leaves formed after one third of the vegetative phase of plant life. Also, other photosynthetic characteristics of leaves, including their responses to irradiance, CO_2 supply etc., are different in leaves of various ages.

References

BAKER, N. R. and LEECH, R. M. (1977). *Pl. Physiol.*, **60**, 640–644
BLENKINSOP, P.G. and DALE, J. E. (1974). *J. exp. Bot.*, **25**, 899–912
ČATSKÝ, J., TICHÁ, I. and SOLÁROVÁ, J. (1976). *Photosynth.*, **10**, 394–402
CLENDENNING, K. A. and GORHAM, P. R. (1950). *Can. J. Res.*, **28C**, 114–139
EHARA, Y. and MISAWA, T. (1975). *Phytopathol. Z.*, **84**, 233–252
HARDWICK, K., WOOD, M. and WOOLHOUSE, H. W. (1968). *New Phytol.*, **67**, 79–86
HAVELANGE, A. (1977). *Physiol. Végét.*, **15**, 723–734
HODÁŇOVÁ, D. (1979). *Photosynth.*, **13**, 376–385
HODÁŇOVÁ, D. (1981). *Biol. Plant.*, **23**, 58–67
HOLOWINSKY, A. W., MOORE, P. B. and TORREY, J. G. (1965). *Protoplasma*, **60**, 94–110
KAMEYA, T. (1972). *J. exp. Bot.*, **23**, 62–64
KERESZTES, Á. and FALUDI-DÁNIEL, Á. (1973). *Acta Biol. Acad. Sci. Hung.*, **24**, 175–189
KIRK, J. T. O. and TILNEY-BASSETT, R. A. E. (1978). *The Plastids*. 2nd edn. Elsevier/North Holland Biomedical Press, Amsterdam
LLOYD, E. J. (1976). *Z. Pfl.-physiol.*, **78**, 1–12
LUDLOW, M. M. and WILSON, G. L. (1971). *Austr. J. biol. Sci.*, **24**, 1077–1087
LYUBIMENKO, V. N. (1910). *Trudy Sanktpeterburgskogo Obshchestva Estestvoznanii*, Ser. III, **20**, 76–84; *Botanicheskaya*, **41**, 1–266
MILLER, M. M. and NOBEL, P. S. (1972). *Pl. Physiol.*, **49**, 535–541
MOKRONOSOV, A. T. (1973). *Trudy Biologo-Pochvennogo Instituta, Novaya Seriya*, **20**, 76–84
MÖLLER, G., STAMP, P. and GEISLER, G. (1977). *Z. Pfl.-ernähr. Bodenk.*, **140**, 481–490
PASSERA, C. (1975). *Riv. Agron.*, **9**, 56–60
RAWSON, H. M. and HACKETT, C. (1974). *Austr. J. Pl. Physiol.*, **1**, 551–560
ŠESTÁK, Z. (1977a). *Photosynth.*, **11**, 367–448
ŠESTÁK, Z. (1977b). *Photosynth.*, **11**, 449–474
ŠESTÁK, Z. (1978a). *Photosynth.*, **12**, 89–109
ŠESTÁK, Z. (1978b). In *Photosynthetic Oxygen Evolution*. Ed. by H. Metzner, pp. 489–493. Academic Press, London
ŠESTÁK, Z. (1979). In *Photosynthesis and Plant Development*. Ed. by R. Marcelle, H. Clijsters and M. Van Poucke, pp. 21–29. Dr. W. Junk, The Hague
ŠESTÁK, Z. and ČATSKÝ, J. (1962). *Biol. Plant.*, **4**, 131–140
ŠESTÁK, Z. and ČATSKÝ, J. (1967). In *Le Chloroplaste, Croissance et Vieillissement*. Ed. by C. Sironval, pp. 213–262. Masson et Cie, Paris
ŠESTÁK, Z., ČATSKÝ, J., SOLÁROVÁ, J., STRNADOVÁ, H. and TICHÁ, I. (1975). In *Genetic Aspects of Photosynthesis*. Ed. by Yu. S. Nasyrov and Z. Šesták, pp. 159–166. Dr. W. Junk, The Hague
ŠESTÁK, Z., ZIMA, J. and STRNADOVÁ, H. (1977). *Photosynth.*, **11**, 282–290
SMILLIE, R. M. (1962). *Pl. Physiol.*, **37**, 716–721

TANAKA, A., FUJITA, K. and KIKUCHI, K. (1974). *Soil Sci. Pl. Nutr.*, **20**, 173–183
TICHÁ, I. (1968). *Photosynth.*, **2**, 167–171
TICHÁ, I. (1976). *Biol. Plant.*, **18**, 237–240
TICHÁ, I. and ČATSKÝ, J. (1977). *Photosynth.*, **11**, 361–366
TICHÁ, I. and ČATSKÝ, J. (1981a). *Photosynth.*, **15**, in press
TICHÁ, I. and ČATSKÝ, J. (1981b). *Photosynth.*, **15**, in press
TICHÁ, I., ČATSKÝ, J. and PEISKER, M. (1980). *Photosynth.*, **14**, 489–496
VÁCLAVÍK, J. (1975). *Biol. Plant.*, **17**, 411–415
WIRTH, E., KELLY, G. J., FISCHBECK, G. and LATZKO, E. (1977). *Z. Pfl.-physiol.*, **82**, 78–87
WOODWARD, R. G. (1976). *Photosynth.*, **10**, 274–279
WOODWARD, R. G. and RAWSON, H. M. (1976). *Austr. J. Pl. Physiol.*, **3**, 257–267
WOOLHOUSE, H. W. (1967). In *Aspects of the Biology of Ageing*, pp. 179–213. Cambridge University Press, Cambridge
WOOLHOUSE, H. W. (1974). *Sci. Progr.*, **61**, 123–147
WOOLHOUSE, H. W. and BATT, T. (1976). In *Perspectives in Experimental Biology*, Vol. 2: *Botany*. Ed. by N. Sunderland, pp. 163–175. Pergamon Press, Oxford
ZIMA, J. and ŠESTÁK, Z. (1979). *Photosynth.*, **13**, 83–106

9

ADAPTATION TO SHADE

HARRY SMITH
Department of Botany, University of Leicester, UK

Shade is an ever-present factor in crop production, be it horticulture, agriculture or forestry. Crop plants shade each other, themselves and weeds—indeed, often the weeds shade the crop plants! If the objective of good crop husbandry is to ensure the achievement of maximum productivity within the theoretical limits of all constraining factors, then clearly shade is an important phenomenon to be understood and, if possible, controlled. The effects of shade on higher plants, particularly with regard to photosynthetic productivity, have been studied in great detail throughout the whole of this century. Many comprehensive reviews exist (e.g. Rabinowitch, 1951; Grime, 1966; Boardman, 1977) and it is not the intention of this author to attempt an encyclopaedic coverage. Rather, this chapter concentrates on the mechanisms whereby plants perceive, and react developmentally, to vegetational shade since the capacity of a crop plant to adapt to shade is a major physiological limitation to productivity.

In the natural—i.e. non-crop—environment, several different strategies in response to shade may be recognized, but these may be conveniently reduced to three major categories:

(1) *shade avoiders*, or obligate sun plants;
(2) *shade tolerators*, or facultative sun/shade plants;
(3) *shade requirers*, or obligate shade plants.

The majority of crop plants—indeed all of those currently grown in significant amounts in the temperature zones—are shade avoiders, as are most herbaceous weeds, although some weeds are shade-tolerant. Shade-requiring, flowering plants are rare and probably absent from the temperate zones, although many lower plants may be in this category. Since shade requirers are rare, I shall restrict myself here to shade avoiders and shade tolerators.

Both strategies may be thought of as adaptations to shade, since in both cases individual plants are more productive in the non-shaded habitats, as long as water and nutrient supplies are not limiting. The responses of both types of plant to shade, therefore, should be seen as strategies for survival, rather than for productivity. In the case of the shade avoiders, the adaptation consists of developmental changes which result, if possible, in the plant growing out of the shade. Changes in the photosynthetic machinery do occur, but they appear to be less important to the overall objective of survival than are the growth changes. The alternative strategy, employed by the shade tolerators, is to modify development such that the plant becomes more capable of photosynthesizing efficiently under the shaded conditions. Growth changes are evident but less striking than in the case of the shade avoiders. The general nature of the two types of adaptation to shade is as follows:

(1) *shade avoiders*: extreme extension growth of stems and petioles, reduced branching (enhanced apical dominance), reduced leaf area, reduced leaf thickness, increased specific leaf area (SLA) due principally to the decrease in leaf thickness (Grime, 1966);

(2) *shade tolerators*: little or no stimulation of extension growth (sometimes inhibition), increased leaf area, increased leaf thickness, increased SLA due to increased area, increased chlorophyll content, increased complexity of photosynthetic machinery (Grime, 1966; Boardman, 1977).

The above adaptational strategies are physiological responses to shade which are initiated whenever an individual plant, or part of a plant, becomes shaded, and may continue for days or weeks. In order for the complex of responses to be initiated, it is obvious that the plant must perceive the fact that it is shaded. The central part of this chapter is concerned with describing the physical nature of canopy shade, and outlining the putative shade perception mechanisms possessed by higher plants.

The Physical Characteristics of Shade and its Experimental Simulation

Although a vast body of research has been carried out in the past on the assumption that vegetational shade is merely a reduction in overall irradiance, this is not the case. Both quantitative *and* qualitative differences exist between unfiltered daylight, and canopy light, as may be readily seen from spectroradiometer scans (*Figure 9.1*). Vegetation selectively absorbs the red and blue wavelengths (due, of course, to the photosynthetic pigments) but transmits some of the green and most of the far-red (i.e. 700–800 nm). Due to the previously mentioned vast body of work based on somewhat false premises, we have long known that plants have a mechanism which allows them to perceive changes in light quantity; in the last few years, however, the existence of a wholly separate mechanism for the perception of light quality changes in the red and far-red wavelength regions has been demonstrated. With these mechanisms higher plants are able to adapt sensitively, accurately and rapidly to a wide range of shaded conditions.

Although, in nature, changes in light quality never occur without simultaneous changes in light quantity, experimentally it is necessary to alter the two variables independently in order to provide evidence on the nature of the perception mechanisms. First of all, however, a simple, arbitrary, parameter describing the typical light quality changes is needed. For cogent reasons explained at length elsewhere (Smith and Holmes, 1977; Smith, 1981), the following parameter ζ (zeta) has been used:

$$\zeta = \frac{\text{photon fluence rate 655–665 nm}}{\text{photon fluence rate 725–735 nm}}$$

This parameter, which is, crudely, the ratio of the red photons to the far-red photons, is a good measure of the light quality in the red/far-red regions.

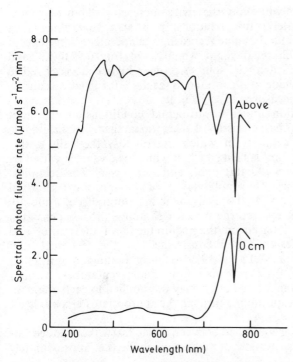

Figure 9.1 *The spectral photon fluence rate distribution of solar radiation above a wheat canopy, and at ground level (0 cm) within the canopy. (Data of M. G. Holmes)*

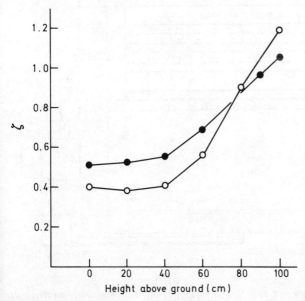

Figure 9.2 *The relationship between the red/far-red photon fluence rate ratio (ζ) and depth within a wheat canopy (height of crop, 90 cm) on overcast (●—●) and clear (○—○) days. (Data of M. G. Holmes)*

Inspection of *Figure 9.1* will show that with increasing depth of shade, ζ decreases proportionately; this relationship is shown graphically in *Figure 9.2*. The highest natural ζ value recorded in temperate regions (except underwater) is that of unfiltered daylight, which ranges from 0.99 to 1.32 and means at 1.03 ± 0.02; only artificial sources provide higher ratios of red to far-red. Within canopy shade, it is common to reach $\zeta=0.1$ in cereals and as low as $\zeta=0.05$ has been measured in sugar beet.

The separate simulation under experimental conditions of the light quantity and light quality characteristics of vegetational shade is a formidable task, requiring growth cabinets in which the photosynthetically active radiation (PAR, 400–700 nm) is held constant while the ζ value is varied by the addition of far-red (>700 nm) light, and vice versa. The technical problems arise from the need to supply very large amounts of extra far-red radiation, in order to depress ζ to the levels found in natural shade, while still providing sufficient PAR to keep the plants well above the compensation point. One such cabinet, based on the requirements of the author and colleagues, has been designed and constructed by the technical staff (Heathcote, Bambridge and McLaren, 1979) at Sutton Bonington. It has four chambers with uniform PAR provided from fluorescent tubes; various quantities of additional far-red radiation may be supplied to each chamber from tungsten-halogen lamps filtered through the appropriate Perspex sheets (*Figure 9.3*).

In this cabinet, ζ may be controlled between about 3.0 (without added far-red) down to values characteristic of vegetation shade (i.e. approximately

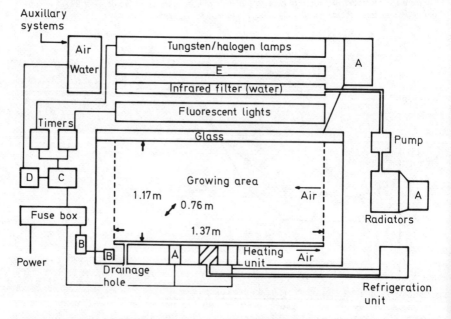

Figure 9.3   *Layout of one compartment in the light quality cabinet. A: Blower units for cooling. B: Temperature sensor and control unit. C: Main isolating relay. D: Safety devices control unit. E: 3.2 mm red 400 plus 3.2 mm green 600 Perspex (ICI). (From Heathcote, Bambridge and McLaren, 1979)*

0.1). Several previously published attempts to test experimentally the effects of varying the red/far-red photon balance on plant growth failed to reveal the full extent of the morphogenetic effects because the amounts of extra far-red applied were too low. Fluorescent lamps, depending on type, give a ζ value of between 2.0 and 9.5; the addition of unfiltered incandescent bulb radiation to the latter type of fluorescent radiation would not depress ζ below that found in ordinary daylight, and thus growth effects similar to those observed in shade could not be expected (Young, 1975).

Table 9.1 THE HABITAT AND COLLECTION SITE FOR EACH OF THE SPECIES USED IN THE EXPERIMENTS DESCRIBED HERE. (FROM MORGAN AND SMITH, 1979)

Species	*Habitat*	*Collection site*
Open habitat		
Chamaenerion angustifolium (L.) Scop.	Rocky places, scree slopes, wood margins, wood clearings, disturbed ground	Arable headland, Sutton Bonington, Notts. (SK 514 261)
Chenopodium album L.	Waste places, cultivated land	Arable land, Sutton Bonington, Notts. (SK 506 262)
Medicago arabica (L.) Huds.	Grassy places, waste ground, light soils near sea	Arable land, Sutton Bonington, Notts. (SK 506 262)
Sinapis alba L.	Arable and waste land	Commercial stock (arable land)
Senecio vulgaris L.	Cultivated land, waste places	Arable land, Sutton Bonington, Notts. (SK 506 262)
Tripleurospermum maritimum ssp, inodorum (L.) Hyl ex Vaarama	Arable and waste land	Arable land, Sutton Bonington, Notts. (SK 506 262)
Intermediate canopy habitat		
Urtica dioica L.	Hedge-banks, woods, grassy places, fens	Hedgerow, Sutton Bonington, Notts. (SK 507 258)
Closed canopy habitat		
Circaea lutetiana L.	Woods, shady places	Closed oak woodland, Swithland Wood, Leics. (SK 537 117)
Geum urbanum L.	Woods, scrub, hedge-banks, shady places	Closed oak woodland, Swithland Wood, Leics. (SK 537 117)
Mercurialis perennis L.	Woods, shady mountain rocks	Closed ash woodlands, nr Bowers Hall, Derbys. (SK 233 652)
Oxalis acetosella L.	Woods, hedge-banks, shady rocks	Closed oak woodland, Swithland Wood, Leics. (SK 537 129)
Silene dioica (L.) Clairv.	Woods, rocky-slopes, hedgerows, cliff ledges	Closed mixed woodland, Domleo Spinney, Notts. (SK 514 263)
Teucrium scorodonia L.	Woods, grassland, heaths, dunes	(a) For survey: Closed oak woodland, Swithland Wood, Leics. (SK 539 122) (b) Ecotypes: Open: short calcareous grassland, Lathkilldale, Derbys. (SK 176 654) Shade: closed oak woodland, Swithland Wood, Leics. (SK 543 118)

164

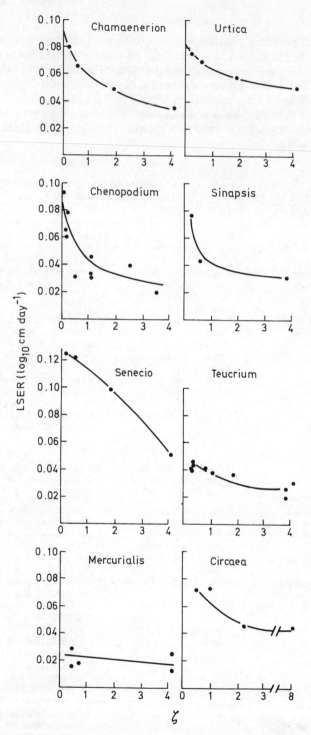

Figure 9.4 *The relationship between logarithmic stem extension rate (LSER) and ζ for a range of seedlings characteristic of open and shaded habitats (see Table 9.1 for details of species and habitats)*

Effects of Shade Light Quality on Development

When a range of plants (*Table 9.1*) were grown in the light quality cabinets, marked effects on development were observed (Morgan and Smith, 1979). Prominent among these effects, particularly with shade avoiders, is extreme internode elongation at low ζ (*Figure 9.4*), but several other developmental parameters also change. In general, the shade avoiders react to low ζ levels by concentrating their growth potential into stem extension; on the other hand, woodland plants, although obviously capable of perceiving the light quality signal, do not react by extreme extension growth.

The dramatic nature of these effects of ζ on extension growth is better seen when the extension growth is monitored continuously with a position-sensitive transducer (*Figure 9.5*). A large increase in growth rate is evident within 10–15 minutes of applying extra far-red light via an interference filter. Thus, plant growth and development are rapidly and extensively responsive, either simply to the far-red fluence rate, or to the relative proportions of far-red and shorter visible wavelengths in the incident radiation.

Effects of Shade Light Quantity on Development

When *Chenopodium album, Urtica dioica* and *Teucreum scorodonia* seedlings—chosen to represent plants from open, hedgerow and woodland habitats,

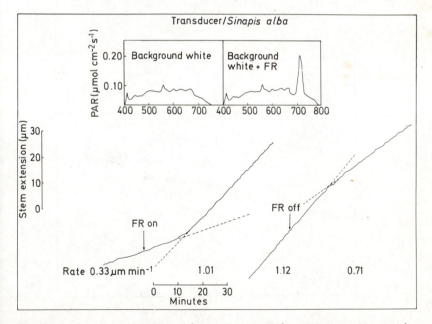

Figure 9.5 *Continuous measurement of stem extension rate by a position–sensitive transducer. A seedling attached to the transducer and exposed to background white fluorescent light was given far-red (FR) light via a fibreoptic probe fitted with a 719 nm interference filter. The far-red source was switched on and off as indicated. The inset diagrams show the spectral photon fluence rate distributions of the background white light with and without added far-red. (Data of D. C. Morgan)*

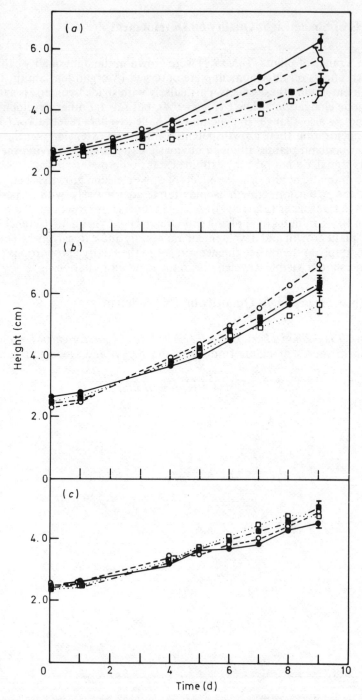

Figure 9.6 *Time courses of the increase in stem length of (a)* Chenopodium album, *(b)* Urtica dioica, *and (c)* Teucreum scorodonia *under various photon fluence rates keeping ζ uniform and high. Photosynthetically active radiation (PAR), in µmol m^{-2} s^{-1}: □ 43; ■ 76; ○ 100; ● 214. (Data of D. C. Morgan)*

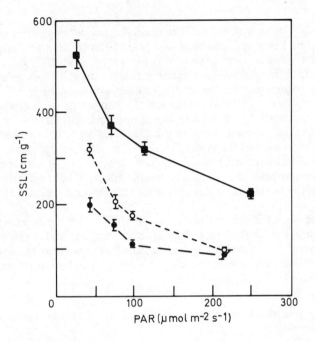

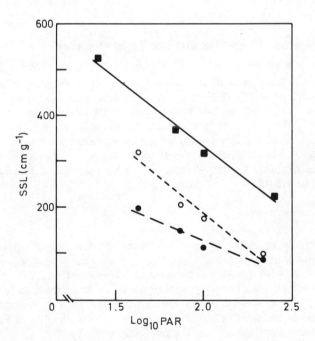

Figure 9.7 *Effect of light quantity on stem development. PAR: Photosynthetically active radiation (total photon fluence rate 400–800 nm). SSL: Specific stem length (stem length/stem dry weight).* ■: Urtica dioica. ○: Chenopodium album. ●: Teucreum scorodonia. *(Data of D. C. Morgan)*

respectively—were transferred from high fluence rates to a range of lower fluence rates, keeping ζ constant and high, little or no effect on extension growth rate was observed (*Figure 9.6*; Morgan and Smith, 1980). Indeed, a reduction in light quantity—simulating shade—actually caused a small but significant *reduction* in extension rate in *Chenopodium album*.

However, more subtle developmental changes do occur when shade-avoiding plants are transferred to low light levels. Although stem extension is little changed, specific stem length (i.e. length per unit weight) is quite markedly increased in the lower light levels in all three species (*Figure 9.7*). In other words, the plants produce the same length of stem with less material. This is an important difference between the effects of light quantity and light quality; in the latter case, clear effects on specific stem length have not been observed.

More obvious differences between the effects of the two environmental variables are seen in relation to leaf development. *Figure 9.8* shows that two parameters of leaf development—the leaf area ratio and the specific leaf area—are affected in one direction by changes in ζ, and oppositely by changes in light quantity simulating shade.

The general conclusions seem to be that:

(1) low ζ results in maximum extension rate at the expense of leaf development;
(2) low PAR results in maximum leaf development at the expense of stem development.

The Perception of Light Quality and Light Quantity

The demonstration that light quality (ζ) and light quantity (PAR) affect development in quite different ways implies that separate perception mechanisms exist for the two environmental variables. The perception of light quality must involve a comparative element, by which the photon fluence rates at two rather widely separated wavelengths may be compared (Smith, 1980). Phytochrome is ideally suited to such a function as it exists in two stable isomeric forms, Pr and Pfr, which are photointerconvertible according to the following scheme:

$$Pr \underset{hv;\,\lambda_{max}730}{\overset{hv;\,\lambda_{max}660}{\rightleftharpoons}} Pfr$$

Pr and Pfr have overlapping absorption spectra below approximately 730 nm, and thus with broad-band visible radiation a photoequilibrium is established at which both photoconversions occur at equivalent rates. At photoequilibrium, the relative proportions of Pr and Pfr (usually described as Pfr/Ptotal, or simply as ϕ) are determined by the spectral photon distribution of incident radiation, the extinction coefficients of Pr and Pfr, and the quantum efficiencies of the photoconversions.

Thus ζ, which is a measure of the red/far-red ratio of the incident radiation, is related to ϕ. Unfortunately, it is currently not possible to estimate ϕ spectrophotometrically in green plants, due to absorbance and fluorescence of chlorophyll. Consequently, we have have been forced to use an indirect

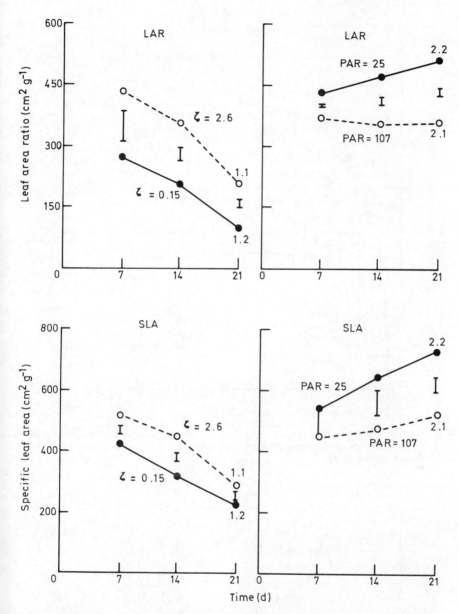

Figure 9.8 *The opposite effects of light quality and light quantity on leaf development in the shade avoider* Chenopodium album. *LAR: Leaf area ratio (leaf area/shoot dry weight). SLA: Specific leaf area (leaf area/leaf dry weight)*

approach. By making many measurements of ϕ in etiolated test materials exposed to a range of natural and artificial light sources, the curve shown in *Figure 9.9* was constructed (Smith and Holmes, 1977). This provides the relationship between ζ, the red/far-red ratio of the incident radiation, and ϕ_e, the photoequilibrium present in etiolated plants.

170

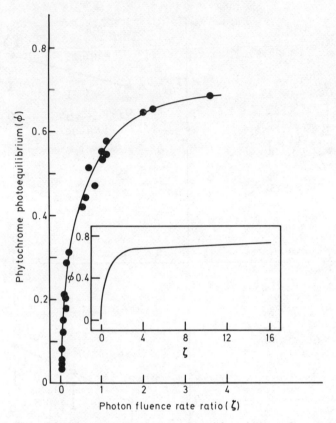

Figure 9.9   *The relationship between the red/far-red photon fluence rate ratio of the incident radiation (ζ) and the phytochrome photoequilibrium established in etiolated test plants (ϕ_e). (From Smith and Holmes, 1977)*

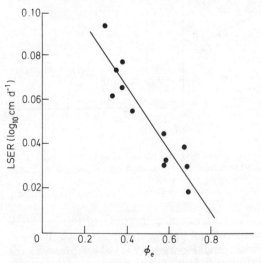

Figure 9.10   *The relationship between ϕ_e, derived from the ζ versus ϕ_e curve in* Figure 9.9, *and the logarithmic stem extension rate (LSER) in* Chenopodium album. *(From Morgan and Smith, 1976)*

We can now use ϕ_e, derived from ζ via *Figure 9.9*, instead of ζ itself; in this way we can plot growth parameters, such as the logarithmic stem extension rate used in *Figure 9.4*, against a parameter related to the phytochrome photoequilibrium expected to be present in the experimental plants, namely ϕ_e. When this transformation is employed, the curves of *Figure 9.4* are converted into linear relationships, as exemplified in *Figure 9.10* (Morgan and Smith, 1976).

It is important to stress at this point that there is no intention whatsoever of using ϕ_e as an 'estimate' of ϕ in the green tissues. Since chlorophyll is distributed throughout the leaf, a gradient of ζ will exist across the leaf, and actual ϕ will vary from cell to cell. At present, the estimation of ϕ within green tissues can only be approached theoretically; such theoretical analysis, however, shows that ϕ at any point within a leaf, although clearly dependent on the optical properties of the leaf, is primarily determined by the spectral photon distribution (and therefore, ζ) of the incident radiation (Holmes and Fukshansky, 1979).

Thus ϕ_e, although not an estimate of ϕ, is systematically related to ϕ. *Figure 9.10*, therefore, may be taken as evidence for a linear relationship between growth rate and phytochrome photoequilibrium, which suggests that phytochrome operates, via the photoequilibrium established by the incident radiation, to perceive the relative amounts of red and far-red radiation. Other possibilities do exist and cannot as yet be eliminated, but space precludes full discussion here (*see* Smith, 1980 for further comments on physiological mechanisms).

The perception of light quantity must involve a photon-counting mechanism of one form or another. One possibility is photosynthesis itself. The light reactions of photosynthesis present a number of opportunities for photon counting (e.g. formation of adenosine triphosphate (ATP) or reduced nicotinamide adenine dinucleotide phosphate (NADPH)); even the accumulation of carbohydrate and other end-products represents ultimately a measure of the number of photons absorbed per unit time. It seems unlikely, however, that photosynthesis itself would be employed to initiate a *qualitative* change in developmental pattern such as is observed. It is easy to see how the availability of ATP, or of reducing power or indeed of photosynthetic assimilates, could modify the overall extent of growth; but building in a selective mechanism is a major conceptual problem. On the other hand, complex feedback controls between the photosynthetic end-products and the processes that lead to the synthesis and construction of the photosynthetic apparatus itself, could certainly be involved in the increased development of chloroplasts in response to low total fluence rates (Björkman *et al.*, 1972).

A further candidate, however, seems more likely at least on *a priori* grounds. Plants possess a blue-light absorbing photoreceptor, responsible for photoperception in phototropism and for a number of blue-light controlled growth phenomena (Smith, 1975). This photoreceptor, which is often known as cryptochrome, is almost certainly one or more flavoproteins (Presti and Delbrük, 1978) and must operate in phototropism by counting photons. Indeed, phototropism in fungi is so sensitive that it is capable of detecting the difference in photon fluence rate on two sides of a sporangiophore, of diameter approximately 0.1 mm (Shropshire, 1963; Bergman *et al.*, 1969). Even allowing for the sporangiophore acting as a lens and thus amplifying

172   *Adaptation to shade*

the signal, this must represent an extremely sensitive transduction mechanism. The effects of blue light on plant growth have been only sporadically studied, in comparison with those of red and far-red light, but nevertheless there is good evidence for responses to blue light (e.g. Gaba and Black, 1979) which are consistent with cryptochrome operating as a perceptor of total light quantity as already proposed by Delbrück (1976). It seems possible, therefore, that shade light quality is perceived by phytochrome and shade light quantity by cryptochrome; this highly tentative statement should not be taken as a definite conclusion, rather as an indication of the direction in which experimentation might be constructively pursued.

Agricultural and Ecological Implications

In the early part of this chapter, attention was drawn to the differing strategies in shade of the shade avoiders and the shade tolerators. If the evidence summarized here may be taken to be indicative of the general situation, then it appears that the shade avoiders, which respond to shade by extreme stem extension at the expense of leaf development, perceive principally the light quality aspect of shade; conversely, the shade tolerators which, in shade, concentrate their effort into leaf development, appear to perceive principally the light quantity aspect of shade. Continuing this trend of thought, it may be that the shade tolerators have evolved to 'take more notice', as it were, of the phytochrome signals they receive, than of the cryptochrome signals, while the shade tolerators behave in the converse fashion. This may be merely trite speculation but, on the other hand, it could point the way to understanding the very marked variation that is known to exist between species in their relative photomorphogenic sensitivities to red/far-red and blue light.

Again, as stated earlier, the majority of temperate zone crop plants, if not all, are probably shade avoiders. Such species react in an 'escapist' fashion when subject to vegetation shade; they tend to use up their available resources in a reaction which aids survival by enabling a proportion, at least, of all individuals to escape from the shade conditions. Since parts of all crop plants are shaded at some point, a fraction of the gross productivity must be utilized in shade avoidance reactions which are agriculturally wasteful. Shade tolerators, on the other hand, increase their photosynthetic efficiency upon shading. Is it perhaps possible, if crop plants were to be developed from shade-tolerating species, that they would be more productive in intensive cultivation than existing crop species?

Acknowledgements

The author is grateful for the collaboration, over several years, with Dr M. G. Holmes, Dr D. C. Morgan, Dr R. Tasker, Dr C. B. Johnson, Dr J. C. McLaren, Mr T. O'Brien, Mr P. Foulkes and Mr R. Child, all of whom have contributed significantly to this work.

References

BERGMAN, K., BURKE, P. V., CEDRA-ALMEDO, E., DAVID, C. N., DELBRÜCK, M., FOSTER, K. W., GOODELL, E. W., HEISENBERG, M., MEISSNER, G., ZALOKAR, M., DENNISON, D. S. and SHROPSHIRE, W. Jr (1969). *Bacteriol. Rev.*, **33**, 99

BJØRKMAN, O., BOARDMAN, N. K., ANDERSON, J. M., THORNE, S. W. and PHYLIOTIS, N. A. (1972). *Carnegie Inst. Yearb.*, **71**, 115

BOARDMAN, N. K. (1977). *Ann. Rev. Pl. Physiol.*, **28**, 355

DELBRÜCK, M. (1976). *Light and Life*. III. Lecture given at the inauguration of the New Carlsberg Laboratory, Copenhagen, 27 September 1976

GABA, V. and BLACK, M. (1979). *Nature*, **278**, 51

GRIME, J. P. (1966). In *Light as an Ecological Factor*, Vol. 1, p. 187. British Ecological Society Symposium No. 6. Ed. by R. Bainbridge, G. C. Evans and O. Rackham. Blackwell Scientific Publications, Oxford

HEATHCOTE, L., BAMBRIDGE, K. R. and MCLAREN, L. S. (1979). *J. exp. Bot.*, **30**, 347

HOLMES, M. G. and FUKSHANSKY, L. (1979). *Pl., Cell Env.*, **2**, 59

MORGAN, D. C. and SMITH, H. (1976). *Nature*, **262**, 210

MORGAN, D. C. and SMITH, H. (1979). *Planta*, **145**, 253

MORGAN, D. C. and SMITH, H. (1981). *New Phytol.*, in press

PRESTI, D. and DELBRÜCK, M. (1978). *Pl., Cell Env.*, **1**, 81

RABINOWITCH, E. I. (1951). *Photosynthesis and Related Processes*, Vol. 2, Part 1. Interscience, New York

SHROPSHIRE, W. Jr (1963). *Physiol. Rev.*, **43**, 38

SMITH, H. (1975). *Phytochrome and Photomorphogenesis*. McGraw-Hill, London

SMITH, H. (1981). In *Plants and the Atmospheric Environment*. British Ecological Society Symposium No. 21. Ed. by P. G. Jarvis. Blackwell Scientific Publications, Oxford.

SMITH, H. and HOLMES, M. G. (1977). *Photochem. Photobiol.*, **25**, 547

YOUNG, J. E. (1975). In *Light as an Ecological factor*, Vol. II, p. 135. British Ecological Society Symposium No. 16. Ed. by G. C. Evans, R. Bainbridge and O. Rackham. Blackwell Scientific Publications, Oxford

10

PHOTOPERIODISM AND CROP PRODUCTION

DAPHNE VINCE-PRUE
K. E. COCKSHULL
Glasshouse Crops Research Institute, Littlehampton, Sussex, UK

The daily photoperiod is a feature of the environment to which the majority of wild species is to some extent adapted. Under natural conditions, responsiveness to photoperiod can confer a substantial advantage on the organism since it locates the response in time and can allow a particular response to be synchronized with advantageous environments, or enable the plant to avoid or minimize the effects of disadvantageous environments. For example, a long day (LD) response in high latitudes can synchronize flowering with a high light integral and thus support the energetic demand of seed production; similarly, a short day (SD) response may be synchronized with the end of a tropical rainy season. Good examples of day-length dependent responses that allow avoidance of disadvantageous environments are the SD-induced formation of resting buds in high latitude trees (avoidance of low temperature damage) and the LD-induced formation of resting buds in the desert liverwort *Lunularia cruciata* (avoidance of water stress; Schwabe and Nachmony-Bascombe, 1963). A photoperiodic response may also enable an organism to fit an ecological niche by occupying a particular time in the habitat. The selective advantage of responsiveness to photoperiod is well illustrated by reference to populations of *Xanthium strumarium*. Plants of American origin with a wide array of photoperiodic types have been successful introductions in diverse parts of the world, showing preadaptation to a range of habitats; on the other hand, those of Eurasian origin, most of which are day-neutral or nearly so, have proved to be at a selective disadvantage in comparison to American introductions even in parts of the old world (McMillan, 1974a, 1974b).

For crop plants, a photoperiod response may be an advantage or a disadvantage. While it is clear that latitudinal distribution may be restricted if cultivars are strongly adapted to photoperiod (as, for example, local cultivars of sorghum in Nigeria; Curtis, 1968), it does not always follow that the best approach is to breed for day neutrality. A cultivar fitted by photoperiod sensitivity to a particular climate confers the advantage of tight seasonal control over a desired response. Furthermore, some horticultural crop plants can be manipulated by the use of artificial photoperiods in precisely scheduled cropping systems. Thus it is necessary for plant breeders, together with plant physiologists, to assess the advantages and disadvantages of photoperiodic responsiveness for particular crops and environmental conditions (cf. Langton and Cockshull, 1976), and to devise appropriate protocols for screening.

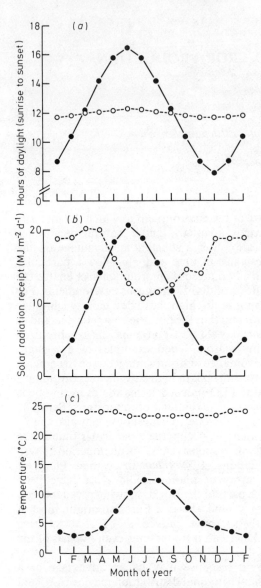

Figure 10.1 *The day length (a), daily integral of solar radiation (b) and average minimum air temperature (c) for low altitude sites (<60 m) at latitudes of 5°N (o) and 50°N (●). Solar radiation and air temperatures at Lymington, UK (50°N) from Efford EHS Annual Reports; solar radiation for Accra, Ghana (5°N) after Black (1956) and air temperature for Andagoya, Colombia (5°N) after Summerfield and Wien (1979)*

The Physical Environment

The daily photoperiod (from sunrise to sunset) varies widely with latitude from a nearly constant 12 hours at the equator to a range of 0–24 hours at the poles (*Figure 10.1*). However, the threshold intensity at which plants become sensitive to light at dawn or lose sensitivity at dusk may vary between species, so that determination of the effective biological photoperiod (or dark period) under natural conditions is not straightforward. It is partly for this reason that results obtained in cabinets with precisely timed light-on and light-off signals may depart from those obtained in the field. For example, in sugar cane, differences in blossoming time indicated a night of some 26 minutes shorter in darkrooms than out-of-doors; 26 minutes represented the twilight time at which the illuminance exceeded 40 lx (Clements, 1968). In Biloxi soya bean, the biological night appeared to begin near the time of astronomical sunset and to end at about the beginning of civil twilight, while in rice the beginning and end of the inductive night coincided approximately with sunrise and sunset (Takimoto and Ikeda, 1961). Clouds, too, may influence the duration of the effective dark period but their effect is likely to be small in species sensitive to low threshold irradiances (Salisbury, 1963); however, a difference of 20–30 minutes in the length of the effective photoperiod between clear and cloudy days has been recorded for *Pharbitis* where dark processes begin at about 200 lx (Takimoto and Ikeda, 1960), and similar differences might be expected in other species with sensitivities of the same order, such as soya bean and rice.

Recently, it has been suggested that photoperiodic timing under natural conditions may be locked to shifts in the ratio of red (R) to far-red (FR) light (and the resultant lowering of the amount of phytochrome present in the biologically active Pfr form) that have been observed at sunrise and sunset (Shropshire, 1973; Holmes and McCartney, 1976). Dusk and dawn signals of this kind might be more precisely timed than would attainment of a threshold irradiance, but the magnitude of the shift in R/FR ratio varies widely according to the density of cloud cover, and in some geographical locations scarcely any change has been detected. Moreover, timing is precisely effected in artificial conditions when plants receive a light-to-dark transition without a change in the ratio of R to FR. Since efficient protocols for photoperiodic screening must be based on the natural signals, the question of how the biological day and night are perceived by a particular plant needs to be determined.

A further point is that days shorten as well as lengthen, and do so at different rates at different times of the year. Although some animals appear to be able to perceive shortening versus lengthening days (for example, the domestic fowl), relatively little attention has been given to this possibility in plants where a particular critical night/day length is usually the primary controlling factor. Nevertheless, in plants of sugar cane, lengthening nights resulted in a faster response, and under *shorter* dark periods, than did constant or shortening nights; even so, the range of effective dark period durations in sugar cane is very narrow (*Figure 10.2*; Clements, 1968, 1975). Results with latitudinally adapted native cultivars of sorghum under natural conditions suggested to Curtis (1968) that they responded to the number of successively shorter days after the summer solstice, irrespective of their absolute length; however, other explanations were not rigorously excluded and the hypothesis remains to be tested experimentally.

Since the majority of plants do nevertheless respond to the absolute length of day, rather than to a changing one (Vince-Prue, 1975), it is axiomatic that they will experience the same inductive photoperiods in spring and autumn.

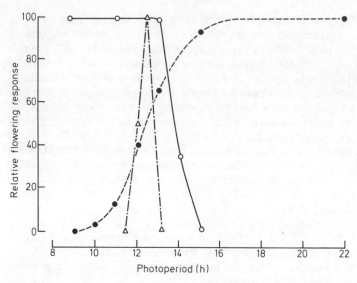

Figure 10.2   *Flowering response curves for plants in different photoperiodic classes.* ○ : Xanthium strumarium *(SDP).* ● : Lolium temulentum *Ceres (LDP).* △ : *Sugar cane H37–1933 (intermediate)*

Confusion between them is normally avoided by having a long juvenile phase when shoots are insensitive to day length, by associated responses to temperature, or by changing day length requirements. In the latter case, successive stages of development may have a narrowing response and require successively shorter (or longer) days; alternatively, the plant may have a dual day length requirement (*see* page 181). Although the underlying rhythm is temperature compensated (Bünning, 1967), temperature interactions with photoperiodic requirements are common (Table 1.1 in Vince-Prue, 1975). A plant can have an absolute photoperiodic requirement at one temperature but only a quantitative one, or none at all, at another; this type of response is well illustrated by the summer fruiting strawberry, which is an absolute SD plant (SDP) for flower induction at temperatures above 16–17°C but which will flower readily in LD at lower temperatures (Guttridge, 1969). Cultivars adapted to marginal growing areas in Scandinavia have an even more pronounced temperature/photoperiod interaction and can form buds in continuous light up to 18°C; at 12°C they are almost indifferent to day length (Heide, 1977). Warm temperatures, particularly at night, can compensate in part for shorter nights in some SDP (as in soya bean; Hamner, 1969) and cool temperatures can similarly counteract longer nights in some LDP (as in *Trifolium subterraneum*; Evans and King, 1975). Thus, temperatures experienced in the field commonly modify the photoperiodic response, and interactions of temperature with day length can lead to considerable errors when predicting varietal adaptation to new enviroments at the same latitude (e.g. in soya bean; Laing, 1974). Temperature as well as day length changes with latitude (*Figure 10.1*) and closely adapted varieties are often sensitive to both factors. A vernalization requirement can also interact with photoperiod and lead to a range of behaviour in natural climates (*see* page 187).

As well as with temperature, effects of natural photoperiod may be confounded with effects of the daily light integral, which changes in parallel with changes in day length throughout the season in all high latitude environments (*Figure 10.1*). In the tropics, cloudy conditions associated with a rainy season may obscure this correlation but the rainy or dry seasons themselves have profound effects on crop growth and may interact with a photoperiod response. In the strongly photoperiodic sugar cane, for example, field blossoming can vary from zero to very heavy depending on the soil moisture tension during the induction period (Clements, 1975).

Thus, although photoperiod may be limiting for a particular response, the magnitude and even the timing of the response under natural conditions can be modified by associated environmental factors such as temperature, light integral and water supply, as well as by plant factors such as age or stage of development.

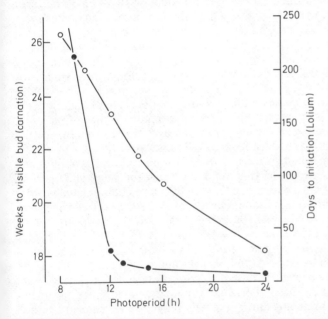

Figure 10.3 *Differences in photoperiodic sensitivity in two long day plants.* ○: Dianthus caryophyllus *(carnation; after Harris and Ashford, 1966).* ●: Lolium temulentum *Ceres. (After Evans, 1969)*

The Range of Responses and Mechanism(s)

Plant responses to day length are numerous. In addition to the initiation and further development of flowers, vegetative growth is influenced in a number of ways by day length; these include the formation of dormant resting buds, leaf fall, the formation of vegetative reproductive organs such as bulbs, tubers and runners, rooting, leaf and stem growth, and plant habit. Any one, or more, of these responses may be important or limiting in particular crops.

The classification of plants according to their photoperiodic responses has been made on the basis of flowering but similar categories might, of course, be set up for other processes. Three main groups are usually considered: long day plants, which flower or show accelerated flowering with increase in length of the daily photoperiod; short day plants, which flower or show accelerated flowering with decrease in length of the daily photoperiod; and day-neutral plants (DNP) with no response to day length (*Figure 10.2*). Other response groups, such as intermediate day plants (with a fairly restricted range of effective day lengths) and plants with dual day length requirements, may be combinations of the major ones. Many plants with absolute photoperiodic requirements show a sharp change from full to zero response as the day length is changed, but in others the response is more quantitative, increasing from zero to maximum over several hours (*Figure 10.3*), thus the *photoperiodic sensitivity* (the change in degree of response per unit change in day length) can vary widely. The absolute value of the critical day length (above or below which the response is suppressed, or at which 50 per cent response occurs) also varies widely; where the response to day length is a quantitative one (as in chrysanthemum flower initiation; Schwabe, 1953; Cockshull, 1976) it is, of course, not completely suppressed at any day length. Absolute values of critical day length can overlap (*Figure 10.2*) so that, under some photoperiods, SD and LD response types are not distinguished; for example, the critical day lengths for *Phleum pratense* (LDP), *Trioda flava* (SDP) and *Sorghastrum mutans* (intermediate) lie between 13 and 15 hours and all may flower at the same time of year depending on time of planting, length of juvenile phase, and the temperature prevailing between the sowing date and completion of the juvenile period (Evans, 1964).

In photoperiodism, plants respond to the duration and timing of light and dark periods in the daily cycle. It is well known that for many photoperiodically sensitive plants, the effects of a long night can be counteracted when it is interrupted (the time of maximum sensitivity varies between species) with a relatively brief exposure to light (Vince-Prue, 1975). However, not all photoperiodically sensitive plants respond strongly to night breaks; some LDP (and even some SDP, e.g. *Fragaria*; Vince-Prue and Guttridge, 1973) respond poorly when given brief night-breaks and may require one or more LD in order to induce a long day type of response (Vince, 1965; Harris, 1968; Ballard, 1969; Friend, 1969). This light-dominant process has particular spectral quality requirements during the long photoperiod requiring both R and FR light and having a changing sensitivity to R and FR during the course of the day (Vince-Prue, 1976). Although unimportant under natural conditions in sunlight, artificial manipulation of the photoperiod requires a knowledge of the response characteristics so that the most effective combination of quality, duration and timing of the artificial light can be selected (Vince, 1972).

Flowering

The regulation of flowering by day length has been studied in great detail and many different types of response are known (Evans, 1969; Vince-Prue, 1975). In general, it is convenient to divide the process into at least two stages, i.e. the initiation of flowers and their development into open blooms, for these may be affected independently. For example, autumn-flowering chrysanthe-mum (*Chrysanthemum morifolium*) will initiate flower buds readily if the dark period exceeds 9.5 hours but they require a longer dark period (10.5–11 hours) if the flower buds are to develop rapidly (Post, 1948; Cathey, 1957). These cultivars are clearly well adapted to flowering in late autumn. The perpetual flowering carnation (*Dianthus caryophyllus*), on the other hand, initiates flowers most rapidly in continuous light but their development is quite unaffected by day length (Harris and Ashford, 1966). Examples are also known in which the day length requirement for rapid flower initiation is the opposite of that for flower development; for example, the strawberry (*Fragaria x ananassa*) initiates flowers in SD but their development may be most rapid in LD (Guttridge, 1969; Canham, 1975; Tafazoli and Vince-Prue, 1978). These responses combine to encourage flowering in late spring under field conditions. Conversely, in china aster (*Callistephus chinensis*) flowers are initiated readily in LD but develop most rapidly in SD (Laurie and Foote, 1935; Hughes and Cockshull, 1965) and flowering occurs in late summer.

In the above examples, the plant response to the natural photoperiod limits flowering to particular times of year, and similar limitations are encountered in many crop plants where the marketable product is either a flower, fruit or seed (Evans, 1975). The limitation can also operate in plants such as sugar cane, where the formation of flowers leads to cane senescence and may also divert assimilates away from the marketable cane (Bull and Glasziou, 1975; Clements, 1975). The potential of photoperiod limitation has been minimized by breeding and selection, by altering the sowing date and by manipulating the day length artificially.

GLASSHOUSE CROPS

Although control of day length has been attempted in the field (Paleg and Aspinall, 1970) it is obviously a more practicable technique when conducted on a smaller scale and under the protection of a structure like a glasshouse. This confers other advantages, such as the ability to heat the air and soil so as to overcome the low temperatures that are associated with winter in high latitudes. It is also possible to enrich the glasshouse atmosphere with CO_2 and so to increase the rate of net photosynthesis at any given irradiance. This goes some considerable way towards off-setting the reduction in net photosynthesis that results from moving plants into a structure which absorbs and reflects 30 per cent of the incident light.

For complete control of the photoperiod it is necessary to simulate SD in summer by drawing an opaque material over the plant every evening and withdrawing it again in the morning when the desired duration of darkness has elapsed. The simulation of LD in winter is readily achieved using light from artificial sources to shorten or interrupt the long dark period, as described earlier.

When the control of day length is combined with selection of genotypes that have a pronounced photoperiodic flowering response, it provides a powerful tool for the regulation of crop production. Flower production can be timed to yield fruits or flowers when the cash returns are highest; flowering can also be synchronized within the crop so that harvesting is concentrated over a short period; plant size prior to flower initiation can be regulated so as to influence eventual yield as well as plant form, and with vegetatively propagated plants it is possible to keep stock plants in a vegetative state. These advantages are well illustrated in the production of flower crops in heated glasshouses.

The chrysanthemum is one of the best examples of a plant whose productivity, in terms of the marketable yield (i.e. number of blooms) produced per unit area per annum is increased substantially by the control of day length. Considerable genetic variation exists in terms of the flowering response to photoperiod (Langton, 1977) and, although summer-flowering types are known, cultivars for controlled year-round cropping have been selected from among the autumn-flowering types which show a marked response to SD. The chrysanthemum is readily propagated from stem cuttings, and the photoperiod response enables the propagator to keep stock plants in a vegetative state in LD and to maintain the cuttings in this state during the rooting period. This method of propagation is also employed with the LDP carnation but it is not necessary to maintain the stock plants in non-inductive SD because the young shoots have a relatively long juvenile phase during which they cannot be induced to flower (Harris and Bradbeer, 1966).

Figure 10.4   Kalanchoe blossfeldiana *cv. Tetra Vulcan grown from seed in long days at 16°C minimum, and transferred to short days after (from left to right) 6, 8, 10, 12, 14 and 16 weeks of long days. (Cockshull and Hand, unpublished data)*

Once the chrysanthemum cutting has rooted, the grower can either produce a small flowering plant by immediately placing it in SD to induce rapid flower formation, or can delay flower initiation with a period of LD so that the flowers are carried on a long leafy stem. The first approach is used when producing flowering pot plants, while the second is used for cut flower production as the longer growing period also results in a heavier plant with a correspondingly heavier weight of flowers (Cockshull and Hughes, 1967). Similar interrelationships between vegetative and reproductive growth were clearly demonstrated in china aster (Hughes and Cockshull, 1965). Exposure of young plants to LD induced early flower initiation, but the plants carried few flowers and their total weight was low. In general, the later flower initiation is induced, the larger is the final plant size and the greater the weight and number of flowers (*Figure 10.4*); with more flowers, however, there is greater competition between them for assimilates and the weight of each individual flower is often less.

The ability to control flower initiation can also be used to ensure that the crop is marketed at a time of high demand and high market returns. Perhaps the best example of this is the poinsettia (*Euphorbia pulcherrima*), which is closely associated with Christmas and is produced at this time by controlling day length (Laurie, Kiplinger and Nelson, 1968). Early attempts at regulating chrysanthemum flowering in the UK were also mainly concerned with delaying flowering until December by means of LD treatments in August and September (Vince, 1953). Today, chrysanthemum flowers are produced in every week of the year by planting rooted cuttings at weekly intervals and giving each batch an appropriate day length treatment (Machin and Scopes, 1978). In this way, the original limitation of production to the autumn period has been completely removed and productivity has been increased by at least 300 per cent.

The use of heated glasshouses and control of the photoperiod does not entirely eliminate seasonal influences, however, for the daily integral of solar radiation also shows a marked seasonal variation (*Figure 10.1*). In December, it falls below 2.5 MJ m^{-2} d^{-1} in the field (in the southern UK) which provides less than 0.75 MJ m^{-2} d^{-1} of visible radiation (400–700 nm) in a glasshouse. Although this has some effect on flower development, its main effect is to delay flower initiation both in chrysanthemum (Cockshull and Hughes, 1971) and in carnation (Harris and Harris, 1962; Bunt, 1973). Low radiation integrals also have a marked influence on the rate of leaf initiation and may double the plastochron of chrysanthemum (Cockshull, 1979). As a result, fewer leaves and internodes will be produced in a given period of LD in winter than in summer and the stem length will be correspondingly shorter. By giving more LD in winter, the onset of flower initiation can be delayed until a sufficient number of internodes have been produced to give the same stem length as in summer.

Further exploitation of the potential of the photoperiodic control of flowering awaits a better understanding of the underlying plant responses. There is evidence that SD stimulate the production of a promoter of flower initiation in the leaves of chrysanthemum (Cathey, 1969) but such evidence is less evident in other SDP such as the strawberry (Guttridge, 1969). In the latter, the greater weight of evidence is that flowering may be regulated by an inhibitor produced in LD. In this plant, therefore, long days have a positive rather than a neutral effect. Further evidence in favour of this comes from studies of the effect of the spectral quality of light used to extend the natural day. Red light in the second half of a 16 hour night inhibited flower initiation in strawberry whereas it was relatively ineffective in the first half; a mixture of R and FR light, on the other hand, was effective in both periods (Vince-Prue and Guttridge, 1973). These responses are similar to those found for the promotion of flower initiation in LDP such as *Lolium* (Vince, 1965) and carnation (Vince, Blake and Spencer, 1964; Harris, 1968). Grafting experiments between different response types also indicate that both promoters and inhibitors of flowering may be produced (Lang, Chailakhyan and Frolova, 1977).

The beneficial effects on the flowering of LDP of light that contains a mixture of R and FR accounts for the effectiveness of tungsten-filament lamps on carnation (Harris, 1968). Further, a night-break is not as effective in promoting flowering of carnation as is continuous irradiation from dusk to dawn (Harris and Ashford, 1966). The greater effectiveness of continuous light as compared with night-break is not uncommon in LDP (e.g. china aster; Cockshull and Hughes, 1969; *see also* Vince, 1972; Vince-Prue, 1975). In chrysanthemum, on the other hand, night-break treatments (of several hours' duration with tungsten-filament lamps) seem fully effective in delaying flowering and the light can be given intermittently throughout the night-break period without apparent loss of effectiveness (Cathey, 1969). Some evidence has been obtained that R is less effective in inhibiting flowering when given in the first half of the night (Cockshull and Vince-Prue, unpublished observations). If this is confirmed it would indicate that chrysanthemum also produces an inhibitor of flowering in LD.

FIELD CROPS

Whatever the environment, different crops are obliged by their structure to accumulate yield at different stages of their life cycle; a photoperiodic response can be used to locate the yield-forming period appropriately in relation to the local climate and in this way can markedly influence productivity. With respect to the proportion of their life span which can be used to form yield, crops fall into three broad phenological classes (Bunting, 1975). In the first, yield consists of or accumulates in vegetative parts and is produced throughout the crop's life; within this group are fodder grasses, silage plants and sugar cane.

Sugar cane is usually classified as an intermediate day plant (*Figure 10.2*), most clones of which have a narrow range of inductive night lengths (usually between 11 and 12 hours); some *spontaneum* varieties, however, behave as LDP (Bull and Glasziou, 1975). In general, profuse flowering in sugar cane is undesirable since yields may be substantially depressed. In one field experiment, in which night-breaks were used to prevent floral initiation, an increase of about 3 tons of sugar per acre was obtained compared with untreated plots which blossomed heavily; most of the effect occurred in the first season (*Table 10.1*). Tassel production also leads to heavy suckering in the next season; these suckers can lodge and weigh down the top-heavy flowering stalks, thus causing poor quality in the latter (Clements, 1975). Because of such effects, varieties that are not heavy tassellers have often been selected for (as, for example, in Hawaii); the use of such low-tasselling

Table 10.1 EFFECT OF BLOSSOMING ON THE YIELD OF SUGAR CANE H37–1933, OAHU 1949–1951. (FROM CLEMENTS, 1975)

Night-break treatment	Yield (tons sugar per acre)	Per cent tasselling	
		First season	Second season
First season	16.9	0.0	10.3
Second season	14.9	26.0	0.8
Both seasons	16.2	0.3	0.8
Neither season	13.9	36.2	16.7

clones, however, poses problems in breeding programmes and methods for inducing flowering have been devised using various light regimes, cultural treatments and nutrient supplies. Good flowering has been obtained in reluctant tassellers in Hawaii by the use of lengthening nights (Clements, 1975) but free-flowering varieties will initiate and develop a normal inflorescence regardless of whether increasing, decreasing or constant night lengths are used (Daniels, Glasziou and Bull, 1965). One interesting factor in the photoperiodic response of sugar cane is that the initiation of branch and spikelet primordia is more sensitive to a night-break given either near the beginning or near the end of a 11.5–12 hour night, than to one given near the middle (Julien, 1970).

All annual species of pasture plants, and also many perennials, rely on regeneration from seed. Their time of flowering, often under the control of day length, is an important feature in their adaptation to climate and has been widely studied in many grasses (as, for example, in *Lolium* species; Cooper, 1959, 1960, 1965). A recent study of the pasture legume, *Stylosanthes*, has similarly considered its day length response in relation to possible adaptation to a range of environments in northern Australian pastures (Cameron and Mannetje, 1977). The agronomic value of a pasture legume depends on its growth, most of which is made before the plant is fully committed to reproduction; on the other hand, regeneration from seed is essential for both annual and perennial forms of *Stylosanthes*. Short day, day-neutral and long day response types thus provide a choice of possible reproductive strategies for the varying climatic conditions of northern Australian pastures. Low temperature may interfere with the regeneration process in higher latitudes (e.g. in *Stylosanthes guianensis* var. *guianensis* cv. Schofield in southern Queensland) and thus accessions with an LD response are likely to be most successful south of the tropic where flowering will be induced by lengthening photoperiod in summer. SD types are characterized by a determinate flowering behaviour and may be used to select plants for late flowering in regions with little variation in length and timing of the growing season. Where there is substantial within- and between-season variability, however, day-neutral or quantitative SD response types are likely to be more successful since early flowering allows some seed production even in very short growing seasons, whereas the capacity for continued vegetative development results in heavy yields during long growing seasons. Crosses between LD and SD types have been made in an attempt to develop a legume with improved adaptation to southern Queensland.

Plants of indeterminate habit constitute the second phenological class, their yield being produced on lateral inflorescences during a greater or smaller fraction of the life of the crops. Many of the grain legumes are included within this group, with a long period of flowering at successive nodes up the stem; some lines, however, have a more determinate habit, which may be associated with higher productivity (Evans and King, 1975). Photoperiod plays a major role in controlling the time and rate of floral initiation and development in grain legumes, which are dependent for heavy yields on abundant flowering, pod set and seed development. Day length responses (as with vernalization, which is limited to genera of the Fabeae and Genisteae) retain a strong correlation with taxonomic grouping; where they retain photoperiodic sensitivity, plants in the Fabeae and Genisteae are LDP, while those in the Phaseoleae are SDP (Evans and King, 1975). Among the Phaseoleae, soya bean (*Glycine max*) is one of the classic SDP on which a great deal of research into the photoperiodic mechanism has been carried out (Hamner, 1969). The effect of photoperiod on flowering in soya bean is so marked in North American cultivars that most are restricted to within about 4 degrees of latitude (480 km) of their adapted area; outside this range, plants mature too early in the south or fail to mature before the frosts in the north (Major and Johnson, 1975). The adaptation to local climate is, however, also strongly influenced by temperature, especially night temperature (Summerfield and Wien, 1979), so that adaptability to climates at the same latitude outside the USA cannot be predicted on the basis of their day length response alone (Laing, 1974). The photoperiod responses of other grain legumes are similarly modified by temperature (Evans and King, 1975).

A feature of the grain legumes is that, almost without exception, flower development has more stringent photoperiodic requirements than does flower initiation; in this they resemble the chrysanthemum (*see* page 181). In soya bean, for example, the requirement for SD becomes more pronounced at successive stages of development and, in some cultivars, the critical day length also becomes progressively shorter (Evans and King, 1975; Shibles, Anderson and Gibson, 1975). The grain legumes, which are LDP, behave similarly and have a more stringent requirement for long days as flowers develop; this is particularly clear in *Vicia faba* where flower initiation occurs in eight-hour photoperiods but the critical day length for flower development is 12–13 hours (Evans and King, 1975). The period of anthesis and seed set is a critical stage in the development of grain legumes and a substantial loss of buds, flowers and immature pods may occur in unfavourable environments. Long days increase flower abscission in the SDP soya bean (Hamner, 1969; Evans and King, 1975; Shibles, Anderson and Gibson, 1975) and *Phaseolus vulgaris* (Zehni, Saad and Morgan, 1970). In the latter, a transmissible inhibitor of flower development (abscission promoter) may be formed in LD since exposing only the lower trifoliate leaf to LD inhibited bud development in the terminal and upper axillary inflorescences, even though there were leaves in SD between them and the LD leaf. The endogenous gibberellin content of the buds increased in LD and the application of cytokinin stimulated the development of flower buds, which would otherwise have abscised in LD (Bentley *et al.*, 1975). In view of the dominant influence of pod setting on the yields of grain legumes, a response to photoperiod at this stage can have a marked effect on productivity.

In the third group of field crops, yield is produced in terminal or late-formed inflorescences as the last phase in the life of an annual crop or of annual shoots of a perennial crop. The dominant members of the group are the cereal crops, and among these are LDP (temperate cereal grains, e.g. wheat and barley) and SDP (maize, sorghum and rice). In these crops, no more leaves are formed once the apical bud has become reproductive and the main sources for grain filling are the latest formed leaves.

The photoperiodic responses of wheat have been widely studied and can serve to illustrate relationships between day length and yield within the cereal crops. Over a wide range of conditions the storage capacity of the ear for assimilates, and in particular grain number, imposes a major limitation on yield (King, Wardlaw and Evans, 1967; Thorne, Ford and Watson, 1968).

The number of grains per ear depends on the spikelets per ear and the number of grains per spikelet; of these, the number of grains per spikelet has been shown to be relatively constant over a range of climatic conditions (Pinthus, 1967; Rawson, 1970; Cackett and Wall, 1971) so that the number of spikelets may be the more important environment-dependent determinant of grain number per ear. Photoperiod and temperature are the most important environmental factors in the control of spikelet number (Halse and Weir, 1970; Rawson, 1970, 1971).

A survey of the responses of 14 cultivars from a wide range of origins (Wall and Cartwright, 1974) confirmed that the most important environmental factor that controls their development was photoperiod and that there was a major difference between temperate cultivars and those adapted to lower latitudes in their photoperiodic sensitivity (*Figure 10.5*). Ear emergence in

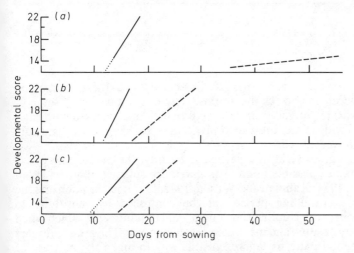

Figure 10.5 *Effect of day length on spikelet development in wheat. Plants were grown in LD (solid line) or SD (dashed line) at 25°C day/20°C night temperatures. (a) Kolibri; (b) Sonora 64; (c) Tokwe. All plants received a main light period for ten hours, and plants in LD received a four-hour day length extension with low intensity light from tungsten-filament lamps. The most advanced spikelet was scored (=developmental score) on the scale of Friend et al. (1963), in which 12 = double ridges, 18 = lemma primordia, 20 = anther primordia. (After Wall, 1972)*

temperate cultivars was delayed indefinitely in SD (10 hours) but they were among the earliest to emerge in LD (14 hours). All of the tropical wheats behaved as quantitative LDP and even in the least sensitive cultivars, where effects on days to heading were small, effects on spikelet number were relatively large and could lead to substantial effects on yield. When vernalized, all of the tropical cultivars grown in 14 hour days were early and formed few spikelets (12–14 per ear). Without vernalization in 14 hour days, the early tropical cultivars remained early with few spikelets per ear, while flowering in the late cultivars was delayed and the ears had many spikelets (about 20); day length had a negligible effect in unvernalized late cultivars (*Table 10.2*). In late tropical cultivars, therefore, maturity seems to be largely controlled by the presence or absence of a 'cold' period, which might effect vernalization; a daily minimum as warm as 13°C appeared partially to

Table 10.2 FINAL NUMBER OF SPIKELETS PER EAR FOR GROUPS OF CULTIVARS OF DIFFERING GEOGRAPHICAL ADAPTATION OR MATURITY CLASS. (FROM WALL AND CARTWRIGHT, 1974)

		Photoperiod (hours)	
		14	10
North temperate	o	14.0	20.3*
	v	14.6	23.6*
Early tropical	o	12.2	17.1
	v	11.8	16.5
Midseason tropical	o	14.4	20.0
	v	12.0	19.1
Late tropical	o	21.6	22.8
	v	12.7	18.5

*Spikelet primordia only.

All plants received ten hours' natural daylight; in the 14 hour treatment, day length was extended for four hours with low irradiance light from 40 W tungsten-filament lamps. Temperatures were 25°C day/20°C night. Seeds were vernalized (v) by keeping part imbibed at 1°C for 28 days. o: Unvernalized control plants.

vernalize some cultivars (e.g. Mexico 120, Cajeme 71). Thus the characteristic high spikelet numbers and relative insensitivity to photoperiod of some of the Mexican wheats derived from Norin 10 (which is associated with high grain yields and wide adaptability in low latitudes) would appear to hold only if they remain unvernalized.

For maximum grain yield in a particular locality, the preheading phase should be as long as possible in order to maximize spikelet numbers (Wall and Cartwright, 1974; Rahman and Wilson, 1977, 1978); its duration and the timing of the grain filling period can be controlled by sensitivity to photoperiod alone, or in combination with a 'cold requirement', which when not met will delay heading and increase spikelet numbers. The use of cultivars with developmental patterns tailored in this way to make the best use of particular seasonal conditions offers an alternative strategy to the breeding of cultivars without photoperiod and temperature sensitivity and, therefore, widely adapted throughout the world (Borlaug *et al.*, 1966).

Vegetative Growth

STEM ELONGATION AND DORMANCY

It is customary to consider day length effects on crop productivity largely in terms of the photoperiodic control of flowering. However, in many crops, there are day length imposed restrictions on aspects of vegetative growth that may be of considerable importance in cropping. We have already made reference (*see* page 183) to an effect on stem elongation in chrysanthemum that arises indirectly; by delaying flowering with LD, stems become acceptably long for cut flower production because more leaves are formed before the terminal bud becomes reproductive. A more direct influence is seen in tulip (Hanks and Rees, 1979) where flower stem length was substantially (20–50 per cent) increased in 16 hour days (using low intensity tungsten-filament day extension for eight hours) compared with eight-hour days. This response suggests that photoperiodic control of stem elongation might be successfully employed in production systems, for example by increasing stem length in early forced tulips or short stemmed cultivars (and vice versa), in attaining suitable stem lengths in unfavourable latitudes, and by reducing excessive stem elongation which leads to the physiological disorder, 'topple'. Photoperiodic control could, therefore, increase the range of tulip cultivars suitable for forcing in a given situation. Since stem length is generally increased in LD compared to SD, especially where tungsten-filament lamps are used to extend the day (Downs, Borthwick and Piringer, 1958; Vince-Prue, 1975, 1977), photoperiodic manipulation of stem elongation might be applicable in other crops where stem length is a quality factor. Photoperiodic control of forced easter lily, where LD leads to taller plants and earlier flowering, is already of considerable commercial importance (de Hertogh, 1974).

Day length has a considerable effect on the growth of many forest trees and can be an important limitation to cropping. In the majority of woody species so far studied, day length influences the duration of extension growth and the time at which buds enter dormancy; in general, the rate of growth is decreased and the onset of dormancy is hastened by SD (Wareing, 1956; Nitsch, 1957; Vince-Prue, 1975). For species of the temperate zone, low winter temperatures are a major unfavourable environmental factor and an increased resistance to below-freezing temperatures accompanies the winter dormant condition (Williams, Pellett and Klein, 1972); thus the time of induction of dormancy and cold hardiness may be crucial for survival. Species with a wide latitudinal distribution frequently show marked ecotypic variation with respect to effects of day length on the time of onset of dormancy. In a study of *Betula pubescens* from several latitudes in Norway, it was found that the critical day length of species increased as they originated from higher latitudes; a very far northern population became dormant even in a day length of more than 20 hours, while the critical day length for more southerly ecotypes was 14–16 hours (Håbjørg, 1972). Low latitude species often continue growing longer if moved to higher latitudes and are damaged by autumn low temperatures; such delayed dormancy and frost damage have been reported for *Robinia* in Russia (Moshkov, 1935) and for *Populus* in Sweden (Sylven, 1942). Similarly, high latitude species grown in shorter day lengths may enter dormancy early and be stunted. However, not all species with a broad latitudinal distribution show a population difference in their response to day length. The broad latitudinal distribution of *Acacia farnesiana*, which occurs from South to North America, seems to depend on a relative indifference to day length (Peacock and McMillan, 1968). In contrast, the wide distribution of *Prosopis juliflora* over a similar latitudinal range has occurred through the selection of ecotypes adapted to photoperiod. Thus, as with many flowering responses, the breeding approach may be to select either for photoperiodic indifference or for closely adapted types. *Prosopis* species are thought to have a high potential for further commercial exploitation since they tolerate conditions found in arid zones of the world and can supply fuel, fodder foliage and tannins, as well as timber (Leakey and Last, 1980). As with other trees, there is evidence to suggest scope for

genetic improvement, and photoperiodic adaptation is one of the variables
that should be considered.

Because the length of day varies much less in equatorial regions than at
higher latitudes, this environmental factor is often dismissed as being of little
or no importance. However, effects of day length on shoot growth have been
demonstrated in a number of tropical trees (Longman, 1969, 1978); in
general, an increase in photoperiod increases the shoot growth rate and some
of the effects are large (e.g. in *Terminalia* and *Chlorophyta*). In *Chlorophyta*,
the rate of production of new leaves was increased in LD by a factor of 4 in
the first month; later the effect was much less but, at the end of the
experiment, shoot dry weight was substantially (60 per cent) greater in plants

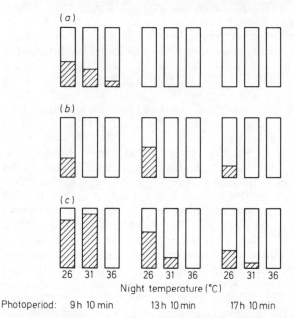

Figure 10.6   *Effects of day length and night temperature on dormancy in some tropical trees: (a)*
Ceiba pentandra; *(b)* Gmelina arborea; *(c)* Bombax buonopozense. *The shaded areas represent
the number of trees which became recognizably dormant during the experimental period (3.5–
4 months). Plants grown at 31°C day temperature and at the night temperatures indicated. (After
Longman, 1969)*

grown in 17 hour than in 10 hour photoperiods (Longman, 1978), although
the former received no additional light energy. Entry into dormancy in
tropical species is favoured by SD and cool nights (*Figure 10.6*), and their
photoperiodic responses seem to be broadly similar to those of many
temperate trees, although their dormancy is probably not so deep (Longman,
1969). Day length may thus be a limiting factor for rapid extension growth,
leaf production and dry weight gain of some trees even in the tropics,
although care must be taken in attempting to extrapolate from results
obtained with controlled environments which often compare widely different
day lengths. However, Njoku (1964) showed that in Nigeria seedlings of
Hildegardia bartei became dormant in 11.5 hour but not in 12.5 hour days.

VEGETATIVE PROPAGATION

Storage organs

Many vegetative storage organs are important food crops and, in several species, their formation depends on or is accelerated by exposure of the leaves to particular photoperiods. With the exception of the formation of bulbs in the genus *Allium*, most photoperiodically induced storage organs are favoured by exposure to SD (Vince-Prue, 1975).

As with many other responses, the formation of storage organs may be quantitatively or qualitatively controlled by photoperiod. Tuberization in potato appears always to be hastened by SD, although cultivars differ considerably in the extent to which they are affected (Wassink and Stolwijk, 1953; Pohjakallio, Salonen and Antila, 1957); some have an absolute SD requirement but most European and North American cultivars form tubers readily in LD. In SD conditions, early tuber formation is associated with early senescence of the haulm and relatively low yields (Pohjakallio, 1953).

Similar considerations apply to onion, where bulb formation is dependent on LD (Brewster, 1977). In lower latitudes, many common cultivars do not form bulbs and the so-called SD bulbing types must be grown; these also have an LD requirement for bulbing but with a much shorter critical day length. The responses of different cultivars to day length remain in the same order over a wide range of photoperiods and, in general, the longer the day the sooner leaf growth ceases and the sooner the bulbs ripen (Austin, 1972). As with many other photoperiodic responses, both day length and temperature are important in controlling bulbing in onion. Higher temperatures favour rapid bulbing, and may decrease the critical day length (Thompson and Smith, 1938), while at lower temperatures (below about 15°C) bulbing is delayed and in some cultivars prevented entirely (Thompson and Smith, 1938; Heath and Holdsworth, 1948). In tropical areas, seasonal temperature differences are particularly important (Abdalla, 1967; Robinson, 1971), but in temperate zones day length is the major determining factor for bulb initiation; even here, temperature modifies the time to maturity in different climates within the same latitude. In longer day lengths, the onset of bulbing is accelerated but yield is low because the emergence of new leaves ceases rapidly and the first formed leaves collapse and senesce with a consequent reduction in photosynthetic capacity (Austin, 1972). Cultivars are best adapted where the minimum temperature and day length requirements occur because, under these conditions, leaf growth and bulb formation are prolonged and heavy yields arise from the large leaf area duration in the bulbing phase (Abe, Katsumata and Nagayoshi, 1955).

Rooting of stem cuttings and runners

Although few agricultural crops are propagated vegetatively, it is not uncommon to find horticultural crops propagated by means of either rooted stem cuttings or runners. In both cases, the natural day length can limit crop production in an indirect manner by affecting the availability of suitable propagating material or its quality. For example, as discussed earlier, photoperiod may influence the onset of bud dormancy and leaf senescence with the result that actively growing stem cuttings can only be taken at certain times of year. Similarly, the formation of runners may be affected as in the strawberry, which forms them in LD (Guttridge, 1960), and *Chlorophytum*, which forms them more readily in SD (Hammer, 1976). The restriction of production to certain times of year may be overcome by modifying the natural photoperiod in an appropriate way. The technique is already widely used to control the quality of cuttings, as with chrysanthemum in which both the stock plants and the cuttings are kept in LD so as to maintain them in a vegetative state (Machin and Scopes, 1978). Long days could similarly be used with *Cornus* to prevent cuttings from becoming dormant and shedding their leaves (Whalley and Cockshull, 1976). Finally, it is known that photoperiod can have a direct effect on root formation, which may be exercised either through the stock plants or the cuttings themselves (Vince-Prue, 1975; Whalley, 1977) and this, too, could be exploited to increase productivity.

THE PRODUCTION OF LEAF SURFACE

As crop growth rate is generally correlated with absorbed radiation (Biscoe and Gallagher, 1977), it is advantageous to develop a high leaf area index early in the growing season and to maintain it for as long as possible while the daily integral of solar radiation is high. Leaf area index is dependent on the separate processes of leaf production, leaf expansion and leaf senescence, all of which are influenced strongly by the daily light integral and by temperature. As already mentioned, in high latitudes these two environmental variables are highly correlated in the field and, in addition, they are also highly correlated with photoperiod (*Figure 10.1*). Using controlled environments, several investigators have attempted to separate effects due to irradiance from those due to temperature but relatively few have examined the effects of photoperiod. This may simply reflect the general opinion that the influence of day length on these processes is relatively minor in the field.

Day length seems to have little *direct* effect on the rate of leaf production except in so far as the latter may be stimulated immediately prior to flower initiation (Thomas, 1961). Nevertheless, there is a major *indirect* effect through the induction of flower initiation at the terminal apical meristem, which prevents further leaf production (cf. chrysanthemum and china aster, page 182); *Stylosanthes*, page 185; and similar effects relating to tuber and bulb formation, page 191). In the field, this problem can be approached by selecting genotypes with a long juvenile phase during which they are incapable of flowering in response to photoperiod (Calder, 1966), or with a vernalization requirement (e.g. wheat, page 187), or by choosing appropriate sowing dates. Under glass, plants can be maintained in non-inductive day lengths until the appropriate size is attained.

In the strawberry, antagonistic effects between vegetative growth and flower initiation are also important in determining final crop yields (Guttridge, 1960). Under natural conditions, day length has correlated effects on leaf size and number of runners (increased by LD) and the initiation of flower trusses (inhibited by LD). The balance between fruiting and vegetative growth is a delicate one and it has been shown that excessive vegetative vigour can depress strawberry yields in the UK; for example, the removal of virus by heat treatment has led to increased vigour and lower yields due to delayed fruit truss formation (Rogers and Fromow, 1958). For this reason, it is suggested that breeding for sharpened photoperiod sensitivity could lead to increased yields because LD would allow vigorous growth capable of supporting large numbers of fruit trusses, while a strong SD response would suppress vigour in autumn and permit rapid truss initiation (Guttridge, 1960). A short cut is to defoliate soon after harvest and, where truss production is a limiting factor for yield in vigorously growing plants, such defoliation treatments can offer a corrective (Guttridge *et al.*, 1961).

Once the formation of a terminal inflorescence has begun, the crop must maintain the production of dry matter with the existing complement of leaves, and leaf senescence becomes an important factor in total dry matter production. Day length does not appear to affect leaf senescence in herbaceous plants. In many woody deciduous species, on the other hand, leaf senescence seems generally to be linked to the induction of bud dormancy and is, therefore, often accelerated by SD.

The third component of leaf surface production is leaf expansion, for which there are examples of both direct and indirect effects of day length. The most common indirect effect is associated with the photoperiodic induction of flower initiation in dicotyledonous plants, which is often accompanied by the production of smaller leaves below the inflorescence (Humphries and Wheeler, 1963). The most usual direct effect is the enhanced expansion of individual leaves in LD (Cockshull, 1966; Vince-Prue, 1975). This response has led to the suggestion that productivity could be increased as a result of increasing leaf area early in the growing season by means of artificial LD treatment (Hughes and Cockshull, 1966; Soffe, Lenton and Milford, 1977); so far, the suggestion has not been widely adopted.

In china aster, individual leaf area was increased by a one-hour night-break from a tungsten-filament source (R+FR) which, in contrast to its effect on flower initiation, was as effective as continuous lighting throughout the night (Cockshull and Hughes, 1969). Both treatments increased surface area by increasing the lamina area per unit weight of leaf. A day length extension with light from a tungsten-filament source (R+FR) had the same effect in sugar beet (Milford and Lenton, 1976); a four-hour extension with FR alone increased leaf area but R alone was relatively ineffective (Milford and Lenton, 1979). This response to day length may be mediated through gibberellins since, in sugar beet, the content of gibberellin in leaves was higher in LD than in SD, and applied gibberellins partly substituted for the LD stimulus (Lenton and Milford, 1977). Sugar beet exhibited a different response to a day length extension when either a *high irradiance* mixed source of fluorescent and tungsten-filament lamps or a *low irradiance* R source was used; in both cases, an increased plant dry weight was largely the consequence of an increase in net assimilation rate (Milford and Lenton, 1976; Lenton and Milford, 1977). Other reports have detailed similar responses to day length extensions (Hurd, 1973).

It is clear that the responses of leaves to day length extensions (and to night-breaks) can be influenced by the quality and irradiance of the light given and that, even at low irradiances, dry matter production may be affected via different pathways.

References

ABDALLA, A. A. (1967). *Exp. Agric.*, **3**, 137–142

ABE, S., KATSUMATA, H. and NAGAYOSHI, H. (1955). *J. hortic. Ass. Jap.*, **24**, 6–16

AUSTIN, R. B. (1972). *J. hortic. Sci.*, **47**, 493–504

BALLARD, L. A. T. (1969). In *The Induction of Flowering*, pp. 376–392. Ed. by L. T. Evans. Macmillan of Australia, Melbourne

BENTLEY, B., MORGAN, C. R., MORGAN, D. G. and SAAD, F. A. (1975). *Nature*, **256**, 121–122

BISCOE, P. V. and GALLAGHER, J. N. (1977). In *Environmental Effects on Crop Physiology*, pp. 75–100. Ed. by J. J. Landsberg and C. V. Cutting. Academic Press, London

BLACK, J. N. (1956). *Arch. Met., Geophys. Bioklimat., B*, **7**, 165–189

BOURLAUG, N. E., ORTEGA, J. C., NARVAEZ, I., GARCIA, A. and RODRIGUES, R. (1966). Hybrid Wheat Seminar Report, pp. 1–19. Crop Quality Council, Minneapolis

BREWSTER, J. L. (1977). *Hortic. Abstr.*, **47**, 17–23, 103–112

BULL, T. A. and GLASZIOU, K. T. (1975). In *Crop Physiology*, pp. 51–72. Ed. by L. T. Evans. Cambridge University Press, Cambridge

BÜNNING, E. (1967). *The Physiological Clock*. Springer-Verlag, New York

BUNT, A. C. (1973). *J. hortic. Sci.*, **48**, 315–325

BUNTING, A. H. (1975). *Weather*, **30**, 312–325

CACKETT, K. E. and WALL, P. C. (1971). *Rhod. J. agric. Res.*, **9**, 107–120

CALDER, D. M. (1966). In *The Growth of Cereals and Grasses*, pp. 59–73. Ed. by F. L. Milthorpe and J. D. Ivins. Butterworth, London

CAMERON, D. F. and MANNETJE, L.'t. (1977). *Austr. J. exp. Agric. Anim. Husb.*, **17**, 417–424

CANHAM, A. E. (1975). Electricity Council Research Report, No. 886

CATHEY, H. M. (1957). *Proc. Am. Soc. hortic. Sci.*, **69**, 485–491

CATHEY, H. M. (1969). In *The Induction of Flowering*, pp. 268–290. Ed. by L. T. Evans. Macmillan of Australia, Melbourne

CLEMENTS, H. F. (1968). *Pl. Physiol.*, **43**, 57–60

CLEMENTS, H. F. (1975). Technical Bulletin, No. 92. University of Hawaii, Hawaii Agricultural Experiment Station

COCKSHULL, K. E. (1966). *Ann. Bot.*, **30**, 791–806

COCKSHULL, K. E. (1976). *J. hortic. Sci.*, **51**, 441–450

COCKSHULL, K. E. (1979). *Ann. Bot.*, **44**, 451–460

COCKSHULL, K. E. and HUGHES, A. P. (1967). *Nature*, **215**, 780–781

COCKSHULL, K. E. and HUGHES, A. P. (1969). *Ann. Bot.*, **33**, 367–379

COCKSHULL, K. E. and HUGHES, A. P. (1971). *Ann. Bot.*, **35**, 915–916

COOPER, J. P. (1959). *Hered. Lond.*, **13**, 317–340, 445–459, 461–479

COOPER, J. P. (1960). *Ann. Bot.*, **24**, 234–246

COOPER, J. P. (1965). In *Essays on Crop Plant Evolution*. Ed. by J. B. Hutchinson. Cambridge University Press, Cambridge

CURTIS, D. L. (1968). *J. appl. Ecol.*, **5**, 215–226

DANIELS, J., GLASZIOU, K. T. and BULL, T. A. (1965). *Proc. int. Soc. Sugar Cane Technol.*, **12**, 1027–1032

DE HERTOGH, A. A. (1974). *Florist*, **8**, 54–58

DOWNS, R. J., BORTHWICK, H. A. and PIRINGER, A. A. (1958). *Proc. Am. Soc. hortic. Sci.*, **71**, 568–578

EVANS, L. T. (1964). In *Grasses and Grasslands*, pp. 126–153. Ed. by C. Barnard. Macmillan, New York

EVANS, L. T. (1969). In *The Induction of Flowering*, pp. 328–349, 457–480. Ed. by L. T. Evans. Macmillan of Australia, Melbourne

EVANS, L. T. (1975). In *Crop Physiology*, pp. 327–355. Ed. by L. T. Evans. Cambridge University Press, Cambridge

EVANS, L. T. and KING, R. W. (1975). In *Report of the TAC Working Group on the Biology of Yield of Grain Legumes*. Publication No. DDDR:IAR/75/2, FAO, Rome

FRIEND, D. J. C. (1969). In *The Induction of Flowering*, pp. 346–375. Ed. by L. T. Evans, Macmillan of Australia, Melbourne

FRIEND, D. J. C., FISHER, J. E. and HELSON, V. A. (1963). *Can. J. Bot.*, **41**, 1663–1674

GUTTRIDGE, C. G. (1960). *Bull. l'Inst. Agron. Stat. Rech. Gembloux, Hors serie* **2**, 941–948

GUTTRIDGE, C. G. (1969). In *The Induction of Flowering*, pp. 247–267. Ed. by L. T. Evans. Macmillan of Australia, Melbourne

GUTTRIDGE, C. G., ANDERSON, M. M., THOMPSON, P. A. and WOOD, C. A. (1961). *J. hortic. Sci.*, **36**, 93–101

HÅBJØRG, A. (1972). *Meld. Norges Landbr.*, **51**, 1–27

HALSE, N. J. and WEIR, R. N. (1970). *Austr. J. agric. Res.*, **21**, 383–393

HAMMER, P. A. (1976). *Hort. Sci.*, **11**, 570–572

HAMNER, K. C. (1969). In *The Induction of Flowering*, pp. 62–89. Ed. by L. T. Evans. Macmillan of Australia, Melbourne

HANKS, G. R. and REES, A. R. (1979). *J. hortic. Sci.*, **54**, 39–46

HARRIS, G. P. (1968). *Ann. Bot.*, **32**, 187–197

HARRIS, G. P. and ASHFORD, M. (1966). *J. hortic. Sci.*, **41**, 397–406

HARRIS, G. P. and BRADBEER, A. P. (1966). *J. hortic. Sci.*, **41**, 73–83

HARRIS, G. P. and HARRIS, J. E. (1962). *J. hortic. Sci.*, **37**, 219–234

HEATH, O. V. S. and HOLDSWORTH, M. (1948). *Symp. Soc. exp. Biol.*, **2**, 326–350

HEIDE, O. M. (1977). *Physiol. Plant.*, **40**, 21–26

HOLMES, M. G. and MCCARTNEY, H. A. (1976). In *Light and Plant Development*, pp. 467–476. Ed. by H. Smith. Butterworth, London

HUGHES, A. P. and COCKSHULL, K. E. (1965). *Ann. Bot.*, **29**, 131–151

HUGHES, A. P. and COCKSHULL, K. E. (1966). *Exp. Hortic.*, **16**, 44–52

HUMPHRIES, E. C. and WHEELER, A. W. (1963). *Ann. Rev. Pl. Physiol.*, **14**, 385–410

HURD, R. G. (1973). *Ann. appl. Biol.*, **73**, 221–228

JULIEN, M. H. R. (1970). PhD Thesis, University of Reading

KING, R. W., WARDLAW, I. F. and EVANS, L. T. (1967). *Planta*, **77**, 261–276

LAING, D. R. (1974). Report No. 8, University of Sydney, Faculty of Agriculture

LANG, A., CHAILAKHYAN, M. Kh. and FROLOVA, I. A. (1977). *Proc. nat. Acad. Sci., USA*, **74**, 2412–2416

LANGTON, F. A. (1977). *Scientia Hortic.*, **7**, 277–289

LANGTON, F. A. and COCKSHULL, K. E. (1976). *Acta Hortic.*, **63**, 165–175

LAURIE, A. and FOOTE, D. (1935). *Ohio Agric. Exp. Stat. Bull.*, **559**, 1–43

LAURIE, A., KIPLINGER, D. C. and NELSON, K. S. (1968). *Commercial Flower Forcing*. McGraw-Hill, New York

LEAKEY, R. R. B. and LAST, F. T. (1980). *J. arid. Env.*, **3**, 9–24

LENTON, J. R. and MILFORD, G. F. J. (1977). *Roth. Exp. Stat. Rep. for 1976*, pp. 42–43

LONGMAN, K. A. (1969). *Symp. Soc. exp. Biol.*, **23**, 471–488

LONGMAN, K. A. (1978). In *Tropical Trees as Living Systems*, pp. 465–495. Ed. by P. B. Tomlinson and M. H. Zimmerman. (Proceedings of the 4th Cabot Symposium.) Cambridge University Press, Cambridge

MACHIN, B. and SCOPES, N. (1978). *Chrysanthemums Year-Round Growing.* Blandford Press, Poole

MAJOR, D. J. and JOHNSON, D. R. (1975). Cited by Summerfield and Wien, 1979

MCMILLAN, C. (1974a). *Bot. Mag., Tokyo*, **87**, 261–269

MCMILLAN, C. (1974b). *Can. J. Bot.*, **52**, 1779–1791

MILFORD, G. F. J. and LENTON, J. R. (1976). *Ann. Bot.*, **40**, 1309–1315

MILFORD, G. F. J. and LENTON, J. R. (1979). *Roth. Exp. Stat. Rep. for 1978*, pp. 46–47

MOSHKOV, B. S. (1935). *Planta*, **23**, 774–803

NITSCH, J. P. (1957). *Proc. Am. Soc. hortic. Sci.*, **70**, 512–525

NJOKU, E. (1964). *J. Ecol.*, **52**, 19–26

PALEG, L. G. and ASPINALL, D. (1970). *Nature*, **228**, 970–973

PEACOCK, J. T. and MCMILLAN, C. (1968). *Am. J. Bot.*, **55**, 153–159

PINTHUS, M. J. (1967). *Euphytica*, **16**, 231–251

POHJAKALLIO, O. (1953). *Physiol. Plant.*, **6**, 140–149

POHJAKALLIO, O., SALONEN, A. and ANTILA, S. (1957). *Acta Agric. Scand.*, **7**, 361–388

POST, K. (1948). *Proc. Am. Soc. hortic. Sci.*, **51**, 590–592

RAHMAN, M. S. and WILSON, J. H. (1977). *Austr. J. agric. Res.*, **28**, 565–574

RAHMAN, M. S. and WILSON, J. H. (1978). *Austr. J. agric. Res.*, **29**, 459–467

RAWSON, H. M. (1970). *Austr. J. biol. Sci.*, **23**, 1–15

RAWSON, H. M. (1971). *Austr. J. agric. Res.*, **22**, 537–546

ROBINSON, J. C. (1971). *Rhod. J. agric. Res.*, **9**, 31–38

ROGERS, W. S. and FROMOW, M. G. (1958). *East Malling Annual Report 1957*, pp. 50–56

SALISBURY, F. B. (1963). *The Flowering Process.* Pergamon Press, Oxford

SCHWABE, W. W. (1953). *XIII International Horticultural Congress, 1952*, Vol. 2, pp. 952–960. Ed. by P. M. Synge. Royal Horticultural Society, London

SCHWABE, W. W. and NACHMONY-BASCOMBE, S. (1963). *J. exp. Bot.*, **14**, 353–378

SHIBLES, R., ANDERSON, I. C. and GIBSON, A. H. (1975). In *Crop Physiology*, pp. 151–189. Ed. by L. T. Evans. Cambridge University Press, Cambridge

SHROPSHIRE, W. (1973). *Sol. Energy*, **15**, 99–105

SOFFE, R. W., LENTON, J. R. and MILFORD, G. F. J. (1977). *Ann. appl. Biol.*, **85**, 411–415

SUMMERFIELD, R. J. and WIEN, H. C. (1979). In *Advances in Legume Science*, pp. 17–36. Ed. by A. H. Bunting and R. J. Summerfield. HMSO, London

SYLVEN, H. (1942). *Svensk Papp.-Tidn*, **43**, 317–324, 332–342

TAFAZOLI, E. and VINCE-PRUE, D. (1978). *J. hortic. Sci.*, **53**, 255–259

TAKIMOTO, A. and IKEDA, K. (1960). *Bot. Mag., Tokyo*, **73**, 175–181

TAKIMOTO, A. and IKEDA, K. (1961). *Pl. Cell Physiol.*, **2**, 213–229

THOMAS, R. G. (1961). *Nature*, **189**, 771–772

THOMPSON, H. C. and SMITH, O. (1938). *Bull. Cornell Agric. Exp. Stat.*, No. 708

THORNE, G. N., FORD, M. A. and WATSON, D. J. (1968). *Ann. Bot.*, **32**, 425–446

VINCE, D. (1953). *Agric.*, **60**, 68–73

VINCE, D. (1965). *Physiol. Plant.*, **18**, 474–482
VINCE, D. (1970). *Proc. 18th Int. Hortic. Congr.*, Vol. 5, pp. 169–180. Ed. by N. Goren and K. Mendel. Tel Aviv, Israel
VINCE, D., BLAKE, J. and SPENCER, R. (1964). *Physiol. Plant.*, **17**, 119–125
VINCE-PRUE, D. (1975). *Photoperiodism in Plants.* McGraw-Hill, London
VINCE-PRUE, D. (1976). In *Light and Plant Development*, pp. 347–369. Ed. by H. Smith. Butterworth, London
VINCE-PRUE, D. (1977). *Planta*, **133**, 149–156
VINCE-PRUE, D. and GUTTRIDGE, C. G. (1973). *Planta*, **110**, 165–172
WALL, P. C. (1972). MSc Thesis, University of Reading
WALL, P. C. and CARTWRIGHT, P. M. (1974). *Ann. appl. Biol.*, **76**, 299–309
WAREING, P. F. (1956). *Ann. Rev. Pl. Physiol.*, **7**, 191–214
WASSINK, E. C. and STOLWIJK, J. A. J. (1953). *Med. Land.-Hoog. Wageningen*, **53**, 99–112
WHALLEY, D. N. (1977). *A.D.A.S. Quart. Rev.*, **25**, 41–62
WHALLEY, D. N. and COCKSHULL, K. E. (1976). *Scientia Hortic.*, **5**, 127–138
WILLIAMS, B. J. Jr, PELLETT, N. E. and KLEIN, R. M. (1972). *Pl. Physiol.*, **50**, 262–265
ZEHNI, M. S., SAAD, F. S. and MORGAN, D. G. (1970). *Nature*, **227**, 628–629

11

WATER STRESS AND CROP GROWTH

W. DAY
Rothamsted Experimental Station, Hertfordshire, UK

Shortage of water restricts crop productivity all over the world—not just in those areas classified as arid or semiarid, but in any area in which the evaporative demand greatly exceeds rainfall during the growing season. Thus in eight years out of ten, irrigation can increase the yield of many crops in the south-east of England. The response of yield to irrigation has been studied, both empirically and analytically, over many years (e.g. work at Rothamsted since 1951 by Penman, 1971; French and Legg, 1979). These studies have produced general relationships between yield and irrigation amounts which are of practical use to the farmer (Ministry of Agriculture, Fisheries and Food, 1974). However, the responses to water stress of specific physiological processes have only been investigated in detail more recently.

In this paper, the author describes some of the ways in which crops respond to water stress. The physiological responses that are observed, and their relative importance for crop productivity, vary with species, soil type, nutrients and climate, but there are general features that can be identified.

Water stress, which results from the withholding of water supply, leads directly to changes in the physical environment of the crops, and these changes may subsequently affect crop physiology. As the soil is dried, the soil water potential decreases and so does the soil hydraulic conductivity. Thus it is more difficult for plants to extract water (Gardner, 1960) and, as a consequence, the plant water potential tends to decrease. This decrease may directly affect the physical aspects of some physiological processes. For example, turgor pressure in the cells will decrease, and turgor forces play a part in the process of leaf expansion (Hsiao and Acevedo, 1974). Loss of turgor can cause leaves to wilt, thus decreasing their light interception and reducing photosynthesis rates. However, under these conditions it is likely that stomatal closure will have a greater effect upon photosynthesis rates than does wilting (Schulze and Hall, this volume).

There will be other direct effects of decreases in plant water potential on physiology, but the indirect effects are likely to be as important. Decreased leaf expansion and stomatal closure both restrict photosynthesis, and as well as slowing dry matter accumulation, this reduction in assimilate supply may affect many physiological processes including the differentiation and expansion of new tissue. Shortage of assimilates at the roots may not only decrease root growth but, as a consequence of this decreased growth, the roots may be less able to utilize all the soil's reserves of water. Other indirect effects of decreased water potential may occur via the action of plant hormones (Mansfield and Wilson, this volume). As well as these effects mediated by the water supply itself, other responses can result from a decrease in nutrient supply. Mobility of ions in the soil decreases as the soil dries, and the ability of roots to take up some ions may also decrease (Dunham and Nye, 1976). The consequent decrease in nutrient uptake may affect a wide range of physiological processes.

These physical, physiological and chemical effects of water stress have been the subject of many reviews in recent years (Hsiao, 1973; and the series of articles edited by Kozlowski, 1968a, 1968b, 1972, 1976). In this paper, a single example is given of crop response to water stress considering many of these effects, and particularly how they combine to affect yield. The crop is spring barley, grown at Rothamsted in 1976 on a field site that was protected from rain by automatic shelters (Legg *et al.*, 1978). Irrigation was applied to individual plots on a predetermined schedule to give a wide range of treatments (*Figure 11.1*) from 'weekly irrigation' to 'no irrigation' from

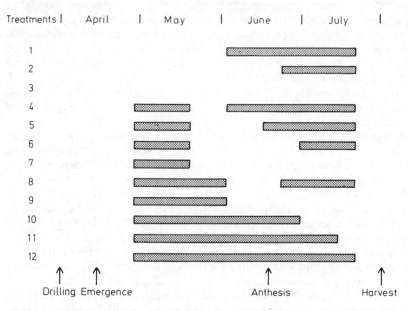

Figure 11.1   *Irrigation treatments for an experiment at Rothamsted in 1976. The bars indicate the period during which each treatment was irrigated*

emergence to harvest (Day *et al.*, 1978). Measurements in the soil, on the plants and their environment, and of the growth of the crop were made on many of the treatments. In this paper, the responses observed are summarized, concluding with a consideration of how these responses modified yield, and how yield responses were related to the drought treatments applied.

The Environment of the Crop, and its Water and Nutrient Status

When the water supply to a crop is stopped, the most immediate effect is that the soil in the rooting zone begins to dry, and eventually transpiration by the

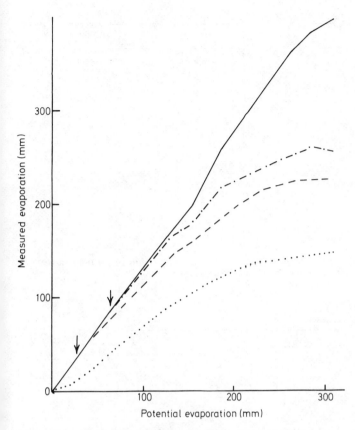

Figure 11.2 *Evaporation from four treatments plotted against the potential evaporation rate (Penman, 1970). The treatments were irrigated weekly throughout the season (——), until the arrow (·—·), (– – –), and unirrigated (· · ·). (After Day* et al., *1978)*

crop will decrease as a consequence of this drying. From our experimental work (*Figure 11.2*), cumulative evaporation from the weekly irrigated crop was almost linearly related to cumulative potential evaporation, as calculated by Penman's (1970) method. When irrigation was withheld in midseason, evaporation continued at the maximum rate for some time and then the rate declined. When irrigation was withheld early in the season, the evaporation was reduced almost immediately; at this stage the plants were small, and a significant proportion of the evaporation from the irrigated crop came from the soil surface—the soil surface of this unirrigated crop dried rapidly and evaporation from the soil surface decreased as a consequence (Day *et al.*, 1978). After irrigation has stopped, the limit to the total evaporation is the amount of water that the roots can extract from the soil profile. This amount will depend on the depth of soil exploited by the roots, and the soil characteristics; the soil environment can thus greatly modify crop response to water stress.

As the soil dries, the soil water potential decreases, so leading to a decrease in plant water potential. In a drying soil, the soil water potential varies down

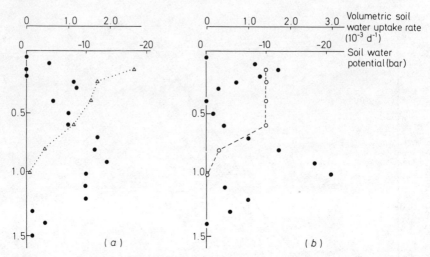

Figure 11.3 *Profiles of soil water potential (○, △) and water uptake rate (●) for two treatments in mid-June after (a) eight weeks', and (b) five weeks' drought*

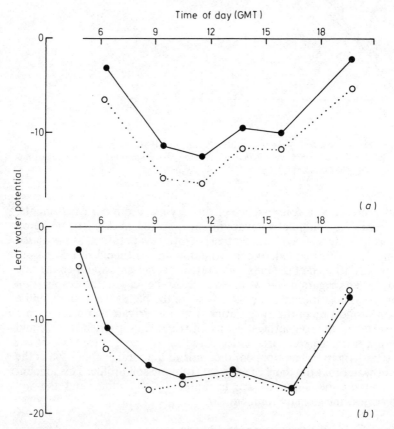

Figure 11.4 *Leaf water potential for weekly irrigated (●) and unirrigated (○) treatments on (a) 27 May, and (b) 8 June*

the soil profile. Thus the water potential (*Figure 11.3*) was as low as -18 bars near the soil surface of the unirrigated treatment just prior to anthesis, but at 1.0 m depth it was little less than -0.1 bar, 'field capacity'. The rate of change of soil water content is also shown in *Figure 11.3*, indicating the depths at which roots were actively extracting water. Clearly, the crop was extracting water from soil at a wide range of water potentials, and it is not possible to give a single value for the soil water potential, nor is it straightforward to define a mean effective soil water potential.

The plant water potential does not depend solely on the soil water potential but also on plant structure and transpiration rate. Although soil water potentials in the unirrigated treatments were low, the leaf area and transpiration rates were also lower than for the irrigated treatment. Leaf water potentials were often little different on irrigated and unirrigated treatments, with maximum differences only about -3 or -4 bar (*Figure 11.4*).

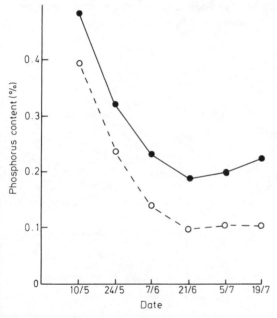

Figure 11.5 *Phosphorus content of the crop dry matter for weekly irrigated (•) and unirrigated (○) treatments*

Water stress modifies aspects of the crop's environment other than its water status. The availability of some nutrients is decreased as the soil dries, although the importance of this decrease depends on the nutrient amounts in the soil. Phosphorus is of particular interest as its availability is decreased (Dunham and Nye, 1976) and most phosphate is held in the top layer of the soil, which is the first to be dried in a period of drought. The phosphorus contents of the plants under the two extreme treatments differed markedly (*Figure 11.5*). In both cases the content declined with time, and it is not possible to decide whether the differences between treatments affected growth and yield. However, comparison with other work at Rothamsted

(Johnston, Warren and Penny, 1970) shows that the final percentage in the unirrigated crop was as low as in other barley crops where lack of phosphorus did limit yield.

Water stress also lead to differences in the aerial environment of stressed and unstressed crops. Because of the differences in the leaf area, the amount of light intercepted differed markedly between crops. As a consequence of decreased transpiration, temperature in a stressed crop is generally somewhat higher and humidity lower than in an unirrigated crop. These differences lead to an increased water vapour pressure deficit in stressed crops, which may have a direct influence on stomatal resistance (Rawson, Begg and Woodward, 1977).

Physiological and Morphological Responses

Having considered the conditions in which the stressed and unstressed crops grew, we can now consider the measured responses of the crop.

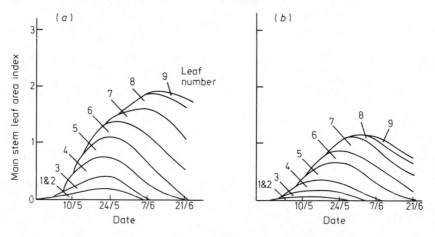

Figure 11.6   *Main stem leaf area index for (a) weekly irrigated, and (b) unirrigated treatments, distinguishing the contribution from each leaf number*

Leaf extension is particularly sensitive to water stress, although responses in the field are generally smaller than those obtained in constant environment experiments. In our experiment, although leaf water potentials did not differ greatly between treatments, there was consistently greater turgor in the leaves of irrigated plants—and leaf growth rates differed markedly (*Figure 11.6*). Leaf area can be influenced by changes in the time of leaf appearance, in the rate and duration of leaf expansion, and in leaf senescence. In this experiment the time of leaf appearance was not affected but maximum leaf size and senescence were.

As well as reducing the area of photosynthetic tissue, water stress can affect photosynthesis itself. Our measurements did not show any systematic differences in either photosynthetic efficiency or in the internal resistance of leaves to CO_2 transfer under different treatments. However, stomatal

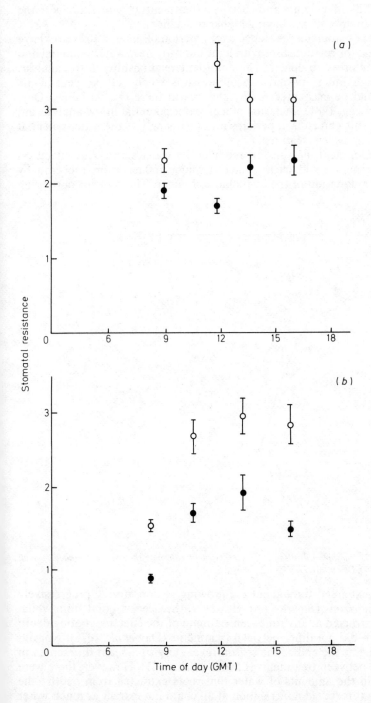

Figure 11.7 *Stomatal resistance of weekly irrigated (●) and unirrigated (○) treatments on (a) 27 May, (b) 8 June*

resistances did differ significantly between treatments (*Figure 11.7*), with the stressed treatments always having higher resistances.

The values of stomatal resistance over a period of about four weeks have been analysed for relationships with light intensity, leaf water potential and water vapour pressure deficit. The dominant relationships in the analysis were with light intensity and vapour pressure deficit—60 per cent of the variance could be accounted for by changes in these two variables (Day, Lawlor and Legg, 1981). Inclusion of leaf water potential in the analysis did not increase this percentage, perhaps not surprisingly as the water potential differed little between treatments.

Crop development may also be modified by water stress. Any effects on root development could clearly have a major influence on the crop by restricting exploitation of the available soil water. The roots of our spring

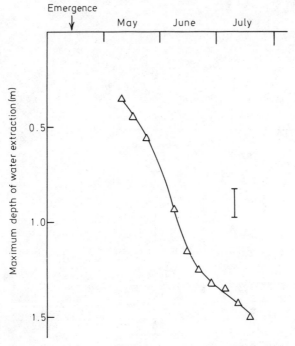

Figure 11.8   *The maximum depth of soil water extraction through the season, averaged over all treatments. (After Day* et al., *1978)*

barley crop extended throughout the growing season, drying progressively deeper soil horizons (*Figure 11.8*). However, the deepest point from which water was extracted at any time—an estimate of the effective rooting depth at that time—did not differ between treatments (Day *et al.*, 1978). Nor did root sampling at the end of the season reveal any significant differences in root density between treatments (Lawlor *et al.*, 1981). However, there were differences in the amounts of water that roots extracted from depth—the treatment that received no irrigation at all could not extract as much water from below 0.5 m as the treatments irrigated early in the season (Day *et al.*, 1978).

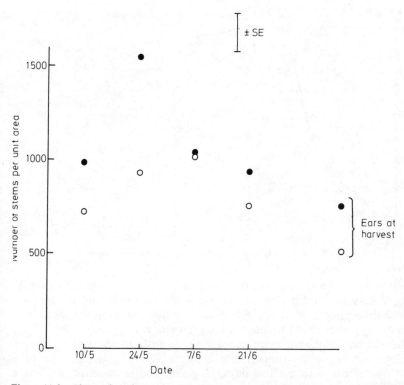

Figure 11.9 *The number of stems per unit ground area through the season, together with the number of ears per unit ground area at harvest, for the weekly irrigated (●) and unirrigated (○) treatments*

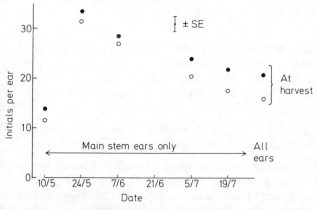

Figure 11.10 *The average number of grain initials on each main stem ear through the season, together with the mean number of grains per ear (main stem and tiller) at final harvest, for the weekly irrigated (●) and unirrigated (○) treatments*

Effects on above-ground development were easier to measure and were quite large. Tiller production was particularly sensitive to stress (*Figure 11.9*), but not all the tillers produced survived to bear ears, and the tiller death phase was also sensitive to water stress. The number of grains in each

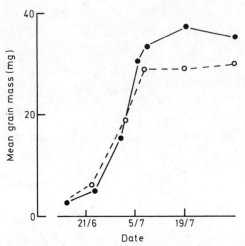

Figure 11.11   *Grain filling for weekly irrigated (●) and unirrigated (○) treatments*

ear was also affected by stress (*Figure 11.10*), as was the process of grain filling (*Figure 11.11*). Final grain size was least in the unirrigated treatment, and the difference in final size was caused by differences in the duration of grain filling rather than the rate of filling (Lawlor *et al.*, 1981).

The stressed and unstressed crops differed little in the timing of their development. Leaf appearance, anthesis and the start of grain filling occurred in all treatments within a few days. However leaf senescence, the end of grain filling, and crop maturity all occurred earlier under water stress.

Dry Matter Accumulation

The final yield of a stressed crop differs from that of an unstressed crop as a result of the integrated effect of many changes in crop physiology. It is important to assess the extent to which each of these changes contributes to loss of yield (Legg *et al.*, 1979). Accumulation of dry matter can be considered as three processes:

$$\begin{array}{c}\text{total dry}\\\text{matter yield}\end{array} = \begin{array}{c}\text{light}\\\text{intercepted}\end{array} \times \begin{array}{c}\text{efficiency of}\\\text{photosynthesis}\end{array} \times \begin{array}{c}\text{fraction remaining}\\\text{after respiration}\end{array}$$

First, light interception: in our experiment the differences in leaf area between treatments were substantial (*Figure 11.6*). The total green area index was decreased by stress both via a shortening of the season and via a smaller maximum green area (*Figure 11.12*). Interception can also be decreased if the leaf angle changes—for example, by leaves wilting—but no significant differences in leaf angle were observed. Over the whole season the unirrigated crop intercepted 40 per cent less light than did the fully irrigated.

The second term is the efficiency with which light energy is converted into chemical energy via carbon fixation. The 'efficiency of photosynthesis' required here is the mean for the whole crop, averaging over all light intensities and all leaves. Our measurements did not distinguish any

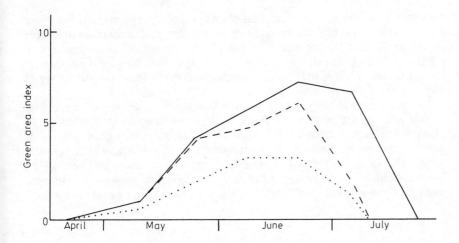

Figure 11.12 *The green area index through the season: treatments which received no irrigation (· · ·), irrigation until 12 May (– – –), and weekly irrigation throughout the season (——–). (After Legg et al., 1979)*

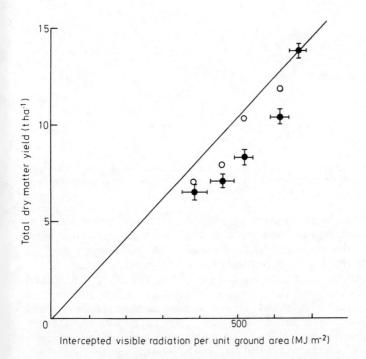

Figure 11.13 *The relationship between total dry matter yield and the amount of radiation intercepted by five of the experimental treatments. The yields estimated (○) from the model of Legg et al. (1979) can be compared with the measured yields (●). (After Legg et al., 1979)*

systematic differences in the internal photosynthetic performance of stressed and unstressed leaves. However, the increased stomatal resistances that were observed also restricted the CO_2 supply to stressed leaves, and hence photosynthesis, particularly for those leaves exposed to high light intensities; the estimated effect of the stomatal closure was to decrease the 'efficiency of photosynthesis' by no more than 7 per cent (Legg *et al.*, 1979).

The third term in the yield equation represents the efficiency of conversion of assimilated carbon into stored dry matter. We made no measurements of plant respiration rates and so cannot comment on any direct influence of water stress on respiration. It is probable that the increased temperature within the stressed crop led to enhanced respiration rates (McCree, 1970). We have calculated the likely differences in respiration between treatments, using our measurements of crop temperature and dry matter accumulation.

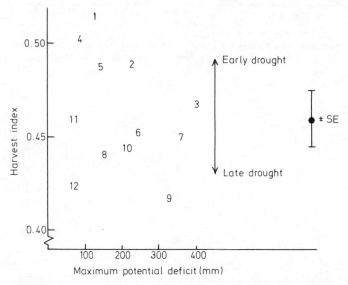

Figure 11.14   *The relationship between harvest index and maximum potential deficit for each treatment. Smaller treatment numbers tended to achieve their maximum deficit early in the season. (After Day et al., 1978)*

Bringing the three terms together, we can compare the estimated decreases in dry matter yield, based on our model and our measurements of the three processes, with the measured yield decreases. The differences in light interception for the five analysed treatments accounted for much of the difference in total dry matter yields (*Figure 11.13*). Yields for the stressed crops were lower than was predicted by the decrease in light interception alone. Some of this decrease could be associated with stomatal closure and changes in crop structure, but not all (*Figure 11.13*; Legg *et al.*, 1979). There is still some difference in yield for which we have not been able to account.

For agriculture, it is the grain yield and not the total dry matter yield which is of primary importance. The harvest index is the fraction of the total dry matter that is in the grain: for cereals in general, where the grain only fills at the end of the crop's life, it is to be expected that late stress will decrease the harvest index more than does early stress (*Figure 11.14*). The effect of late stress on grain filling (shown in *Figure 11.11*) was not as great as its effect (via loss of leaf area, and stomatal closure) on crop photosynthesis in the period of grain filling. This may indicate that the plants were able to compensate to some extent for adverse conditions at the time of grain filling (Gallagher, Biscoe and Scott, 1975; Lawlor *et al.*, 1981).

Yield Decreases and the Degree of Drought

Consideration of the degree of water stress, or drought, is not only important when considering the response of specific physiological processes to stress (Hsiao, 1973) but also for the response of yield to the overall stress treatment.

Plant water potential is a measure of water stress, but its importance may be complicated by changes in the component potentials (osmotic and turgor) and by changes in crop structure. Osmotic adjustment—decreases in the osmotic potential under stress, which lead to the maintenance of turgor at lower water potentials—is a mechanism whereby plants adapt to water stress conditions (Turner, Begg and Tonnet, 1978). For our spring barley, the differences in osmotic potential were small and any osmotic adjustment was minimal. Neither were the differences in leaf water potential large (*Figure 11.4*), although lower leaf water potentials generally correlated with lower yields. Alternatively, crop response can be related to the imposed environmental conditions—the soil water supply or the atmospheric demand—and such analysis is more straightforward, although results will be influenced by aspects of the environment other than the water balance, e.g. soil type (French and Legg, 1979). This type of analysis is particularly useful in studying yield response to water stress, as in many experiments detailed plant measurements are not made but the relevant environmental variables are measured.

The results of our experiment were analysed in a number of ways. We used the mean soil water deficit—the difference between the measured water content of the profile and its content at 'field capacity'—averaged over periods of a few weeks, in an analysis of the effects of stress on the components of grain yield (Day *et al.*, 1978). This analysis showed that each component was affected by stress in a particular phase of the crop's growth. The number of grains per ear was related to the deficit in the first six weeks of growth when the ear was developing; the number of ears to the deficit in the next period, up to anthesis, when tiller numbers were declining; and grain mass was related principally to the deficit in the grain filling period (*Figure 11.15*). In particular, it is noteworthy that although the tillering phase was sensitive to stress, with large differences between treatments in the number of tillers after six weeks of growth (*Figure 11.9*), it was the response to stress in the next three weeks, when tillers were dying, that affected the final number of tillers and hence the yield.

The evaporative demand during a period of drought can also give a measure of the stress experienced by a crop, without the need for detailed measurements of soil water content. So a second analysis of the response of yield used the maximum potential deficit for each treatment. In general, this maximum potential deficit incorporates both the level of stress and its duration; and yield is often simply related to it (French and Legg, 1979). In

212

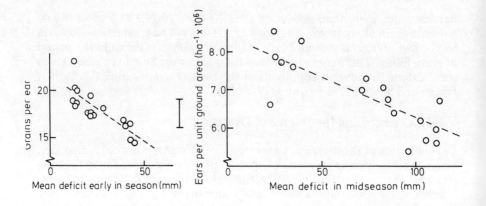

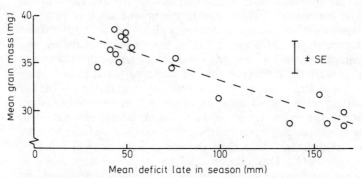

Figure 11.15 *Components of yield plotted against the mean soil water deficit in one of three periods: period 1, tillering and ear formation; period 2, stem elongation and tiller death; period 3, grain filling. For each component the linear regression shown accounted for most of the variance in that component, independent of the deficit in the other two periods. (After Day et al., 1978)*

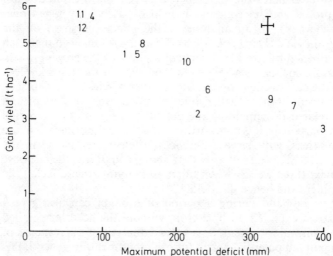

Figure 11.16 *Grain yield as a function of maximum potential deficit for each treatment. (After Day et al., 1978)*

our experiment the maximum potential deficit amounted to the integral of the potential evaporation rate over the period of drought; in other experiments, where water supply is not withheld in just one period, a rather more complicated calculation must be made (Penman, 1971). Our results indicate that there was no difference between the effect of drought at different times of the season (*Figure 11.16*; Day *et al.*, 1978), with the possible exception of treatment 2, which only received irrigation after anthesis.

Finally, we considered the effect of drought by relating yield to the water used by the crop. There is an upper limit to the amount of water that plants can extract from the soil, and when this limit is approached transpiration is decreased by loss of leaf area and stomatal closure. These responses will also decrease assimilation. A single linear relationship between grain yield and water use largely covers all treatments in our experiment (*Figure 11.17*). When stress decreased water use, it also led to a decrease in grain yield. Even treatment 2 is within the general trend—late irrigation did not greatly stimulate grain yield nor did it greatly increase water use.

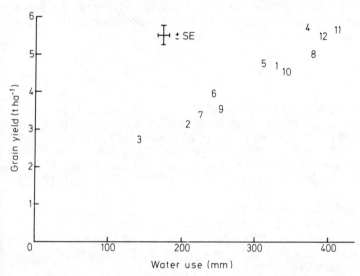

Figure 11.17   *Grain yield as a function of water used under each treatment. (After Day* et al.. *1978)*

Summary

There are many physiological processes that are modified by water stress and that may contribute to loss of productivity. Of these processes, the production and maintenance of leaf area were of particular importance in our experiment. This importance was evidenced by the magnitude of the differences in leaf area between treatments, and quantified by consideration of the utilization of incident light energy on the different treatments.

Relating the response of the crop to the applied water stress gave information both on the stress-sensitive aspects of development and on how yield was affected. Thus, although barley grains only fill towards the end of the crop's life, we found a response of yield to stress at all stages in the growing season. This response appeared in different components of the yield depending on the timing of the stress, but the magnitude of the yield response was independent of the timing.

There was, in particular, a nearly linear relationship between the grain yield of each treatment and the water used under that treatment. When shortage of water decreased transpiration there was a proportional decrease in grain yield. For the crop to produce more yield under these conditions would require either an increase in the amount of water available—by an increased depth of rooting or more efficient extraction—or a decrease in the amount of water used to produce unit assimilate, i.e. an increase in the water use efficiency.

Acknowledgements

The work discussed in this paper was done in conjunction with B. J. Legg, D. W. Lawlor, K. J. Parkinson and A. E. Johnston. My thanks are due to them for many valuable discussions prior to the presentation of this paper, and to B. J. Legg for comments on the manuscript.

References

DAY, W., LAWLOR, D. W. and LEGG, B. J. (1981). *J. agric. Sci.*, in press

DAY, W., LEGG, B. J., FRENCH, B. K., JOHNSTON, A. E., LAWLOR, D. W. and JEFFERS, W. de C. (1978). *J. agric. Sci.* (*Cambridge*), **91**, 599–623

DUNHAM, R. J. and NYE, P. H. (1976). *J. appl. Ecol.*, **13**, 967–984

FRENCH, B. K. and LEGG, B. J. (1979). *J. agric. Sci.* (*Cambridge*), **92**, 15–37

GALLAGHER, J. N., BISCOE, P. V. and SCOTT, R. K. (1975). *J. appl. Ecol.*, **12**, 319–336

GARDNER, W. R. (1960). *Soil Sci.*, **89**, 63–67

HSIAO, T. C. (1973). *Ann. Rev. Pl. Physiol.*, **24**, 519–570

HSIAO, T. C. and ACEVEDO, E. (1974). *Agric. Meteorol.*, **14**, 59–84

JOHNSTON, A. E., WARREN, R. G. and PENNY, A. (1970). *Roth. Exp. Stat. Rep. for 1969*, Part 2, pp. 39–68

KOZLOWSKI, T. T. (1968a). *Water Deficits and Plant Growth*, Vol. I: *Development, Control and Measurement*. Academic Press, New York

KOZLOWSKI, T. T. (1968b). *Water Deficits and Plant Growth*, Vol. II: *Plant Water Consumption and Response*. Academic Press, New York

KOZLOWSKI, T. T. (1972). *Water Deficits and Plant Growth*. Vol. III: *Plant Responses and Control of Water Balance*. Academic Press, New York

KOZLOWSKI, T. T. (1976). *Water Deficits and Plant Growth*, Vol. IV: *Soil Water Measurement, Plant Responses and Breeding for Drought Resistance*. Academic Press, New York

LAWLOR, D. W., DAY, W., JOHNSTON, A. E., LEGG, B. J. and PARKINSON, K. J. (1981). *J. agric. Sci.*, in press

LEGG, B. J., DAY, W., BROWN, N. J. and SMITH, G. J. (1978). *J. agric. Sci.* (*Cambridge*), **91**, 321–336

LEGG, B. J., DAY, W., LAWLOR, D. W. and PARKINSON, K. J. (1979). *J. agric. Sci.* (*Cambridge*), **92**, 703–716

MCCREE, K. J. (1970). *Prediction and Measurement of Photosynthetic Productivity*, pp. 221–230. Ed. by I. Setlick. Centre for Agricultural Publishing and Documentation, Wageningen

MINISTRY OF AGRICULTURE, FISHERIES AND FOOD (1974). Bulletin No. 138. HMSO, London

PENMAN, H. L. (1970). *J. agric. Sci.*, **75**, 69–73

PENMAN, H. L. (1971). *Roth. Exp. Stat. Rep. for 1970*, Part 2, pp. 147–170

RAWSON, H. M., BEGG, J. E. and WOODWARD, R. G. (1977). *Planta*, **134**, 5–10

TURNER, N. C., BEGG, J. E. and TONNET, M. L. (1978). *Austr. J. Pl. Physiol.*, **5**, 597–608

12

SHORT-TERM AND LONG-TERM EFFECTS OF DROUGHT ON STEADY-STATE AND TIME-INTEGRATED PLANT PROCESSES

E.-D. SCHULZE
A. E. HALL*
Lehrstuhl Pflanzenökologie Universität Bayreuth, West Germany

Introduction

Approximately one third of the earth's land surface is covered with vegetation types such as desert, savannahs and steppes where water deficits limit productivity. In addition, there are many other habitats where drought, which we define as the occurrence of substantial water deficits in soil, plant or atmosphere, significantly influences plant performance and survival. Even the arctic and alpine vegetation is to a large extent determined by the effects of drought due to cold soils or frozen tissues. Drought limitations on plant productivity also occur in agricultural systems because irrigation is only possible on a relatively small area of the earth's surface and high evaporative demands can cause plant water deficits even with abundant water in the soil. The extent to which we are able to understand the influence of environmental drought on plant processes will strongly influence management of cultivated and natural vegetations.

Despite a long history of ecological and physiological research, present understanding of whole plant responses to drought is very limited. Much research has emphasized attempts to relate the physiological response of a plant to its water status at the time at which this process occurs. The time dimension of drought effects has only been studied for biomass production and yield, whereas the long-term influence of drought on momentary events such as stomatal conductances has been neglected (Hall, Schulze and Lange, 1976). Instead, emphasis has been given to dynamic models which relate stomatal response to changes in plant water status operating with a half time of seconds or minutes.

This paper examines the following contrasting hypotheses for the effects of drought on plant processes. First, there are unique relationships between the rates of plant processes and the water relationships of the plant at the time at which these processes occur. We define these as the instantaneous or short-term effects of drought. Second, the previous history of the water relationships of the plant have to be taken into account and plants respond to some integral of drought over time. We define this as the long-term effects of drought. The influences of drought on steady-state stomatal conductance and CO_2 assimilation, and on the time-integrated processes of biomass production and seed yield, are considered.

*Department of Botany and Plant Science, University of California, Riverside, CA 92521, USA.

Long-term and Short-term Effects of Drought on Transpiration and Assimilation

THE RELATIONSHIP BETWEEN PLANT WATER STATUS AND DIURNAL COURSES OF GAS EXCHANGE

Drought causes significant changes in the diurnal courses of transpiration and CO_2 assimilation. During the course of a year, one-peaked curves of CO_2 uptake and water loss change with increasing atmospheric and soil drought into two-peaked patterns (Stocker, 1960; Schulze *et al.*, 1975a). In irrigated and non-irrigated treatments of the apricot *Prunus armeniaca*, grown in a desert environment (*Figure 12.1*), diurnal patterns were one-peaked with

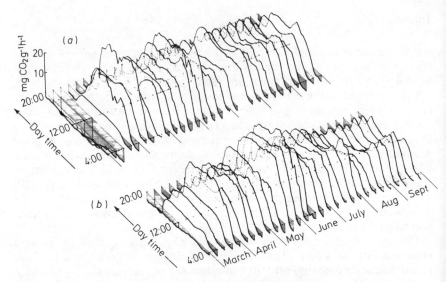

Figure 12.1   *Seasonal changes of diurnal courses of CO_2 assimilation of (a) irrigated, and (b) non-irrigated* Prunus armeniaca *trees growing in the run-off farm Avdat, Israel. The days of lowest air humidity are depicted for weekly periods. (Schulze* et al., *1980)*

high rates of photosynthesis until early July except on days with cloud cover or sand storms. Two-peaked curves were observed in both treatments in late summer, but the dry treatment exhibited a much stronger noon depression and lower rates of photosynthesis. It is important to note (*Figure 12.2*) that the minimal daily leaf water potentials (Ψ_{min}) of both treatments reached a very low level in June when diurnal courses were still one-peaked, and Ψ_{min} did not change when diurnal courses became two-peaked with the proceeding dry season. The non-irrigated apricot is strongly avoiding drought. Ψ_{min} was only 3–4 bar lower than in the irrigated treatment, but leaf biomass was strongly reduced (Evenari *et al.*, 1977). These seasonal and diurnal patterns of CO_2 exchange indicate no obvious relationship with the concurrent plant water status. Consequently, available information on both short-term and long-term effects of drought on stomatal conductance and carbon assimilation is examined.

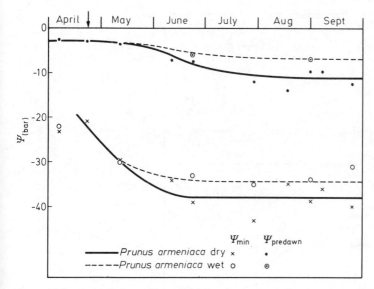

Figure 12.2   *Seasonal change of minimal daily water potentials* (Ψ_{min}) *and predawn water potentials* ($\Psi_{predawn}$) *of irrigated and non-irrigated* Prunus armeniaca *trees. (After Evenari et al., 1977)*

SHORT-TERM EFFECTS OF DROUGHT ON STOMATAL CONDUCTANCE

In the field, environmental and plant internal water factors vary simultane-ously, and leaf water potential is closely related to the transpiration rate (Kaufmann, 1976). Consequently, it is difficult to separate the stomatal effects of leaf water status and of the environmental factors which also influence evaporative demand. Dynamic and steady-state responses should be separately considered.

Transient stomatal responses to abrupt changes in water supply have been described by Raschke (1970). Similar dynamic responses are also observed after excising leaves or twigs. Oscillations are another type of dynamic stomatal response due to a hydraulic feedback loop (Cowan, 1977). With respect to an understanding of plant responses under field conditions it must be realized, however, that leaf conductance usually operates in a quasi-steady-state and oscillations are not generally observed. Also, studies in which leaves or twigs are cut or in which roots are treated and osmotically shocked or excised are necessary in a causal analysis but may not be relevant to explain stomatal responses in more natural environments, as was pointed out by Hall and Hoffmann (1976). Many complicated interactions may occur with such treatments and the findings of Raschke (1975) on the interaction of hormonal levels and stomatal responses may serve as an example for the manifold alterations which may be induced with such treatment.

Under field conditions with intact plants, stomatal conductance was found to be constant or to vary without an obvious relation to short-term changes of xylem pressure potential, Ψ (Jordan and Ritchie, 1971; Schulze *et al.*, 1975b). In contrast, it is well known that stomata may close at some threshold level of water stress (Turner, 1974). Therefore, the former results could be

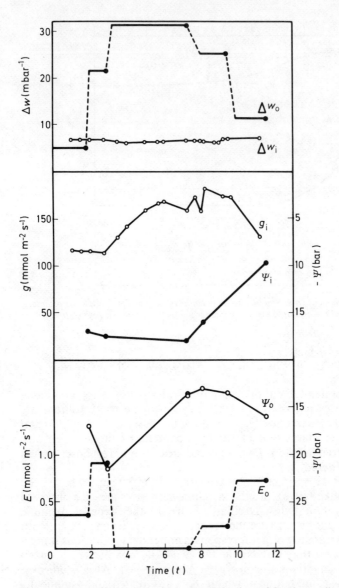

Figure 12.3   *Leaf evaporation (E$_o$) of a whole plant of* Corylus avellana *exposed in a growth chamber, and the associated leaf water potential (Ψ_o) during a change in air humidity (Δw_o). At the same time, water vapour conductance of leaves on a single test twig (g$_i$) are measured at constant humidity in a cuvette (Δw_i) in response to changes in water potential in the cuvette (Ψ_i). (After Schulze and Küppers, 1979)*

due to the occurrence of extremely low threshold levels of plant water status for stomatal closure, which were not reached in these field environments. But these findings also indicate that threshold correlations between stomatal conductance and Ψ, which are based on diurnal curves, may not represent cause–effect relationships. Stomatal responses during the day may be due to some other environmental factor such as low humidity (Schulze *et al.*, 1974).

Correlations between minimal diurnal water potential and stomatal conductance and its change over a period of days may implicitly contain a long-term stress component (Turner *et al.*, 1978).

Schulze and Küppers (1979) were able to investigate separately the short-term effects of plant internal and plant external factors on the stomatal response with special, controlled environment facilities. An apparatus was used which consisted of a controlled environment growth chamber, in which short-term changes in the plant water status of intact plants could be created, with a controlled environment cuvette inside the chamber, in which the effect on stomatal conductance of a test leaf could be determined under constant ambient conditions. For example, short-term changes of the water potential of the test leaf, Ψ_i, were created by changing the evaporation of whole plants using different levels of humidity in the growth chamber (*Figure 12.3*). With a time-lag, leaf water potential in the cuvette (Ψ_i) followed the

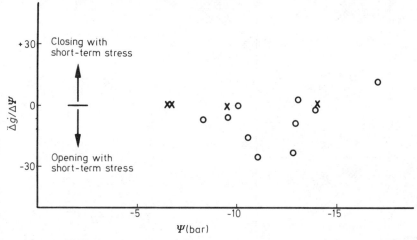

Figure 12.4 *Changes of stomatal conductance of* Corylus avellana *with short-term changes of water potential* $(\Delta g/\Delta\Psi)$ *as related to average conditions of water potential* (Ψ). *Positive values indicate closing response of stomata with short-term stress. Negative values indicate opening response of stomata with short-term stress.* Δw: *mbar bar^{-1} water vapour difference between leaf and air during the experiments.* $\circ$:Δw=5.5 *mbar bar^{-1}.* $\times$:Δw=30 *mbar bar^{-1}.* (*After Schulze and Küppers, 1979*)

water status of the plant outside (Ψ_o). But, unexpectedly, the test leaves in the cuvette had maximal stomatal conductance when Ψ_i was lowest and a short-term increase in Ψ_i occurred only when the stomata closed. This experiment was repeated at various stages of plant water stress during a drying experiment (*Figure 12.4*). A negative feedback control of the stomata, mediated by instantaneous changes of whole leaf water potential, should result in positive values of $\Delta g/\Delta\Psi$, where Δg represents a change in stomatal conductance, g, which is correlated with a change in water potential. This was not found. The experiments show greater stomatal conductance at decreased Ψ (negative values of $\Delta g/\Delta\Psi$), which indicates that Ψ is purely a function of the evaporation rate and that a feedback loop is not effective. It is possible that stomata of hazel close at a threshold plant water potential which is lower than that which can be achieved with experiments of the type

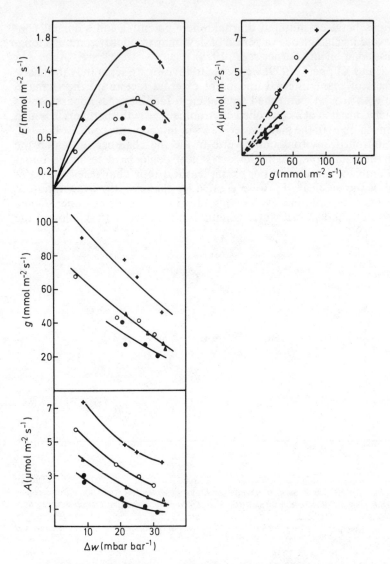

Figure 12.5   *Leaf evaporation (E), leaf conductance (g) and CO$_2$ assimilation (A) of* Corylus avellana *as related to changes in relative vapour pressure difference between leaf and air (Δw) and as influenced by long-term water stress, as well as CO$_2$ assimilation as related to leaf conductance at various levels of long-term plant water deficits.* + : Ψ = − 15 ± 1.3 *bar.* ○ : Ψ = − 18 ± 0.3 *bar.* ▲ : Ψ = − 19 ± 0.4 *bar.* ● : Ψ = − 23 ± 1.4 *bar. (After Schulze and Küppers, 1979)*

explained in *Figure 12.3*. It is not expected that this threshold level would occur precisely at a bulk leaf turgor of zero, since it is known that stomata have independent osmotic relationships. At least for plants which respond to air humidity, the ecological significance of threshold stomatal responses to short-term changes in plant water status must be questioned if they are rarely experienced by intact plants even under extreme environments except by irreversible treatments.

LONG-TERM EFFECTS OF DROUGHT ON STOMATAL RESPONSE

Long-term effects of plant water deficits on stomatal response and photosynthetic activity are apparent from field studies where diurnal variations in leaf conductance were correlated with changes in soil water potential (Morrow and Mooney, 1974; Havranek and Benecke, 1978) or with predawn leaf water status (Running, 1976). Examples of long-term drought effects during a drying cycle are shown in *Figure 12.5*. At high rates of evaporation *Corylus avellana*, a woody plant, had a water potential of − 12 to − 15 bar even when well irrigated. Transpiration rate decreased despite a higher evaporative demand above a relative vapour pressure difference between leaf and air of 27 mbar⁻¹. With increasing drought, maximal

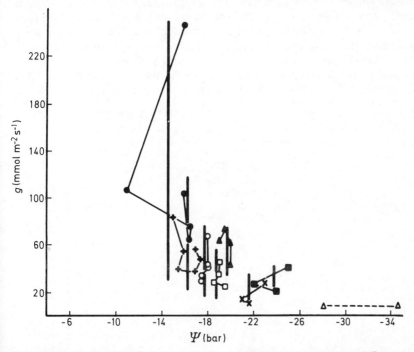

Figure 12.6 *Leaf conductance (g) as related to average levels of water potential (Ψ) in* Corylus avellana. *Thick lines indicate average conditions of long-term stress. Thin lines indicate conditions during individual experiments, denoted by different symbols. (After Schulze and Küppers, 1979)*

stomatal conductance was progressively reduced and transpiration peaked at smaller vapour pressure differences. Assimilation rate was reduced due to both a decrease in stomatal conductance and subsequently also to changes in the relationship between CO_2 assimilation and stomatal conductance.

The changes of leaf conductance during humidity response experiments (*Figure 12.6*, thin lines) at various levels of long-term drought were related to the average level of Ψ, which occurred as short-term changes of plant water status during each of these experiments (*Figure 12.6*, thick lines). At each drought level, stomata could be opened and closed within a certain range by changes in air humidity. With increased long-term drought the range of the response to humidity was decreased, but stomatal response to humidity was still obvious when leaves visibly wilted. Threshold responses of stomata to long-term water stress were not observed. Similar findings were made by Havranek and Benecke (1978), who used light as a factor for changing stomatal conductance at various levels of plant drought.

SHORT-TERM AND LONG-TERM EFFECTS OF DROUGHT ON ASSIMILATION

The effects of drought on assimilation have been extensively studied. Earlier findings were discussed by Slatyer (1973), who found for several species no change of residual internal resistance with decreasing relative water content of the tissue, except for cotton plants and maize where internal resistance increased. Boyer (1976) has reported several studies in which drought affected photosystem function but more recent work by Mooney, Björkman and Collatz (1977) has demonstrated that drought effects on photosynthesis may depend strongly on the plant history. Field plants of *Larrea divaricata* in Death Valley showed no inhibition of quantum yield with severe drought (-50 bar) whereas laboratory grown plants were significantly affected when water potentials fell below -20 bar. Under field conditions the major effects

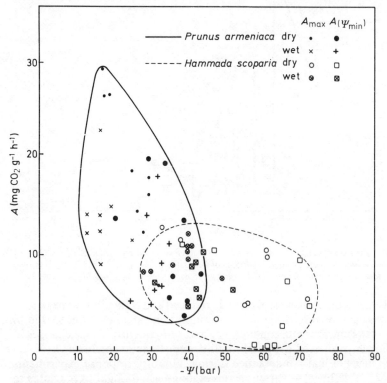

Figure 12.7   *CO_2 assimilation of irrigated and non-irrigated species of* Prunus armeniaca *and* Hammada scoparia *as related to simultaneous measurements of water potential (Ψ). Maximal rates of assimilation (A_{max}) during selected diurnal courses of a drying period and the according water potential, as well as assimilation rates at the time when diurnal water potentials were minimal ($A_{\Psi_{min}}$) are shown. (Schulze* et al., *1980)*

of drought on assimilation were due to stomatal closure, similar to the observations of Slatyer (1973). Working with plants which lack stomatal control, namely lichens, Lange et al. (1975) demonstrated that CO_2 assimilation was affected only at very severe levels of desiccation.

Possible relationships between CO_2 assimilation and concurrent leaf water potential are examined in *Figure 12.7*, where data from diurnal courses are used. Maximal rates of CO_2 assimilation and the concurrent measured value of Ψ, as well as assimilation rates which were reached when Ψ was at its daily minimum, are shown for irrigated and non-irrigated plants on representative days of a dry season. There is no unique relationship between assimilation and Ψ for the mesophytic leaves of *Prunus armeniaca* or for the xeromorphic stems of the desert shrub *Hammada scoparia*. Certainly, for each individual day, a correlation between assimilation and water potential may be drawn but such correlation is mediated by other environmental factors, as was shown by Schulze et al. (1975b).

However, there is no doubt that drought does influence CO_2 assimilation, as shown in *Figures 12.1* and *12.5*. The difficulty exists of how to describe quantitatively the long-term effects of drought on CO_2 assimilation. It is necessary to normalize the data due to effects on assimilation other than drought. In this case, the ratio of assimilation by droughted and well-irrigated plants was used. Also, the data were initialized by using relative rates considering assimilation to be unity at the beginning of the dry season. At the present stage of knowledge it is still unclear how long-term effects of drought may be quantified. In the following example cumulative predawn water potential of the droughted plants was used since weekly irrigations assured that the cumulative predawn water potential of the irrigated plants was low at any time for the wet treatments. The effect of long-term water stress on assimilation was estimated according to equation (12.1):

$$\frac{A_{d_{(t)}}}{A_{d_{(t_0)}}} \Bigg/ \frac{A_{w_{(t)}}}{A_{w_{(t_0)}}} = \int_{t_0}^{t} \left(\sum \Psi_{\text{predawn}} \right) \tag{12.1}$$

where A_d and A_w indicate CO_2 assimilation of droughted and irrigated plants, respectively, at the time t of observation and at the time t_0 of the last rainfall on droughted treatments.

Figure 12.8 shows this relationship for a number of species which are extremely different in life form and leaf morphology: *Prunus armeniaca*, a mesophytic plant grown in a run-off farm system ($\Psi_{\min} = -43$ bar); *Artemesia herba-alba*, a desert shrub with soft, deeply dissected mesophytic leaves in spring and rather xeromorphic leaf scales at the end of the dry season ($\Psi_{\min} = -111$ bar); *Reaumuria negevensis*, a desert shrub with small, succulent salt-excreting leaves ($\Psi_{\min} = -96$ bar); and *Hammada scoparia*, an aphyllous, xeromorphic shrub ($\Psi_{\min} = -96$ bar). All these species exhibited the same curvilinear decline of normalized assimilation with predawn bar days. It is remarkable that the data for mesophytic leaves of *Prunus armeniaca* and *Artemesia herba-alba* and the succulent leaves of *Reaumuria negevensis* are on the same curve. Only *Hammada scoparia* which has C_4 metabolism, was different, showing a slower initial decline. The linear integration of Ψ, implicit in equation (12.1), is biologically questionable. For example, a change of predawn water potential from -60 to -70 bar may have a

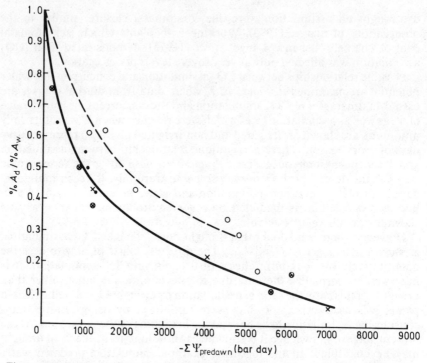

Figure 12.8   *The change of normalized maximal assimilation rates (% A_d/% A_w) of dry and wet plants during the course of a dry season under desert conditions (Negev, Israel). The assimilation rates were initialized to be 100 per cent at the time of last rainfall for the dry treatments. Horizontal axis: cumulative predawn water potentials ($\sum \Psi_{predawn}$). Experimental plants are cultivated apricot trees* (Prunus armeniaca) *growing in the run-off farm Avdat, Israel, and plants of the natural desert vegetation* (Artemisia herba-alba, Reaumuria negevensis *and* Hammada scoparia), *(Schulze et al., 1980)*

different effect on metabolism compared to a change from -5 to -15 bar. However, the characteristics of the function in equation (12.1) could only be reliably established when more data sets are available to relate CO_2 assimilation to long-term drought in different environments and with different species.

The results may be summarized to indicate that for a great range of conditions with intact plants there is no unique relationship of stomatal conductance and CO_2 assimilation to concurrent water potential, but both appeared to be related to long-term drought in an integrated manner.

Quantifying the Effects of Environmental Drought on Biomass Production and Seed Yield using Plant Parameters

Despite considerable research on plant water relationships, few concepts and data are available concerning the quantitative relationships between biomass production or yield and plant *water deficits*. In contrast, considerable information is available, mainly from agricultural experiments, on the relationship between biomass production or seed yield and plant *water use*.

These two approaches of using either water deficits or water use as a basis for quantifying the effects of drought on plant performance will be compared. The discussion will be restricted to studies with annual species because quantifying the effects of drought on the productivity of perennials is more difficult due to carry-over effects from previous years, as was shown for example for apricot trees by Spiegel-Roy, Evenari and Manzing (1971).

CHOICE OF PARAMETERS

It is most important to know which parameters best describe the impact of environmental drought. Plant water deficits may influence biomass production through changes in plant turgor or osmotic potential (Hall, Foster and Waines, 1979) but it is not known whether the average turgor or osmotic potential of a plant or organ is the correct parameter, or whether the water status of specific tissues, cells or organelles determines the plant response to drought. Practical considerations indicate that it may be expedient to use pressure chamber values as a measure of plant water status. But some species avoid drought to such an extent (cf. *Figure 12.2*) that plant water potentials change only to a small degree (Turk and Hall, 1980a), even though the plant performance is substantially affected by environmental drought (Turk, Hall and Asbell, 1980). Therefore, additional parameters are needed to describe these effects.

Plants mainly avoid drought by reducing rates of water loss. This may be achieved through a decrease in leaf area or leaf conductance. Both responses also decrease CO_2 assimilation and thus biomass production. Consequently, some measure of transpiration may provide a parameter for describing the impact of environmental drought on plants, which takes into consideration the detrimental effects of drought avoidance on productivity. Drought-induced changes in biomass production have been correlated to integrated values of transpiration (De Wit, 1958), or evapotranspiration (Hanks, Gardner and Florian, 1969).

It is possible that hormonal balance may provide an integrated measure of the drought experience of the plant (Quarrie and Jones, 1977), but data on endogenous hormone levels are difficult to obtain and to interpret.

TIME INTEGRATION OF PLANT PARAMETERS

For most cases where environmental drought influences plant performance it may be assumed that it is the drought experience of the plant over several days or weeks that is responsible for the particular effect on biomass production or seed yield. But consideration must be given to the time period over which the plant parameter should be integrated. Where reduction in transpiration is a measure of reduction in CO_2 assimilation it may be anticipated that transpiration must be integrated over a considerable period to correlate well with biomass production. Therefore, seasonal transpiration and evapotranspiration have frequently been used as parameters in studies of biomass productivity. But for correlation with *seed yield*, transpiration during the period of flowering or seed development may be more appropriate than total transpiration (Passioura, 1976). The most appropriate method for integrating plant water status parameters for yield correlations is not obvious.

Integrating diurnal curves of water potential or its components throughout the season of growth does not appear to be practical at this time. Maximum or minimum daily pressure chamber values could be determined, but it is not known whether plant performance will exhibit better correlations with one or the other or with a parameter that combines both values.

It is conceivable that on rare occasions such as drought during tasselling with corn, the plant water status over a specific short period of time may account for much of the reduction in seed yield due to drought. This led to the use of sensitivity coefficients, as an attempt to account for changes in plant susceptibility to environmental drought. For example, controlled environment studies by Fischer (1973) defined the sensitivity of seed yield in wheat to drought during flowering and seed filling. Hiler and Clark (1971) applied models for quantifying yield relationships with drought, which included sensitivity coefficients to account for variations in drought susceptibility at different developmental stages. But there are additional complexities. It has been shown that the effect on seed yield of drought at specific developmental stages can be influenced by subsequent weather conditions (Turk, Hall and Asbell, 1980). The extent to which relationships between biomass production or seed yield and plant water status or transpiration are dependent upon the developmental stage at which drought occurs, has not been adequately established.

Different approaches may be needed for quantifying the combined effects of drought on yield at different stages. Jensen (1968) proposed that for indeterminate plants the effect of drought (D) at different stages (i) should be multiplied by a specific sensitivity coefficient (S) and then summed to obtain the overall effect on yield (Y), as described in equation (12.2):

$$Y = \sum_{i=1}^{n} (S_i D_i) \tag{12.2}$$

This model assumes that drought at different stages has independent and additive effects on yield.

For determinate plants, Jensen (1968) proposed that the effect of drought at different stages should be multiplied to obtain the overall effect on yield with the sensitivity coefficient as an exponent, equation (12.3):

$$Y = \prod_{i=1}^{n} (D_i)^{S_i} \tag{12.3}$$

This model assumes that the effects of drought on yield at one stage influence yield response to drought at other stages in a multiplicative manner.

GENERALIZED MODELS

Normalizing the yield and plant response parameters generalizes the models permitting comparisons of drought effects on plants during different seasons, as shown earlier for CO_2 assimilation, and at different locations. Yield is frequently normalized by taking Y/Y_w, where Y_w is the yield of plants with optimal water supplies. This approach assumes that a specific drought level, as described by equation (12.2) or (12.3), will have a proportional effect on yield irrespective of the yield level. Water use rate (E) may be normalized by a similar procedure using E/E_w or by using E/E_p, where E_p is an environmental measure of potential evaporation. Normalization with E_p produces a more general function than normalization with E_w but use of E_p has disadvantages due to the difficulties involved in its measurement. The following equations are obtained:

$$Y/Y_w = \sum_{i=1}^{n} S_i \left(\int_{t_0}^{t} \frac{E}{E_w \text{ or } E_p} \right) \tag{12.4}$$

$$Y/Y_w = \prod_{i=1}^{n} \left(\int_{t_0}^{t} \frac{E}{E_w \text{ or } E_p} \right)^{S_i} \tag{12.5}$$

When using plant water status parameters they should be normalized whenever yield is normalized because even well-watered plants exhibit plant water deficits. A measure of plant water potential (Ψ) may be normalized by taking Ψ_w / Ψ. This procedure has the advantage of producing values between 0 and 1 analogous to E/E_w and necessary when using equations with coefficients as exponents such as equations (12.3), (12.5) and (12.7). The following equations are obtained.

$$Y/Y_w = \sum_{i=1}^{n} S_i \left(\int_{t_0}^{t} \frac{\Psi_w}{\Psi} \right) \tag{12.6}$$

$$Y/Y_w = \prod_{i=1}^{n} \left(\int_{t_0}^{t} \frac{\Psi_w}{\Psi} \right)^{S_i} \tag{12.7}$$

When quantifying the responses of a particular species to environmental drought, it is necessary to determine plant responses to different levels of drought at different stages of development and in different environments. This should be followed by empirical studies to determine the parameters, and methods of normalization and integration which provide the most accurate and general predictions of relationships between yield and environmental drought. Some examples are presented to illustrate the relationships between biomass production or seed yield and the effects of environmental drought that have been observed.

Drought responses of an indeterminate species (*Vigna unguiculata*)

Biomass and seed yield of cowpeas (*Vigna unguiculata*) were correlated with water use (*Figures 12.9* and *12.10*). In similar studies but with a smaller number of drought treatments, biomass and seed yield were linearly correlated with transpiration (Shouse *et al.*, 1977). The correlation between biomass and water use did not appear to be dependent on the developmental stage at which drought occurred. In contrast, for *seed yield* the influence of drought was dependent on the developmental stage at which it occurs.

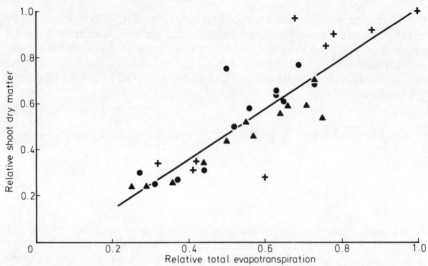

Figure 12.9   *Shoot dry matter and total evapotranspiration of cowpeas* (Vigna unguiculata) *70 days after emergence (after Turk and Hall, 1980b, 1980c). Data are from experiments during two years with two similar cultivars. The treatments consisted of different levels of water supplied weekly (I_0) or where water was withheld during the vegetative (I_1) or flowering stages (I_2). Shoot dry matter and evapotranspiration values were normalized by dividing by the values obtained on treatments which were irrigated weekly at potential levels. The regression equation was:*

relative shoot dry matter = $-0.08 + 1.09 \times$ relative evapotranspiration

and had an r^2 of 0.82. $+ : I_0.$ • $: I_1.$ ▲ $: I_2$

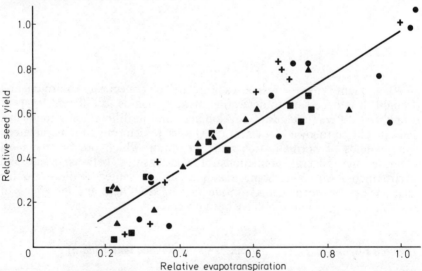

Figure 12.10   *Seed yield and evapotranspiration of cowpeas* (Vigna unguiculata) *(after Turk, Hall and Asbell, 1980; Turk and Hall, 1980c). Data are from experiments during two years with two similar cultivars. The treatments consisted of different levels of water supplied weekly (I_0) and where water was withheld during the vegetative (I_1), flowering (I_2) or pod filling (I_3) stages. Evapotranspiration values are for the period from the end of the vegetative stage to harvest. Seed yield and evapotranspiration values were normalized by dividing by the values obtained on treatments, which were irrigated weekly at potential levels. The regression equation was:*

relative seed yield = $-0.07 + 1.03 \times$ relative evapotranspiration

and had an r^2 of 0.84. $+ : I_0.$ • $: I_1.$ ▲ $: I_2.$ ■ $: I_3$

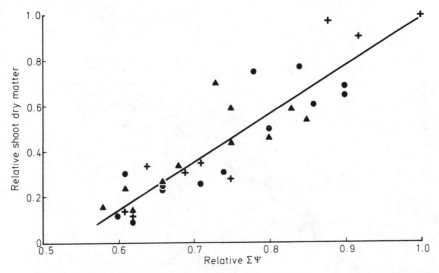

Figure 12.11 *Shoot dry matter and cumulative daily pressure chamber values of cowpeas* (Vigna unguiculata) *(after Turk and Hall, 1980a, 1980b). Pressure chamber values were the average of predawn and early afternoon values from 33 to 69 days after emergence. Data are from the experiments and treatments described in the legend of* Figure 12.9. I_0 *are weekly irrigated treatments, I_1 are treatments with a vegetative stage drought, and I_2 are treatments where water was withheld during flowering. The data were normalized as in Figures 12.9 and 12.10, except that the cumulative pressure chamber values of well-watered treatments were divided by the cumulative pressure chamber values of the other treatments. The regression equation was:*

relative shoot dry matter $= -1.11 + 2.10 \times$ *relative* $\Sigma \Psi$

and had an r^2 of 0.86. $+ : I_0$. $\bullet : I_1$. $\blacktriangle : I_2$.

Drought during the vegetative stage reduced water use but had little influence on seed yield, providing the subsequent environment was conducive for rapid recovery, whereas drought during pod filling substantially reduced seed yield but with only a small reduction in water use (Hall, Dancette and Turk, 1977). An attempt was made to incorporate these effects by including only water used by the plants between the beginning of flowering and harvest (*Figure 12.10*). Alternatively, equation (12.4) could be used but with a small value for the sensitivity coefficient during the vegetative stage and larger sensitivity coefficient for the pod filling stage.

The biomass and seed yield of the experiments described in *Figures 12.9 and 12.10* were also correlated with long-term water stress experience (*Figures 12.11 and 12.12*). Water stress experience was estimated by daily cumulation of the average of predawn and early afternoon pressure chamber values. Correlations with biomass production and seed yield were also observed with either cumulative predawn or cumulative afternoon values (Turk and Hall, 1980a, 1980b). For these, the correlations of biomass and seed yield to cumulative water potentials, with separate values for different developmental stages and using sensitivity coefficients and summation (equation (12.3)), did not appear advantageous. Hiler *et al.* (1972) related seed yield to afternoon pressure chamber values in glasshouse studies of drought effects on potted cowpea plants. They used an equation that is similar to equation (12.2) and incorporated sensitivity coefficients for

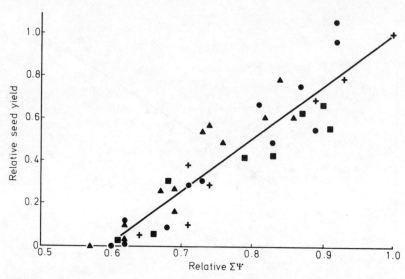

Figure 12.12 *Seed yield and cumulative daily pressure chamber values of cowpeas* (Vigna unguiculata) *(after Turk, Hall and Asbell, 1980; Turk and Hall, 1980a) Pressure chamber values were the average of predawn and early afternoon values from 33 to 78 days after emergence. Data are from the experiments and treatments described in the legend of Figure 12.10.* I_0 *are weekly irrigated treatments,* I_1 *are treatments with a vegetative stage drought, and* I_2 *and* I_3 *are treatments where water was withheld during flowering and pod filling, respectively. The data were normalized as described for Figure 12.11, The regression equation was:*

  relative seed yield $= -1.46 + 2.46 \times$ *relative* $\sum \Psi$

and had an r^2 *of 0.90.* $+: I_0$. $\bullet: I_1$. $\blacktriangle: I_2$. $\blacksquare: I_3$

different stages, but the relevance of their model to cowpea responses to drought under field conditions has not been established.

The cowpea exhibits extreme drought avoidance and the drought-induced changes in pressure chamber values are small (Turk and Hall, 1980a). In contrast, drought-induced changes in transpiration are large (Hall and Schulze, 1980) and may provide an indication of the effects of environmental drought on biomass production and yield which are easier to detect. However, pressure chamber values have a lower requirement for normalization than transpiration and are less subject to errors associated with normalization.

COMPARISON OF DROUGHT RESPONSES OF DETERMINATE WITH INDETERMINATE SPECIES

Biomass and grain yield of corn (Hanks *et al.*, 1978) and biomass of several determinate crops (Hanks, Gardner and Florian, 1969) were linearly correlated with seasonal evapotranspiration. For biomass this response to drought may be generally applicable. However, for grain yield responses to drought it may be expected that the response may be more complex, depending on the stage at which drought occurs and its intensity. Jensen (1968) described grain yield responses to drought using the product equation (equation (12.5)) and substantially different sensitivity coefficients for different developmental stages. Generally, the early flowering stage is most sensitive to drought and the vegetative stage is least sensitive (Hiler and Clark, 1971).

The sensitivity of determinate crops to drought has a phenological explanation. Drought during early stages of flowering may determine the maximum number of grains that may develop. Drought during grain filling determines the extent to which these grains accumulate carbohydrates, and grain yield is the product of the two quantities: grain number and grain size (Fischer and Turner, 1978). Modern cultivars of determinate crops usually have little ability to overcome the detrimental effects of a desiccated inflorescence by producing secondary inflorescences, but more primitive cultivars may still retain morphological plasticity of this type (Hall, Foster and Waines, 1979). Extreme sensitivity to drought at specific times during the season is more pronounced where the cultivar is genetically uniform and low-tillering, or has the ability to produce only one head per culm. In contrast, indeterminate species can produce a new set of flowers and leaves after drought is relieved by rain. A remarkable example of this response was described for cowpea by Turk, Hall and Asbell (1980). During one year, drought at flowering caused total abscission of flowers and young pods. But when the drought was relieved by irrigation, the cowpeas produced more flowers and leaves and achieved seed yields which were 71 per cent of well-irrigated plants. This recovery response only required 15 days in addition to the 97 day growing season of the well-irrigated plants. The short time required by this crop for pod filling, 20 days compared with 50 days for soya bean, may be partially responsible for this strong recovery.

Additional complexities are frequently present when attempting to relate plant performance to drought. For example, soil drought may influence mineral nutrition, plant sensitivity to temperature, disease resistance and susceptibility to herbivory. Some of these interactions may be subject to elucidation by multifactorial experiments but field experiments of this type, which include drought treatments, are often difficult to execute and to interpret. Quantifying the effects of drought on yield of perennials will be particularly difficult because of carry-over effects from previous seasons, but it is possible that these may be treated as having multiplicative effects on yield.

References

BOYER, J. S. (1976). *Phil. trans. R. Soc. Lond. B*, **273**, 501–512

COWAN, I. R. (1977). *Adv. Bot. Res.*, **4**, 117–228

DE WIT, C. T. (1958). *Inst. Biol. Scheik. Onderzoek. Landb., Wageningen*, **59**, 1–88

EVENARI, M., LANGE, O. L., SCHULZE, E.-D., KAPPEN, L. and BUSCHBOM, U. (1977). *Flora*, **166**, 383–414

FISCHER, R. A. (1973). In *Plant Responses to Climatic Factors*, pp. 233–242. Ed. by R. O. Slatyer. UNESCO, Paris

FISCHER, R. A. and TURNER, N. C. (1978). *Ann. Rev. Pl. Physiol.*, **29**, 277–317

HALL, A. E., DANCETTE, C. L. and TURK, K. J. (1977). *Proc. Int. Symp. on Rainfed Agriculture in Semi-Arid Regions*, pp. 398–418. University of California at Riverside

HALL, A. E., FOSTER, K. W. and WAINES, J. G. (1979). In *Agriculture in Semi-Arid Environments*. Ed. by A. E. Hall, G. H. Cannell and H. Lawton. Springer-Verlag, Berlin

HALL, A. E. and HOFFMAN, G. J. (1976). *Agron. J.*, **68**, 876–881

HALL, A. E. and SCHULZE, E.-D. (1980). *Austr. J. Pl. Physiol.*, **7**, 141–147

HALL, A. E., SCHULZE, E.-D. and LANGE, O. L. (1976). In *Ecological Studies. Analysis and Synthesis. Water and Plant Life*, pp. 169–188. Ed. by O. L. Lange, L. Kappen and E.-D. Schulze. Springer-Verlag, Berlin

HANKS, R. J., ASHCROFT, B. L., RASMUSSEN, V. P. and WILSON, G. D. (1978). *Irrig. Sci.*, **1**, 47–59

HANKS, R. J., GARDNER, H. R. and FLORIAN, R. L. (1969). *Agron, J.*, **61**, 30–34

HAVRANEK, W. M. and BENECKE, U. (1978). *Pl. Soil.*, **49**, 91–103

HILER, E. A. and CLARK, R. N. (1971). *Am. Soc. agric. Engs. Trans.*, **14**, 757–761

HILER, E. A., VAN BAVEL, C. H. M., HOSSAIN, M. M. and JORDAN, W. R. (1972). *Agron. J.*, **64**, 60–64

JENSEN, M. E. (1968). In *Water Deficits and Plant Growth*, Vol. 2, pp. 1–22. Ed. by T. T. Kozlowski. Academic Press, New York

JORDAN, W. R. and RITCHIE, J. T. (1971). *Pl. Physiol.*, **48**, 783–788

KAUFMANN, M. R. (1976). In *Transport and Transfer Processes in Plants*, pp. 313–327. Ed. by I. F. Wardlaw and J. B. Passioura. Academic Press, New York

LANGE, O. L., SCHULZE, E.-D., KAPPEN, L., BUSCHBOM, K. and EVENARI, M. (1975). In *Environmental Physiology of Desert Organisms*, pp. 20–37. Ed. by N. F. Hadley. Dowden, Hutchinson, Ross: Stroudsburg, Penn.

MOONEY, H. A., BJÖRKMAN, O. and COLLATZ, G. J. (1977). *Carnegie Inst. Yearb.*, **76**, 328–334

MORROW, P. A. and MOONEY, H. A. (1974). *Oecologia (Berl.)*, **15**, 205–222

PASSIOURA, J. B. (1976). *Austr. J. Pl. Physiol.*, **3**, 559–565

QUARRIE, S. A. and JONES, H. G. (1977). *J. exp. Bot.*, **28**, 192–203

RASCHKE, K. (1970). *Pl. Physiol.*, **45**, 415–423

RASCHKE, K. (1975). *Planta*, **125**, 243–259

RUNNING, S. W. (1976). *Can. J. Forest. Res.*, **6**, 104–112

SCHULZE, E.-D., HALL, A. E., LANGE, O. L., EVENARI, M., KAPPEN, L. and BUSCHBOM, U. (1980). *Oecologia (Berl.)*, **45**, 11–18

SCHULZE, E.-D. and KÜPPERS, M. (1979). *Planta (Biol.)*, **146**, 319–326

SCHULZE, E.-D., LANGE, O. L., EVENARI, M., KAPPEN, L. and BUSCHBOM, U. (1974). *Oecologia (Berl.)*, **17**, 159–170

SCHULZE, E.-D., LANGE, O. L., EVENARI, M., KAPPEN, L. and BUSCHBOM, U. (1980). *Oecologia (Berl.)*, **45**, 19–25

SCHULZE, E.-D., LANGE, O. L., KAPPEN, L., EVENARI, M. and BUSCHBOM, U. (1975a). In *Photosynthesis and Productivity in Different Environments*, pp. 107–119. Ed. by J. P. Cooper. Cambridge University Press, Cambridge

SCHULZE, E.-D., LANGE, O. L., KAPPEN, L. EVENARI, M. and BUSCHBOM, U. (1975b). *Oecologia (Berl.)*, **18**, 219–233

SHOUSE, P., DASBERG, S. D., JURY, W. A. and STOLZY, L. H. (1977). *Proc. Int. Symp. on Rainfed Agriculture in Semi-Arid Regions*, pp. 424–440. University of California at Riverside

SLATYER, R. O. (1973). In *Plant Responses to Climatic Factors*, pp. 271–276. Ed. by R. O. Slatyer. UNESCO, Paris

SPIEGEL-ROY, P., EVENARI, M. and MAZING, D. (1971). *J. Am. Soc. hortic. Sci.*, **96**, 696–701

STOCKER, O. (1960). In *Encyclopedia of Plant Physiology*, Vol. 5, pp. 460–491. Ed. by A. Pirson. Springer-Verlag, Berlin

TURK, K. J. and HALL, A. E. (1980a). *Agron. J.*, **72**, 421–427

TURK, K. J. and HALL, A. E. (1980b). *Agron. J.*, **72**, 428–433

TURK, K. J. and HALL, A. E. (1980c). *Agron. J.*, **72**, 434–439

TURK, K. J., HALL, A. E. and ASBELL, C. W. (1980). *Agron. J.*, **72**, 413–420

TURNER, N. C. (1974). *Pl. Physiol.*, **53**, 360–365

TURNER, N. C., BEGG, J. E., RAWSON, H. M., ENGLISH, S. D. and HEARN, A. B. (1978). *Austr. J. Pl. Physiol.*, **5**, 179–194

13

REGULATION OF GAS EXCHANGE IN WATER-STRESSED PLANTS

T. A. MANSFIELD
J. A. WILSON
Department of Biological Sciences, University of Lancaster, UK

Introduction

The control of transpiration by the movements of stomata has an obvious bearing on crop productivity, because as stomata close to conserve water they increasingly restrict the rate of entry of CO_2 into the leaf for photosynthesis. Water conservation is not only necessary for plants that are suffering measurable water stress. It must be looked upon as a continuous process that requires responses of the stomata to many factors, which together control the rate of transpiration even in plants which may apparently have an abundant supply of soil water. The occasions are rare when a plant can support the transpiration that occurs through fully open stomata, which allow a rate of evaporation nearly as great as that from a free water surface of the same area as the leaf. So little water is held in reserve in mesophytes that a leaf which transpires at a rate greater than the replacement of water via the xylem may not survive more than a few minutes, as can be demonstrated by the use of simple treatments to cause stomata to open to abnormally wide apertures (Mansfield and Davies, 1981).

The subject of this paper has to be considered in the context of all the factors that have been recognized as important in regulating the apertures of stomata. We have elsewhere suggested that it is convenient to look upon the responses of stomata as representing different lines of defence (Davies, Mansfield and Orton, 1978; Mansfield and Davies, 1981). The first lines of defence were considered to be the responses to factors of the aerial environment, especially water vapour pressure deficit (VPD) and CO_2 concentration, which can bring about the adjustment of stomatal apertures to a level appropriate for the prevailing conditions. The responses to VPD may be especially important in this context, and in many species, particularly those of dry habitats, they can play a part in water conservation just as significant as the mechanisms we shall discuss in detail below. The second lines of defence are those which come into operation when the responses of the stomata to the immediate above-ground environment are unable to exert sufficient control over transpiration. It is these with which we shall be mainly concerned in this paper.

It is important to emphasize that in referring to 'first' and 'second' lines of defence we are not implying a strict temporal separation between the two. Indeed, we believe that all these mechanisms are superimposed on one another in plants growing in most natural environments. The rationale for a separation into 'first' and 'second' is based on the knowledge that it is possible to grow plants under favourable conditions (for example, well-watered greenhouse crops) with the result that the 'second' lines of defence are absent or barely detectable, although the stomata are sensitive to factors such as light, VPD and CO_2 concentration.

Stomatal Behaviour in Water-stressed Plants

When the first lines of defence are inadequate, further closure of stomata becomes necessary to reduce transpiration and so protect those organelles, such as chloroplasts, of which functioning and integrity are affected by water stress (Boyer, 1976). There is evidence that the necessary control is achieved by the transmission of a 'distress signal' from the mesophyll chloroplasts to the guard cells. It is with the nature and mode of operation of this signal that we shall be mainly concerned. Evidence for the existence of a factor capable of causing long-term closure of stomata in plants that had experienced wilting was known for many years (e.g. Glover, 1959) before any clues as to its nature were forthcoming. It was the work of Wright (1969) and Wright and Hiron (1969) which was a stimulus to the research that has brought about our present understanding of this topic. They found that after wheat leaves had been allowed to wilt, they contained greatly increased amounts of the growth regulator, abscisic acid (ABA). Much work that followed established a clear quantitative relationship between water stress and endogenous ABA in a variety of species. It was also found that externally applied ABA could cause stomata to close and initiate a pattern of behaviour similar to that in plants that had raised levels of ABA because of wilting. There was prolonged closure of stomata lasting up to nine days after a single application of ABA (Jones and Mansfield, 1970, 1972). This paralleled the behaviour of stomata after wilting, for they were found to remain closed for many days even though full turgor had been restored (Heath and Mansfield, 1962; Allaway and Mansfield, 1970; Fischer, 1970).

   We are now able to postulate a sequence of events which occurs in leaves as they experience a loss of turgor, with the objective of understanding the mechanisms involved in the control of stomata by abscisic acid.

THE FORMATION OF ABSCISIC ACID AND ITS RELEASE FROM MESOPHYLL CHLOROPLASTS

Some experiments of Loveys (1977) provided information of central importance to our understanding of the role of ABA as a stress hormone. He showed that the bulk of the ABA which appears in the epidermis of water-stressed plants is not formed there but is apparently transported from the mesophyll. There was no detectable increase in the ABA content of isolated epidermal tissue after exposure to an osmotic stress sufficient to induce its formation in the intact leaf. However, when epidermis was removed from leaves which had previously been stressed it was found to contain large quantities of ABA.

   Prior to this discovery, Milborrow and Robinson (1973) and Milborrow

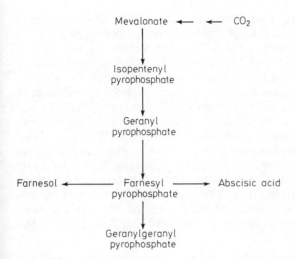

Figure 13.1 *Part of the pathway of terpenoid biosynthesis to show the point of origin of farnesol and abscisic acid*

(1974) had come to the conclusion that chloroplasts are the main sites of ABA formation in green leaves. One route for the biosynthesis of ABA is known to involve the conversion of mevalonic acid (MVA) to isopentenyl pyrophosphate (IPP), three units of which are incorporated into the C-15 precursor of sesquiterpenoids, all-*trans* farnesyl pyrophosphate. The relevant parts of the terpenoid biosynthetic pathway are shown in *Figure 13.1*.

Another observation of Loveys (1977) suggests, however, that attention should not only be paid to the control of the synthesis of ABA in the chloroplasts. He showed that nearly all of the ABA in the leaves of well-watered plants is contained in the chloroplasts, but that after four hours of water stress the total ABA rose 11-fold while the amount in the chloroplasts only doubled. This draws attention to the importance of the mechanism of release of ABA from mesophyll chloroplasts, and it could provide an explanation for the discrepancies observed between the total amount of ABA in leaves and the apertures of stomata. For example, Beardsell and Cohen (1975) found in maize that as water stress developed, stomatal closure began before the level of ABA had risen appreciably in the leaves. This could be due to a redistribution of ABA that was already present, but contained within the mesophyll chloroplasts because it could not penetrate their envelope membranes.

The permeability of plastid envelopes towards ABA was studied by Wellburn and Hampp (1976), who found that etioplast envelope membranes were only moderately permeable to ABA, but that permeability increased considerably after greening. This demonstrated a capacity of the membrane system to undergo permeability changes which could be used in a regulatory manner. If such a change occurred during the early stages of water stress then a rapid response of stomata, mediated by ABA, could occur without any new synthesis of the hormone. We could also postulate that control over the rate of synthesis of ABA in the chloroplasts is, in whole or in part, determined by the rate of its release into the cytoplasm.

Since we must regard the *speed of reaction* of the stomata to water stress as an essential factor in the protection of a leaf against physical injury, attention is focussed on the mechanism of release of ABA already present prior to stress. An understanding of this process is seen to be of the utmost importance in our appreciation of the role of ABA as a stress hormone.

Extracts from water-stressed plants of sorghum revealed the presence of another sesquiterpenoid in addition to ABA (Ogunkanmi, Wellburn and Mansfield, 1974). This was identified as all-*trans* farnesol (Wellburn *et al.*, 1974), and further experiments by Fenton, Davies and Mansfield (1977) showed that it was capable of inducing stomata to close when applied to non-stressed leaves of sorghum. It was initially considered to be a compound with a separate function as an endogenous 'antitranspirant', but another suggestion was made by Mansfield, Wellburn and Moreira (1978). They postulated that farnesol could have a role in determining the release of ABA contained within chloroplasts. Several pieces of evidence lend support to this hypothesis:

(1) farnesol has been found to accumulate, along with ABA, in water-stressed sorghum leaves (Ogunkanmi, Wellburn and Mansfield, 1974);
(2) although farnesol was able to close stomata when applied to isolated epidermis, its effects were always destructive as it caused disintegration of the guard cell chloroplasts (Fenton, Mansfield and Wellburn, 1976);
(3) it could cause non-destructive and reversible closure of stomata if applied to whole leaves of sorghum, but the nature of the treatment required did not suggest a *direct* effect on the guard cells. Fine emulsions of farnesol sprayed on to the leaf surfaces did not affect the stomata, but if leaves were immersed in the emulsion for 30 minutes, partial closure occurred followed by a slow recovery over several days;
(4) all-*trans* farnesol was found to be capable of inhibiting the metabolic functions of isolated chloroplasts (Fenton, Mansfield and Wellburn, 1976). This appeared to be the outcome of its effects on membranes, which were easily disrupted by higher concentrations or more prolonged treatments. This property was not a general one among other homologous isoprenoid alcohols, for geraniol (C-10) was inactive, and geranylgeraniol (C-20) showed only a small activity. There was, however, high activity shown by nerolidol (C-15), which is the tertiary alcohol isomer of farnesol.

Although there is no conclusive evidence in support of the hypothesis, the citations above together amount to strong circumstantial evidence, especially when considered alongside the findings of Loveys (1977). The failure of farnesol to close stomata on isolated epidermis, except at concentrations sufficient to disrupt the chloroplasts of the guard cells, could be explained in terms of the absence of ABA from these chloroplasts and their inability to manufacture it. The chloroplasts of guard cells are structurally and biochemically different from those of the mesophyll (Mansfield, 1981) and their inability to manufacture ABA is not unexpected; they may be unable to reduce CO_2 photosynthetically (Raschke and Dittrich, 1977) and might possibily lack MVA, the essential precursor for terpenoid biosynthesis (*Figure 13.1*). The guard cell chloroplasts are not, however, so different from those of the mesophyll that their membranes are unaffected by farnesol.

In summary, then, it is suggested that when the water potential (or the turgor) of the chloroplast falls, there is a block in the conversion of farnesyl pyrophosphate to geranylgeranyl pyrophosphate (*Figure 13.1*), and some production of farnesol occurs. This alters the permeability of the chloroplast envelope membrane, and there is a release of ABA into the cytoplasm. More ABA formation then occurs in the chloroplast because of the raised level of the precursor, farnesyl pyrophosphate, and because the continued release of ABA into the cytoplasm prevents end-product inhibition. When the stress on the chloroplast is relieved, the conversion of farnesyl to geranylgeranyl pyrophosphate returns to normal and the rate of formation of ABA and farnesol declines.

Our most recent studies (Wilson and Davies, 1979) have consisted of an examination of some cultivars of sorghum and maize known to have different levels of tolerance to water stress. Extracts from stressed and non-stressed plants were purified and then bioassayed for antitranspirant activity using the system of Ogunkanmi, Tucker and Mansfield (1973), which employs pieces of isolated epidermis of *Commelina communis*. Compounds capable of closing stomata were detected, and there was some correlation with stress pretreatment and the known stress tolerance of the cultivars. There was not, however, a clear relationship between farnesol production and water stress as had appeared to be the case with the sorghum cultivar 'Piper' (Ogunkanmi, Wellburn and Mansfield, 1974). Chromatograms showed a distinct zone of antitranspirant activity which resembled that of farnesol, but further analysis revealed that this was principally due to the presence of unsaturated fatty acids.

An indication that unsaturated C-18 fatty acids could affect stomata in a manner comparable to that of farnesol had been obtained in our previous studies (Fenton, 1976; Fenton, Mansfield and Wellburn, 1976). There have also been reports that levels of short-chain fatty acids increase with water stress and that they, too, are capable of inducing stomata to close (Willmer, Don and Parker, 1978).

The effect of fatty acids on guard cells is closely similar to that of farnesol. Concentrations of 10^{-5} to 10^{-4} M are required to produce a visible response, and stomatal closure is accompanied by a lysis of the chloroplasts and cell death. It is, therefore, difficult to envisage an antitranspirant role for endogenous fatty acids which involves a *direct* effect on the guard cells.

In considering the likely action of fatty acids during water stress it has been profitable to draw parallels with other fields of research where their involvement has been clearly established, namely the ageing of chloroplasts and the study of chilling injury. This latter topic is particularly relevant since a predominant component of the stress experienced during chilling may be a loss of turgor (Wilson, 1976).

As a result of both chilling injury and chloroplast ageing, ultrastructural changes within the chloroplasts occur comparable to those seen in guard cells when epidermal strips are treated with fatty acids (e.g. Siegenthaler, 1969a, 1969b). These changes are associated with an inhibition of electron flow and energy-linked reactions (Siegenthaler, 1972), an alteration of the physico-chemical properties of membranes (Wintermans, 1966; Helmsing, 1969) and conformational changes of proteins. Many of these changes can be induced if some of the fatty acids normally found in chloroplast membranes

(Siegenthaler, 1972) are applied to freshly isolated chloroplasts. Indeed, it has recently been shown that linolenic acid, the acid mainly associated with the 'antitranspirant activity' in extracts from maize and sorghum during and after water stress (Wilson and Davies, 1979), can be bound by isolated chloroplasts (Okamoto and Katoh, 1977; Okamoto, Katoh and Murakami, 1977).

There have been several reports of changes in chloroplast ultrastructure, comparable to those discussed above, being induced by water stress (Giles, Beardsell and Cohen, 1974; Vieira da Silva, Naylor and Kramer, 1974; Freeman and Duysen, 1975). In the first study cited, these changes were attributed to an increase of enzymatic activity on the chloroplast lipids, causing a release of free fatty acids. It has also been shown that there are alterations in electron transport (Boyer and Bowen, 1970; Keck and Boyer, 1974), photophosphorylation (Nir and Poljakoff-Mayber, 1967; Keck and Boyer, 1974) and carbon fixation (Plaut, 1971; Plaut and Bravdo, 1973) of chloroplasts isolated from leaves with low water potentials.

These gross ultrastructural and biochemical changes are generally observed in tissue that has been subjected to a severe water stress. More information is needed on any such events that occur *during* water stress, as opposed to before or after stress, before we can realistically assess the involvement of fatty acids in any changes that do occur. We must recognize that if fatty acids have a physiological role of the type proposed, they are likely to act in a more subtle manner than that observed in severely stressed tissue.

The increase in the unsaturated C-18 fatty acid, linolenic acid, which we have found to be associated with water stress in maize and sorghum, might point to an increase in the unsaturation of membranes leading to permeability changes. These, if they occur in the chloroplast envelopes, could effect the release of ABA.

Thus we have two distinct classes of compounds, C-15 isoprenoid alcohols and fatty acids (both short-chain and C-18 unsaturated) which could be involved in an essential step in the regulation of stomata by ABA, namely its release from its sites of formation. The close metabolic link between farnesol and ABA makes the hypothesis concerning its involvement a neat and attractive one, but our recent studies have not always provided confirmation of an association between water stress and farnesol accumulation. It is likely, however, that membrane permeability changes could be induced by changes in the amounts of farnesol not readily detected. The changes in fatty acids during water stress are convincing, and future studies of their involvement could be rewarding.

THE TRANSPORT OF ABSCISIC ACID TO THE GUARD CELLS

Although many naturally occurring compounds with the ability to regulate growth and other phenomena have been isolated from plants, there have been comparatively few cases in which they have been shown to move from a site of synthesis to another location in which they cannot be synthesized, and where they exert their main effect. This is the classic concept of hormonal action, and there is much evidence to suggest that the control of guard cell turgor by ABA may be a clear example of it. The route by which ABA is transported from mesophyll chloroplasts to the guard cells is of some significance, because it could provide control over the rate at which molecules arrive at the sites of action and thereby influence the speed and magnitude of response.

Weyers and Hillman (1979) have shown that radioactively labelled ABA accumulates quickly in the stomatal complex when applied to isolated epidermal tissue. They were able to estimate that no more than 6 fmol ABA were present in each stomatal complex at the time of closure, but uptake of the hormone continued. Thus ABA is accumulated by guard cells of both open and closed stomata, and irrespective of whether or not its control function in the guard cells has been achieved. This suggests that no regulatory mechanism over uptake exists in the guard cells themselves.

The route for transport from mesophyll to guard cells is largely a matter for speculation. Weyers and Hillman (1979) point out that ABA could be released into the apoplast, but this would involve little control over movement since fluxes of materials dissolved in the apoplastic solution are largely dependent on flow to sites of evaporation. A symplastic route through plasmodesmata from cell to cell is an alternative possibility, and Mansfield, Wellburn and Moreira (1978) showed that this need only involve three or four cells. The final step in the sequence, from the subsidiary cells into the guard cells, is unlikely to be symplastic because there are no plasmodesmatal connections at this point (Carr, 1976; Weyers and Hillman, 1979). Although movement along the shortest linear route cannot be assumed without further exploration, it seems unlikely that more complicated routes involving xylem or phloem would be involved to any great extent, because there would still be a need for transport along a pathway of several epidermal cells. The fact that ABA is found in xylem and in phloem (Lenton, Bowen and Saunders, 1968; Hoad, 1973) may be related to its other regulatory functions in the plant rather than to its control of stomata.

The observations of Weyers and Hillman outlined above tend to suggest that there may be a lack of control in the epidermis of the amounts of ABA reaching the guard cells. Externally applied ABA moves readily through the xylem and reaches the guard cells of fully turgid leaves (Mansfield, 1976). The picture we have, therefore, is of guard cells accumulating ABA indiscriminately, just as they do many other substances (Weyers and Hillman, 1979). It remains a possibility, however, that guard cells have a way of storing ABA by compartmentalization or sequestration. This could enable controlled release to occur to sites of action, and such a mechanism would explain the prolonged effect of externally applied ABA, which can suppress stomatal opening for many days (Jones and Mansfield, 1972).

THE CONTROL OF GUARD CELL TURGOR BY ABSCISIC ACID

Large changes in solute content of the guard cells have been shown to accompany the movements of stomata, and there is much to suggest that potassium ions play a major part in bringing about these changes (Fujino, 1967; Fischer, 1968; Humble and Raschke, 1971; Penny and Bowling, 1974). Electroneutrality in the guard cells is maintained either by the internal generation of malate (Allaway, 1973; Travis and Mansfield, 1977) or by the intake of chloride from without (Schnabl and Ziegler, 1977; Vankirk and Raschke, 1978). The well-known loss of starch from the chloroplasts of guard cells, which normally accompanies stomatal opening, may be due to its conversion to malate (Willmer and Rutter, 1977).

The regulation of stomata by ABA could be achieved if it were to interfere with the ionic changes that are the essential basis for changes in guard cell turgor. The outcome of the action of ABA is to inhibit both the uptake of potassium ions and the disappearance of guard cell starch that normally accompany stomatal opening (Mansfield and Jones, 1971).

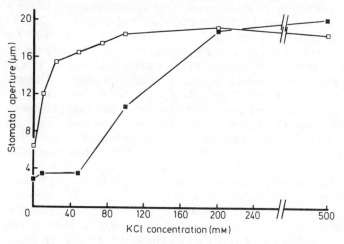

Figure 13.2   *Apertures of 'isolated' stomata of* Commelina communis *incubated for 2.5 h in a range of KCl concentrations, with (■) or without (□) abscisic acid. Each point is the mean of 30 measurements. Standard errors were within the limits of the insignia. Before being transferred to the final incubation media, stomata were wide open (17.0 ± 0.3 µm). (From Wilson, Ogunkanmi and Mansfield, 1978)*

Raschke (1975) has concluded that ABA acts by inhibiting a hydrogen ion expulsion mechanism in the plasmalemma of the guard cells. The great rapidity of the response to low doses of ABA, for example within five minutes to a 10^{-7} M solution (Cummins, Kende and Raschke, 1971), lends support to the view that effects on the plasmalemma are involved. Wilson, Ogunkanmi and Mansfield (1978) found that the response of isolated stomata to ABA could be counteracted by high levels of potassium ions in the incubation medium (*Figure 13.2*). There was evidence from this work that a potassium concentration of around 50 mM produced a maximum response to ABA. These findings could be interpreted in terms of an effect of ABA blocking 'channels' in the plasmalemma for potassium ions entering the guard cells. Thus, although ABA does ultimately affect hydrogen ion expulsion from the guard cells, this could result from a linkage between this process and the number of potassium ions entering. It is not possible, therefore, to say at which point the inhibitory effect of ABA operates, but a close connection with ion transport through the plasmalemma seems to be indicated.

Raschke (1975) found that the stomata of some species have a low sensitivity to CO_2 when the leaves have a high water potential, and considered that this was because the ABA level in the guard cells was low. This and other evidence (Mansfield, 1976) have suggested some interrelationship between the responses of stomata to CO_2 and to ABA, but whether these are of practical consequence in plants growing in normal conditions is doubtful. Dubbe, Farquhar and Raschke (1978) have emphasized that it is imperative to take great care over the pretreatment of plants, avoiding even mild water stress if this interrelationship between CO_2 and ABA is to be fully recognized.

The question of whether ABA interacts with CO_2 or not is of some import in our understanding of the control of stomata in plants under stress. Dubbe, Farquhar and Raschke (1978) have argued that there is a highly sophisticated control of gas exchange, with ABA affecting the gain of a feedback loop that involves intercellular CO_2 concentration and stomata. They took precautions to avoid interference from endogenous ABA, and applied 5×10^{-6} M (+)-ABA via the transpiration stream to detached leaves. This is in itself a dynamic situation, because ABA must continuously accumulate in the guard cells as the observations are made. From experiments of this sort it is therefore arguably hazardous to draw any firm conclusions about the role of ABA in the determination of 'physiological gains' in relation to CO_2, simply because there is not a 'steady state' with respect to ABA.

We have recently carried out an experiment to compare the responses of stomata on detached epidermis to ABA at two levels of CO_2. The epidermis was taken from well-watered plants of *Commelina communis* and incubated in ABA for three hours. The results in *Figure 13.3* show no evidence of an increased sensitivity to CO_2 in the presence of ABA; over the range of ABA concentrations which produced a maximum response, the two lines are virtually parallel. This and previously presented evidence (Mansfield, 1976) has led us to the view that effects of ABA and CO_2 are essentially additive. We should emphasize that an adequate picture of stomatal responses to ABA may not be achieved from single applications of the hormone to well-watered plants, because there is evidence of responses differing in magnitude if the hormone is repeatedly applied (Davies, 1978).

The analysis carried out by Dubbe, Farquhar and Raschke (1978) led them to view the control of stomata by ABA as an elegant optimization of gas exchange when leaves are in the stressed state. Apart from the criticism above we do not dispute the excellence of their analysis of the situation in leaves of experimental plants, carefully treated to avoid stress, and presented with an external supply of ABA. Data are to be found in the literature, however, which suggest that ABA can act in a manner that amounts to a crude and arbitrary control in relation to external conditions. This is how we are inclined to view the often reported after-effect of water stress on stomata (e.g. Allaway and Mansfield, 1970; Fischer, 1970). The stomata remain closed for varying periods of time, up to several days, after turgor has been restored. During this period they can obstruct photosynthesis under totally favourable external conditions. The advantages of such a prolonged action of ABA can perhaps be appreciated when considered in ecological terms (Mansfield, Wellburn and Moreira, 1978) but it is not possible to regard this rather arbitrary imposition of a 'ceiling' on the extent of stomatal opening in

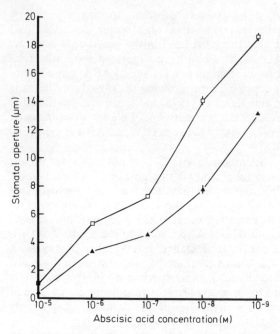

Figure 13.3   *Apertures of stomata on detached epidermis taken from well-watered plants of* Commelina communis *after incubation for 3.0 h in a range of ABA concentrations with zero* CO_2 (□) *or 330 ppm* CO_2 (▲). *Stomata were not 'isolated' as in* Figure 13.2. *Light intensity 80 W* m^{-2}; *temperature 25°C. Standard errors are shown as vertical bars where they are larger than the insignia*

the same light as an elegant control system related to the plant's immediate physiological state.

Endogenous Antitranspirants and Plant Productivity

The relationship between plant productivity and the function of ABA as an endogenous antitranspirant is not open to simple evaluation. This would be difficult to assess even if the whole range of physiological functions of ABA were understood. The fact is, however, that the role of ABA in controlling stomata is better understood than its other activities, many of which are suspected rather than firmly established. There are indications of other processes which may be controlled or affected by ABA, and which could assist in maintaining turgor during times of water stress.

THE ROLE OF ABSCISIC ACID IN MAINTAINING TURGOR

Glinka and Reinhold (1971, 1972) and Collins and Kerrigan (1973) discovered that ABA increased the apparent permeability of root tissue to water. As Clarkson (1974) pointed out, it is not easy to ascertain whether this effect is due to an increase in hydraulic conductivity to water, or to an increase in active solute transport into the xylem. However, there is a reduction in the real or the apparent resistance to radial water flow across the root, and this could be important in making more soil water available to the shoot. Although it is known that transport of ABA can occur from the leaves to the roots (Hocking, Hillman and Wilkins, 1972), and that the ABA produced in stressed leaves moves into the phloem for transport to other parts of the plant (Hoad, 1973), we are not aware of any attempts to relate the timing and quantities of ABA moved during the early stages of water stress to changes in the apparent permeability of the roots. It is an attractive idea that ABA might simultaneously reduce water consumption by causing stomata to close, and increase water intake through the roots, but no one to our knowledge has yet undertaken the difficult task of evaluating the possibility of such integrated responses. It is worthy of note that Tal and Nevo (1973), working with three 'wilty' mutants of tomato (which have a low endogenous ABA content and which cannot maintain turgor because their stomata remain open), found a greater resistance to water flow across the roots than in normal tomato plants. This suggests a root factor closely linked to events controlled by ABA in the leaves.

Malek and Baker (1978) found that ABA inhibits the active proton flux considered to be involved in the co-transport of sucrose during phloem loading in *Ricinus*. A decrease in the rate of transport of photosynthates out of the leaf would enable the mesophyll cells to retain more solutes for the maintenance of turgor. This phenomenon of osmoregulation is of considerable importance during water stress (Osonubi and Davies, 1978). It may be necessary to preserve the integrity of cell ultrastructure, and it is essential if cell expansion and growth of tissues are to continue.

CONSEQUENCES OF THE ACTION OF ABSCISIC ACID ON PLANT PRODUCTION

It is difficult to envisage the events in the plant caused by the hormonal action of ABA having any net effect other than to reduce plant production. The defensive mechanisms which have to be called upon during times of water stress have an inevitable cost in terms of net photosynthesis, and may involve a diversion of photosynthates for protective functions such as osmoregulation. Thus the growth of organs of economic importance may show a larger proportional reduction than net photosynthesis in the leaves.

Our discussion here is concerned mainly with stomatal effects, and with the influence these may have on net photosynthesis in the leaves. We must first distinguish between immediate and long-term consequences of the protection given by partial stomatal closure. When we speak of an inhibition of photosynthesis caused by such closure, this represents an assessment of the immediate consequences during the period of time in which ABA is active on the stomata. An assessment in the longer term will recognize that without the reduction in net CO_2 gain which accompanies the action of ABA, after a restoration of turgor at a point in the future, photosynthesis might be severely hampered because of damage to organelles (or their absence altogether, for the plant may be dead). Thus the reduction in net photosynthesis in the short term must be looked upon as the inevitable consequence of a protection of the cellular equipment required for that photosynthesis which is to be carried out in the future. Any attempts to manipulate the hormonal control of stomatal behaviour in the interests of crop production must be undertaken with an understanding of the above basic considerations.

CAN WE OPTIMIZE THE CONTROL OF TRANSPIRATION BY ABSCISIC ACID?

In the foregoing discussion of the mechanisms which underly the control of stomatal aperture by ABA, we have recognized three essential steps: the production and release of ABA by mesophyll chloroplasts, its transport to the guard cells, and its action to reduce the turgor of the guard cells. All three of these clearly make a contribution to the efficiency of the overall process of hormonal control. We are only in a position to discuss the first and the last in any detail because too little is known about the transport processes involved. A consideration of these in the present context may not, in any case, be very profitable because they are unlikely to be readily manipulated in the interests of plant production.

Speed of reaction to water stress

The processes connected with the release of any ABA which is 'bound' within the mesophyll chloroplasts prior to stress, and those involved in the *de novo* synthesis of ABA, must be regarded as being of prime importance in relation to the speed with which stomata react to a fall in the bulk water potential of the plant. A tardy response of the stomata to the onset of water stress could be the outcome of a delayed appearance of ABA in the guard cells. The fact that stomata do not seem able to manufacture ABA themselves (Loveys, 1977) should cause us, in future research, to place great emphasis on gaining an understanding of the mechanism of release of ABA from the mesophyll. The possibility should be considered of using this as a characteristic to be sought when selecting crop varieties with drought resistance. Until the present time, the ability to form ABA has been looked upon as a desirable characteristic only in rather imprecise terms. A correlation has been sought between drought resistance and the bulk tissue level of ABA, whereas we here argue that the main priority is the speed of arrival of ABA in the guard cells. This may be achieved before any significant amount of new ABA has been synthesized.

The possibility should also be explored of manipulating stomatal aperture in the field by inducing the release of that ABA bound within mesophyll chloroplasts. This could be a way of reducing the consumption of water at times when the onset of water stress can be anticipated several days in advance. A better understanding of the processes behind the release of ABA should ideally precede any studies in this area, but the indications of involvement of certain fatty acids suggest that immediate studies of compounds known to be capable of inducing changes in fatty-acid composition would be worthwhile. There are precedents for such studies, as it has been possible to alter resistance to chilling and frost hardening by inducing artificial changes in linolenic acid composition of seedlings (St. John and Christiansen, 1976; Willemot, 1977).

Action of abscisic acid on stomata

Apart from gaining a better understanding of the mechanism by which ABA closes stomata (page 243), we need to give some thought to the value of the prolonged action that occurs in many species. This has the effect of imposing a ceiling on stomatal aperture for a period of time, from a few days (Allaway and Mansfield, 1970; Fischer, 1970) to a number of weeks (Glover, 1959), even though full turgidity of the plant has been restored.

The failure of the stomata to reach their full opening potential for some time after rewatering has been looked upon as a 'safety mechanism', which enables the recovery of full turgor to be achieved more rapidly (Dörffling *et al.*, 1977). Where the after-effect is of short duration this interpretation seems a reasonable one, but it is not an adequate explanation in those cases in which a recovery to a normal pattern of stomatal behaviour only occurs several days after the leaves have recovered their full turgor. Mansfield, Wellburn and Moreira (1978) suggested that sufficient ecological advantages may accrue to justify a prolonged protective mechanism of this kind for plants growing in the wild. It is, however, arguably a pattern of behaviour which is not ideal when we are seeking to maximize crop productivity. If the stomata fail to recover their full opening potential quickly after the restoration of turgor in the plant, they must act as an obstruction to photosynthesis. Partially closed stomata may sometimes increase 'water use efficiency' (Mansfield, Wellburn and Moreira, 1978), but there is nevertheless a decline in the rate of absolute photosynthesis which can affect overall production.

We mentioned above the rather imprecise terms in which correlations between ABA formation and drought resistance have been sought. Taking into account the arguments advanced in this section, we would suggest that the following pattern of behaviour should be sought:

(1) a quick release of any ABA bound within the mesophyll chloroplasts, followed by a rapid synthesis of new ABA, as soon as water stress is experienced;
(2) an early recovery to a normal pattern of stomatal behaviour after the turgor of the plant has been restored.

Attention has perhaps been misdirected in the past to the amount of ABA produced under stress, based upon the supposition that a greater quantity should be indicative of drought resistance. If our assessment is correct, we would envisage the rapid release of a quantity just sufficient to cause an adequate drop in transpiration as being a more desirable characteristic in crops than an accumulation of sufficient ABA to close the stomata for a prolonged period of time.

Acknowledgements

We are grateful to the Science Research Council for financial support for J.A.W., and to Dr W. J. Davies and Dr A. R. Wellburn for helpful discussion on several aspects of this paper.

References

ALLAWAY, W. G. (1973). *Planta*, **110**, 63–70
ALLAWAY, W. G. and MANSFIELD, T. A. (1970). *Can. J. Bot.*, **48**, 513–521
BEARDSELL, M. F. and COHEN, D. (1975). *Pl. Physiol.*, **56**, 207–212
BOYER, J. S. (1976). *Phil. Trans. R. Soc. Lond. B*, **273**, 501–512
BOYER, J. S. and BOWEN, B. L. (1970). *Pl. Physiol.*, **45**, 612–615
CARR, D. J. (1976). In *Intercellular Communication in Plants: Studies on Plasmodesmata*, pp. 243–289. Ed. by B. E. S. Gunning and A. W. Robards. Springer-Verlag, Berlin
CLARKSON, D. (1974). *Ion Transport and Cell Structure in Plants*. McGraw-Hill, London
COLLINS, J. C. and KERRIGAN, A. P. (1973). In *Ion Transport in Plants*, pp. 589–594. Ed. by W. P. Anderson. Academic Press, London
CUMMINS, W. R., KENDE, H. and RASCHKE, K. (1971). *Planta*, **99**, 347–351
DAVIES, W. J. (1978). *J. exp. Bot.*, **29**, 175–182
DAVIES, W. J., MANSFIELD, T. A. and ORTON, P. J. (1978). Proc. Joint BCPC and BPGRG Symp.—Opportunities for Chemical Plant Growth Regulation
DÖRFFLING, K., STREICH, J., KRUSE, W. and MUXFELDT, B. (1977). *Z. Pfl.-physiol.*, **31**, 43–56
DUBBE, D. R., FARQUHAR, G. D. and RASCHKE, K. (1978). *Pl. Physiol.*, **62**, 413–417
FENTON, R. (1976). PhD thesis, University of Lancaster
FENTON, R., DAVIES, W. J. and MANSFIELD, T. A. (1977). *J. exp. Bot.*, **28**, 1043–1053
FENTON, R., MANSFIELD, T. A. and WELLBURN, A. R. (1976). *J. exp. Bot.*, **27**, 1206–1214
FISCHER, R. A. (1968). *Science N.Y.*, **160**, 784–785
FISCHER, R. A. (1970). *J. exp. Bot.*, **21**, 386–404
FREEMAN, T. P. and DUYSEN, M. E. (1975). *Protoplasma*, **83**, 131–145
FUJINO, M. (1967). *Sci. Bull. Fac. Ed., Nagasaki Uni.*, **18**, 1–47
GILES, K. L., BEARDSELL, M. F. and COHEN, D. (1974). *Pl. Physiol.*, **54**, 208–212
GLINKA, Z. and REINHOLD, L. (1971). *Pl. Physiol.*, **48**, 103–105
GLINKA, Z. and REINHOLD, L. (1972). *Pl. Physiol.*, **49**, 602–606
GLOVER, J. (1959). *J. agric. Sci., Camb.*, **53**, 412–416
HEATH, O. V. S. and MANSFIELD, T. A. (1962). *Proc. R. Soc. Lond. B*, **156**, 1–13
HELMSING, P. J. (1969). *Biochim. biophys. Acta*, **178**, 519–533
HOAD, G. V. (1973). *Planta*, **113**, 367–372
HOCKING, T. J., HILLMAN, J. R. and WILKINS, M. B. (1972). *Nature, New Biol.*, **235**, 124–125
HUMBLE, G. D. and RASCHKE, K. (1971). *Pl. Physiol.*, **48**, 447–453
JONES, R. J. and MANSFIELD, T. A. (1970). *J. exp. Bot.*, **21**, 714–719
JONES, R. J. and MANSFIELD, T. A. (1972). *Physiol. Plant.*, **26**, 321–327
KECK, R. W. and BOYER, J. S. (1974). *Pl. Physiol.*, **53**, 474–479
LENTON, J. R., BOWEN, M. R. and SAUNDERS, P. F. (1968). *Nature*, **220**, 86–87
LOVEYS, B. R. (1977). *Physiol. Plant.*, **40**, 6–10
MALEK, T. and BAKER, D. A. (1978). *Pl. Sci. Lett.*, **11**, 233–239
MANSFIELD, T. A. (1976). *J. exp. Bot.*, **27**, 559–564
MANSFIELD, T. A. (1981). In *Plant Physiology*, Vol. VII. Ed. by J. F. Sutcliffe and F. C. Steward. Academic Press, New York

MANSFIELD, T. A. and DAVIES, W. J. (1981). In *Physiology and Biochemistry of Drought Resistance*. Ed. by D. Aspinall and L. G. Paleg. Academic Press, London
MANSFIELD, T. A. and JONES, R. J. (1971). *Planta*, **101**, 147–158
MANSFIELD, T. A., WELLBURN, A. R. and MOREIRA, T. J. S. (1978). *Phil. Trans. R. Soc. Lond. B*, **284**, 471–482
MILBORROW, B. V. (1974). *Phytochem.*, **13**, 131–136
MILBORROW, B. V. and ROBINSON, D. R. (1973). *J. exp. Bot.*, **24**, 537–548
NIR, I. and POLJAKOFF-MAYBER, A. (1967). *Nature*, **213**, 418–419
OGUNKANMI, A. B., TUCKER, D. J. and MANSFIELD, T. A. (1973). *New Phytol.*, **72**, 277–282
OGUNKANMI, A. B., WELLBURN, A. R. and MANSFIELD, T. A. (1974). *Planta*, **117**, 293–302
OKAMOTO, T. and KATOH, S. (1977). *Pl. Cell Physiol.*, **18**, 539–550
OKAMOTO, T., KATOH, S. and MURAKAMI, S. (1977). *Pl. Cell Physiol.*, **18**, 551–560
OSONUBI, O. and DAVIES, W. J. (1978). *Oecologia*, **32**, 323–332
PENNY, M. G. and BOWLING, D. J. F. (1974). *Planta*, **119**, 17–25
PLAUT, Z. (1971). *Pl. Physiol.*, **48**, 591–595
PLAUT, Z. and BRAVDO, B. (1973). *Pl. Physiol.*, **52**, 28–32
RASCHKE, K. (1975). *Ann. Rev. Pl. Physiol.*, **26**, 309–340
RASCHKE, K. and DITTRICH, P. (1977). *Planta*, **134**, 69–75
SCHNABL, H. and ZIEGLER, H. (1977). *Planta*, **136**, 37–43
SIEGENTHALER, P. A. (1969a). *Pl. Cell Physiol.*, **10**, 801–810
SIEGENTHALER, P. A. (1969b). *Pl. Cell Physiol.*, **10**, 810–820
SIEGENTHALER, P. A. (1972). *Biochim. Biophys. Acta*, **275**, 182–191
ST. JOHN, J. B. and CHRISTIANSEN, M. N. (1976). *Pl. Physiol.*, **57**, 257–259
TAL, M. and NEVO, Y. (1973). *Biochem. Genet.*, **8**, 291–300
TRAVIS, A. J. and MANSFIELD, T. A. (1977). *New Phytol.*, **78**, 541–546
VANKIRK, C. A. and RASCHKE, K. (1978). *Pl. Physiol.*, **61**, 361–364
VIEIRA DE SILVA, J., NAYLOR, A. W. and KRAMER, P. J. (1974). *Proc. Natl Acad. Sci., USA*, **71**, 3243–3247
WELLBURN, A. R. and HAMPP, R. (1976). *Planta*, **131**, 95–96
WELLBURN, A. R., OGUNKANMI, A. B., FENTON, R. and MANSFIELD, T. A. (1974). *Planta*, **120**, 255–263
WEYERS, J. D. B. and HILLMAN, J. R. (1979). *Planta*, **144**, 167–172
WILLEMOT, C. (1977). *Pl. Physiol.*, **60**, 1–4
WILLMER, C. M., DON, R. and PARKER, W. (1978). *Planta*, **139**, 281–287
WILLMER, C. M. and RUTTER, J. C. (1977). *Nature*, **269**, 327–328
WILSON, J. A. and DAVIES, W. J. (1979). *Pl. Cell Env.*, **2**, 49–57
WILSON, J. A., OGUNKANMI, A. B. and MANSFIELD, T. A. (1978). *Pl. Cell Env.*, **1**, 199–201
WILSON, J. M. (1976). *New Phytol.*, **76**, 257–270
WINTERMANS, J. F. G. M. (1966). In *Biochemistry of Chloroplasts*, Vol. I, p. 115. Ed. by T. W. Goodwin. Academic Press, New York
WRIGHT, S. T. C. (1969). *Planta*, **36**, 10–20
WRIGHT, S. T. C. and HIRON, R. W. P. (1969). *Nature*, **224**, 719–720

14

EFFECTS OF LOW TEMPERATURE STRESS AND FROST INJURY ON PLANT PRODUCTIVITY

WALTER LARCHER
Institute of Botany, Universität Innsbruck, Austria

Introduction

Low temperatures present an important limiting factor for plant production and plant distribution in large areas of the temperate regions, but also in warmer climates. The northward extension of high-yielding varieties of wheat, and the cultivation of C_4-forage grasses in Australia and New Zealand are primarily restricted by low winter temperatures (Dilley *et al.*, 1975; Rowley, Tunnicliffe and Taylor, 1975). Low air and soil temperatures at planting time restrict the production period and yields of maize in high latitudes and altitudes (Mock and Skrdla, 1978). In Japan, rice yields may decrease by 35 per cent in years with abnormally cool summers as are expected in the next decades (Toriyama, 1976). In the tropical Andes frost is the major limiting factor for potato production (Li and Palta, 1978). Spring frosts present a regular risk for fruit plantations and horticulture; they may also cause severe impediment in reforestation (Aussenac, 1973). Extremely severe winters which are expected to occur in the northern hemisphere on average every ten years, cause considerable damage in orchards (Kemmer and Schulz, 1955; Schnelle, 1963; Quamme, 1976), in olive and citrus plantations (Morettini, 1961; Yelenosky and Young, 1978) and even for the sugar cane industry in marginal areas (Miller, 1976).

Stress Effects on Productivity and Yield, and Susceptibility of Plants to Chilling and Freezing

Low temperature stress impairs metabolic processes and dry matter production reversibly if moderate; long-lasting severe stress causes irreversible decrease in productivity, and yield losses. Härtel (1976) summarizes schematically the effects of environmental stress on the productivity of plants (*Figure 14.1*).

During a temporary action of non-injurious environmental stress the reactive responses of a plant remain within the physiological pattern; after stress the plant will soon recover completely. The yield, however, is shortened because of the lower metabolic activity during stress. The amount of yield loss after reversible depression of the metabolism depends primarily on the duration of the stress. By severe stress, life functions may be inhibited

253

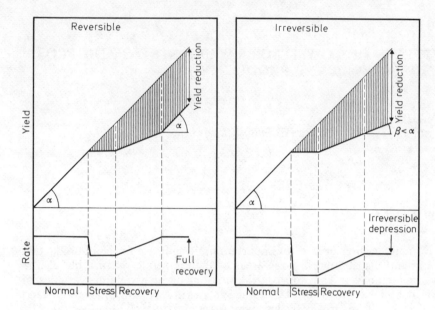

Figure 14.1   *Stress effects on productivity rate and yield. (After Härtel, 1976)*

irreversibly, and the plants are therefore not able to reach their original production capacity after returning to favourable living conditions. In this case, the yield is further reduced by the degree and duration of the after-effects.

Readily reversible disturbances at decreased temperature are: interruption of growth, especially of elongation growth; transient depression of photosynthesis and respiration; stagnation of assimilate translocation; and decreased uptake of water and minerals from cold soils. Examples of pathologic disorders are: irreversibly reduced photosynthetic capacity; premature senescence of leaves; developmental defects; and loss of fertility.

Excessive stress leads to lethal tissue injuries and loss of organs. As a consequence, yield is reduced depending on the proportion of injured plant parts important for productivity.

Three plant groups can be distinguished according to their sensitivity to low temperatures and frost, depending on the mechanisms of injury involved (Larcher, 1973, 1980; Larcher and Bauer, 1982):

(1) chilling-sensitive plants, which suffer severe injury even at positive temperatures below 10–15°C;
(2) freezing-sensitive plants, which are killed if frozen;
(3) freezing-tolerant plants, which are able to survive in the frozen state until a specific characteristic resistance limit is reached.

In chilling-sensitive plants, injuries are induced by low temperatures without freezing. Symptoms are increased biomembrane permeability and metabolic dysfunctions which finally lead to cell necrosis. The functional disorders are primarily caused by a phase-transition of biomembrane lipids at characteristic temperatures and by alterations of the lipid-protein complex (Lyons, 1973).

Although rapid killing after a cold shock of few hours' duration has been reported, as a rule chilling injuries develop chronically and become visible only after several days or weeks. Descriptions of typical chilling injuries are reported by Biebl (1962), Lutz and Hardenburg (1968), Levitt (1972) and Lyons (1973).

Tissue freezing occurs in the intercellular spaces and proceeds on the cell surfaces, and/or by ice nucleation inside the protoplasts. Intracellular freezing under natural conditions always kills the cells. Extracellular ice formation gradually withdraws water from the protoplast as a consequence of the vapour pressure gradient between the external ice phase and the internal liquid phase, thus eventually causing injury by dehydration. Removal of water by extracellular freezing is assumed to cause membrane defects and conformational changes in proteins. The process of ice formation in plant tissues has been investigated for more than a century; many problems, however, are still not resolved (Levitt, 1978). Thorough descriptions of freezing processes in plant tissue are given by Asahina (1956, 1978), Levitt (1957, 1966, 1972), Mazur (1969), Mittelstädt (1969) and Samygin (1974). A thermodynamic analysis of the freezing process is presented by Krasavtsev (1972) and Olien (1973, 1974, 1978); biochemical effects are discussed by Levitt (1972) and by Heber and Santarius (1973).

Effect of Chilling and Freezing on Carbon Dioxide Exchange

RESPONSES OF CHILLING-SENSITIVE PLANTS TO LOW TEMPERATURE STRESS

Photosynthesis

In chilling-sensitive plants, net photosynthesis decreases remarkably below 10–12°C, and at 5–10°C CO_2 uptake ceases completely (Ludlow and Wilson, 1971). When the temperature drops below a critical threshold at about 10°C, the energy of activation (E_a) for the terminal electron transfer processes and for enzymes involved in photosynthesis increases abruptly (Shneyour, Raison and Smillie, 1973). In chilling-sensitive C_4 grasses the E_a for phosphoenol pyruvate (PEP)-carboxylase increases 2–4 times (*Figure 14.2*; Phillips and McWilliam, 1971; McWilliam and Ferrar, 1974).

In *Zea mays* and *Sorghum bicolor* the activity of the light-activated pyruvate-phosphodikinase becomes strongly reduced at temperatures below 12°C (Taylor, Slack and McPherson, 1974; Shirahashi, Hayakawa and Sugiyama, 1978). In sorghum, the translocation of intermediate products of the C_4 metabolism is impaired at low temperatures (Taylor, Jepsen and Christeller, 1972); in maize kernels and avocado cotyledons starch synthesis is inhibited (Downton and Hawker, 1975).

Exposure to low temperatures for several days leads to an irreversible depression of photosynthesis: after rewarming, the CO_2 uptake remains low or completely inhibited (*Figure 14.3*; Kislyuk, 1964a; Taylor and Rowley, 1971; Pasternak and Wilson, 1972; Tschäpe, 1972; Crookston *et al.*, 1974; Drake and Raschke, 1974; Kishitani and Tsunoda, 1974; McWilliam and Ferrar, 1974).

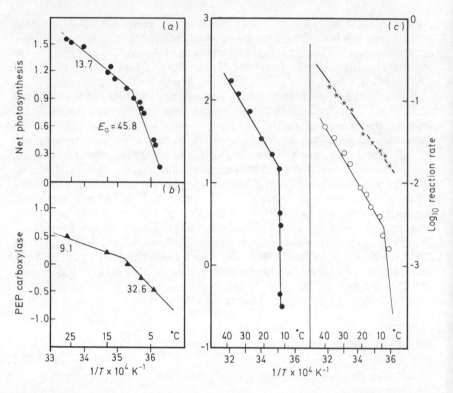

Figure 14.2   *Arrhenius plots of photosynthetic, respiratory and enzyme activities of chilling-sensitive plants. (a) Net photosynthesis of sorghum (after McWilliam and Ferrar, 1974): (b) PEP carboxylase activity of maize (after Phillips and McWilliam, 1971). (c) Oxygen consumption (●—●), alcohol dehydrogenase (*—*) and isocitrate dehydrogenase (○—○) activity of embryonic axes from imbibed soya bean seeds (after Duke, Schrader and Miller, 1977)*

Chloroplasts seem to be extremely sensitive to chilling stress (Ilker *et al.*, 1976); they appear swollen with dense thylakoid structures (Taylor and Craig, 1971; Kimball and Salisbury, 1973). The permeability of the chloroplast membrane is greatly increased at low temperatures (Nobel, 1974). High radiation together with low temperatures causes photo-oxidative chlorophyll degradation in photolabile cucumber leaves (Kislyuk, 1964a, 1964b; van Hasselt, 1972; van Hasselt and Strikwerda, 1976) but, when cooled in darkness, no such defect occurs (Sochanovicz and Kaniuga, 1979). The CO_2 gas exchange may also be affected by after-effects of chilling on stomatal behaviour, i.e. by increased diffusion resistance (Tschäpe, 1972; Ivory and Whiteman, 1978) and by the sensitization of guard cells to changes in the internal CO_2 concentration (Drake and Raschke, 1974).

Respiration

Associated with membrane phase transitions below 10–15°C in chilling-sensitive plants, at low temperatures, much higher energy for activation of

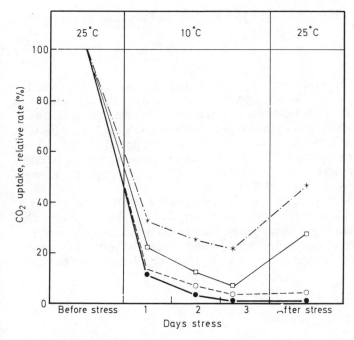

Figure 14.3 *Effect and after-effect of chilling stress (+ 10°C) on net photosynthesis of tropical C₄-grasses and soya bean. * = Paspalum dilatatum. □ = Glycine max. ○ = Sorghum bicolor, ● = Pennisetum typhoides. (Data of Taylor and Rowley, 1971)*

enzymes involved with respiratory processes is needed (E_a of various dehydrogenases of cotton embryos between 0 and 10°C: 88–160 kJ mol^{-1}; between 20 and 30°C: 33–46 kJ mol^{-1}; Duke, Schrader and Miller, 1977). Thus, in the Arrhenius plot a discontinuity appears within the critical temperature range (Lyons and Raison, 1970; cf. also *Figure 14.2*). The disproportions among reactions catalysed by chilling-sensitive membrane-bound mitochondrial enzymes and less susceptible enzymes in the cytosol may result in an accumulation of acetaldehyde and ethanol (Murata, 1969). Shifting of the respiratory metabolism in favour of anaerobic phases (e.g. glycolysis) causes progressive intoxication and cell damage similar to anoxia (Lyons, 1973).

If the chilling stress is of short duration, then in most cases the respiratory dysfunctions are completely reversible. After rewarming, respiration may react with an overshoot (cacao seeds: Casas, Redshaw and Ibáñez, 1965; corn seedlings: Creencia and Bramlage, 1971; Leaves of *Episcia reptans*, but not bean leaves: Wilson, 1978; cool-stored fruits: Eaks, 1960, 1976; Murata, 1969; Kozukue and Ogata, 1972; Fukushima, Yamazaki and Tsugiyama, 1977), and the RQ may decrease temporarily (in cotton seedlings: Christiansen, 1971; in *Episcia*: Wilson, 1978). Even with irreparable chilling injury a short increase in respiratory rate can sometimes be observed after rewarming; respiration ceases soon, however (Amin, 1969).

RESPONSES OF CHILLING-TOLERANT PLANTS TO LOW AND SUBFREEZING
TEMPERATURES

Photosynthesis

Chilling-tolerant plants are capable of CO_2 uptake at low positive
temperatures. At 5°C most herbaceous and woody plants of temperate
regions still achieve photosynthetic rates of 30–50 per cent of the maximal.
Cold-adapted provenances of Sitka spruce from Alaska and Queen Charlotte
Island were able to assimilate at 5°C 60–70 per cent, and at 0°C 30 per cent
as much CO_2 as at optimum temperatures, while provenances from the north
coast of Washington could assimilate 33 and 12 per cent, respectively
(Neilson, Ludlow and Jarvis, 1972).

In vascular plants CO_2 uptake is inhibited, as a rule, when ice appears in
the leaves (Pisek, Larcher and Unterholzner, 1967; Pisek, 1973). This
indicates that chilling-tolerant plants may be photosynthetically active as
long as they are not frozen. In freezing-tolerant vascular plants CO_2 uptake
is interrupted by freezing; after thawing, however, photosynthetic activity

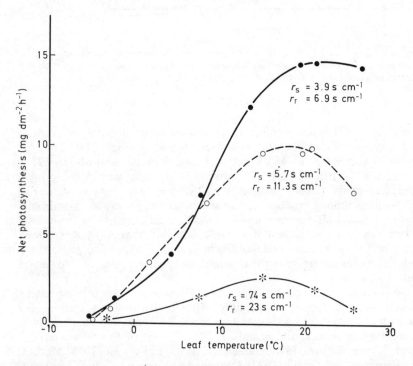

Figure 14.4   *Net photosynthesis at various leaf temperatures, and stomatal (r_s) and residual (r_r)
transfer resistances to CO_2 uptake (at 20°C) of unhardened Sitka spruce before and after 20 min
exposure to −5°C. During this treatment the leaves remained subcooled; the photosynthetic activity
was depressed the day after frost but recovered after two days. Shoots exposed to −5°C for six hours
froze and died. ● = Without frost. * = After frost. ○ = After recovery. (After Neilson, Ludlow and
Jarvis, 1972)*

will be regained as long as no lethal injury occurs. During the growing season the low temperature limit of net photosynthesis of these plants lies within a range of -1 to about $-5°C$; during winter at about -5 to $-8°C$ (data lists: Bauer, Larcher and Walker, 1975; Larcher, 1980).

For dry matter production during periods with repeated night frosts a prompt and complete reactivation of the photosynthetic capacity after rewarming is of special importance. Post-stress depression of photosynthesis can be induced by temperatures just above the tissue freezing point (*Figure 14.4*; Neilson, Ludlow and Jarvis, 1972). In apple leaves with a freezing temperature of $-1.6°C$ no after-effect has been revealed until $+2.5°C$, but after an exposure to $-1.3°C$ photosynthesis decreased to a third of the normal level (Seeley and Kammereck, 1977).

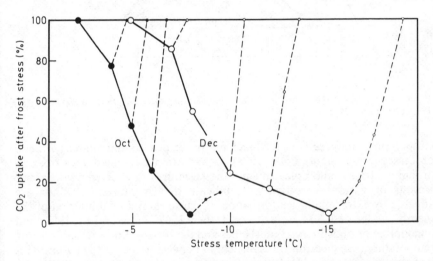

Figure 14.5 *Depression of net photosynthesis after exposure of fir twigs to various degrees of frost stress in autumn and winter. Solid lines = CO_2 uptake immediately after stress. Dotted lines = recovery during the following days. (After Pisek and Kemnitzer, 1968)*

In freezing-tolerant leaves CO_2 uptake after thawing is diminished depending on the intensity, the frequency and the duration of the subfreezing temperatures (*Figure 14.5*; Pavletič and Lieth, 1958; Weise, 1961; Polster and Fuchs, 1963; Pisek and Kemnitzer, 1968; Larcher, 1969; Pharis, Hellmers and Schuurmans, 1970).

In wheat plants, CO_2 uptake is depressed after frost exposure of the aerial parts as well as after soil frost (Koh, Kumura and Murata, 1978a, 1978b). The onset of CO_2 uptake in the morning, therefore, is considerably delayed even after moderate night frost (*Figure 14.6*). A series of consecutive frosty nights may restrict the production capacity of a plant drastically (Tranquillini, 1957; Pisek and Winkler, 1958; Parker, 1961; Keller, 1965; Kozlowski and Keller, 1966; Negisi, 1966; Zelawski and Kucharska, 1967).

There are various possible explanations for the after-effects of low temperatures and frost on the photosynthetic activity of chilling-tolerant plants. Accumulation of soluble carbohydrates, as regularly observed at low

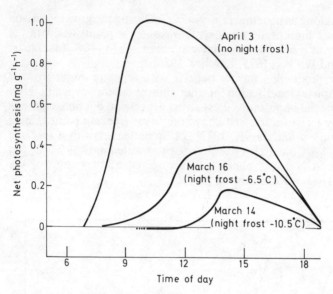

Figure 14.6   *Effect of night frosts on the CO_2 uptake of pine twigs during the following day. (After Polster and Fuchs, 1963)*

temperatures, may lead to a feedback inhibition of secondary photosynthetic processes (Peoples and Koch, 1978). Dehydration associated with freezing probably strains chloroplast membrane structures to a deleterious extent. Results of various investigations indicate that membrane proteins are affected by extracellular ice formation, to cause reversible inhibition of the membrane-located ATPase activity (Heber, 1967; Palta and Li, 1978) and uncoupling of phosphorylation (Heber and Santarius, 1964). At subfreezing temperatures the electron transport chain between photosystems I and II is interrupted (Martin, Mårtensson and Öquist, 1978; Öquist, 1980). As isolated chloroplasts are more freezing-tolerant compared with *in situ* frozen chloroplasts, Senser and Beck (1977) propose that an adverse interference of toxic compounds or lytic enzymes liberated from freeze-injured cell compartments may be effective. The functional effects are paralleled by structural transformations: shape and structure of the chloroplasts change during exposure to low temperatures and frost, and they remain altered for some time after recovery (Kwiatkowska, 1970a; Salisbury *et al.*, 1973; Smillie *et al.*, 1978).

Respiration

Respiratory rates decline exponentially with decreasing temperatures. At low temperatures respiration supplies less energy, less reduction power and fewer intermediary metabolites. Little evidence is available for low temperature effects on photorespiration (Mächler, Nösberger and Erismann, 1977; Sosinska, Maleszewski and Kacperska-Palacz, 1977; Mächler and Nösberger, 1978). Therefore the following information refers merely to dark respiration.

There are no detailed investigations on respiration rates of higher plants at very low subfreezing temperatures. As revealed by measurements with lichens and poikilothermic animals, mitochondrial respiration can still be sustained at negative temperatures. In insects, respiration activity decreases gradually as long as the cells remain subcooled; at freezing, a sharp break in the Arrhenius plot occurs indicating that respiratory gas exchange is interrupted (Asahina, 1966; Kanwisher, 1966). With progressive dehydration by freezing respiratory metabolism is depressed. Frozen tissues are in a latent, anabiotic state. Mitochondria change their shape near 0°C and slightly below (in winter wheat at -3°C: Kwiatkowska, 1970a, 1970b; at $+2$°C: Pomeroy, 1977); however, they retain their functional capacity after freezing (in rye: Singh, de la Roche and Siminovitch, 1977).

After thawing a temporary rise in respiratory rate is frequently observed in vascular plants (Semikhatova, 1953; Pisek and Winkler, 1958; Rakitina, 1960; Larcher, 1961; Weise and Polster, 1962; Pisek and Kemnitzer, 1968; Bauer, Huter and Larcher, 1969). Respiration increases within hours to between two and five times the normal level depending on the frost stress and on the hardening state of the plant; after a few days it decreases gradually. In sensitive plants the extent of the post-thawing respiratory flush is more pronounced than in frost-hardened plants. The transient CO_2 outburst after frost is to be considered as a repair effect, as a consequence of pathological disorders during freezing; the rise in respiratory activity after exposure to low positive temperatures is a physiological response (Semikhatova, 1962; Chabot and Billings, 1972; Stewart and Bannister, 1974; Seeley and Kammereck, 1977; Larcher, 1979). This enhancement is possibly due to increased substrate supply because of the accumulation of metabolites at low temperatures. This improved energy supply together with changes in the mitochondrial structures and the enzyme pattern at low temperatures is interpreted as a capacity adaptation (Precht, 1973). In wheat plants with pronounced ability for hardening, alternative electron transport pathways (antimycin A-resistant respiration) are promoted at low temperatures (Voinikov, 1978).

Excessive respiration causes additional losses for the carbohydrate budget. Together with the restriction of the photosynthetic activity at low temperatures and after frost, high respiration rates may contribute to impair the carbohydrate budget of the plant at prolonged cold weather periods.

Survival and Productivity after Severe Damage

Lethal tissue injury caused by chilling and freezing weakens the plant and diminishes its production capacity. The degree of the final losses depends on the localization and the extent of the injuries, and on the possibility of a restitution of killed plant parts. The individual organs and tissues of a plant usually vary in their susceptibility to low temperature stress (Larcher, 1970; 1973; Larcher and Bauer, 1982); they are also differently endangered depending on the occurrence of microclimatic temperature gradients. In most cases partial damage will not lead to an elimination of the plant in its natural distribution range or cultivation area as long as regeneration buds remain undamaged. However, when a partial damage extends further or becomes irreparable, the total loss of the plant will be the consequence. High resistance of tissues which are important for regeneration therefore represents an essential chance for survival after severe damage.

DAMAGE OF ORGANS IMPORTANT FOR FRUIT SET

Flowers and likewise the flower primordia within the buds are the most susceptible plant parts. In chilling-sensitive plants like rice and sorghum night temperatures of +13 to +16°C during the late period of the pollen mother cell development cause floret sterility (Lin and Peterson, 1975; Brooking, 1976). Spring frosts, which occur frequently in temperate climates, injure blossoms of deciduous and coniferous trees and certain herbaceous crops at temperatures between −1 and −5°C (strawberries: Ourecky and Reich, 1976; fruit trees: Rogers, 1952; Kemmer and Schulz, 1955; Modlibowska, 1956, 1962; Pisek, 1958; Kohn, 1959; Proebsting and Mills, 1978; forest trees: Pankratova, 1956; Till, 1956; Parker, 1963; Timmis, 1977). Young fruits are as frost-sensitive as the blossoms, sometimes even more (strawberries: Ourecky and Reich, 1976). In winter buds of trees and shrubs the primordia of pistils and anthers are less frost resistant than the bud scales (Till, 1956; Larcher, 1970; Bittenbender and Howell, 1976); after severe winters sterile blossoms may come to bloom (Modlibowska and Montgomery, 1947). In wheat plants, too, the floral apex is damaged at first (Szabolcs, 1967).

Chilling and freezing injury of flower primordia and flower organs cause remarkable losses in yield (Toriyama, 1976), and in any case a reduced marketability of the deformed or low-quality fruits which are formed (Modlibowska, 1962).

REDUCED PRODUCTIVITY BY DEFOLIATION

During the growing season, leaves of most herbaceous and woody species are killed at −2 to −6°C; during winter the shoots of herbaceous perennials survive −15 to −30°C, evergreen leaves of woody plants in subtropical and warm temperate regions are damaged below −5 to −25°C, in cold winter regions below −30 to −90°C (for data lists *see* Larcher and Bauer, 1982). The effect of a partial loss of leaf area on the production capacity of plants depends on the extent and localization of freezing injuries. Defoliation experiments provide information on growth reduction in woody plants after loss of foliage. In *Populus tremuloides* the radial growth was reduced to 27 per cent of the normal year ring width after a damage by late frost (Strain, 1966). In evergreen trees the cambial growth decreases for about 20–30 per cent, when 50–70 per cent of the leaf area is lost (*Pinus banksiana*: O'Neil, 1962; *Pseudotsuga menziesii*: Mitscherlich and Gadow, 1968; *Quercus suber*: Magnoler and Cambini, 1968, 1970). When the entire foliage is lost, in many cases no annual ring is formed in the following season; at best, the growth ring will attain 50 per cent of the normal width; complete defoliation, however, may result in tree death (Monange, 1961; O'Neil, 1962; Magnoler and Cambini, 1968; Jahnel, 1969; Kozlowski, 1971). Also shoot elongation and formation of new leaves is poor after extensive injury to the foliage

Leaf area:

1957 { 16.1 cm²

1956 {

Total: 13.0 cm²
Living: 8.5 cm²

1955 {

Figure 14.7 *Reduced needle growth (1956) and full recovery (1957) after a partial frost injury during winter 1955/56 to* Abies nordmanniana. *(Photography: September 1957 by W. Larcher)*

(*Figure 14.7*). The effect of defoliation on dry matter production and regrowth depends also on the season when damage occurs (Craighead, 1940; Kramer and Wetmore, 1943; Buttrose, 1966; Howell, Stergios and Stackhouse, 1978).

Incomplete restitution of the foliage and of the shoot system resulting from heavy frost injury is frequently due to bud killing. Normal refoliation is only to be expected if the buds are more resistant than the leaves. Growth from partially injured buds, however, is still possible, although retarded, as long as some of the bud meristem remains intact (fruit trees: Kemmer and Schulz, 1955; Pisek and Eggarter, 1959; barley: Olien, 1974; Olien and Marchetti, 1976).

SEVERE INJURIES TO SUPPORTING ORGANS

Freezing of the main roots or the lower stem and rootstock causes dieback of whole trees (Chandler, 1954; Kemmer and Schulz, 1955; Karnatz, 1956a,

1956b; Schnelle, 1963; Sakai, 1968). After a partial injury to the shoot system trees may die in subsequent years if the damage is irreparable and eventually spreads. Recovery is to be expected if the cambium remains intact or largely undamaged. Observations after severe winters and freezing experiments with fruit trees have revealed that the prognosis for restitution is doubtful, if approximately 50 per cent of the cambium has been killed (Kemmer and Schulz, 1955; Pisek and Eggarter, 1959; Lapins, 1965). Recovery experiments with 2–3-year-old seedlings of *Abies alba* and *Larix decidua* carried out by Bendetta (1972) have shown that the probability of restitution and thus the long-term survival of artificially frost-damaged plants varies depending largely on seasonal differences in the localization of frost injuries: after frost in spring and summer (when the cambium is the most sensitive shoot tissue) the plants were killed if this tissue was damaged to an extent of 50 per cent or more. During winter when the cambium was extremely frost-tolerant (living even at −80°C), the xylem and the needles of fir were injured first (at −30 to −36°C). Plants with a 50 per cent xylem damage were able to survive depending on the amount of intact leaves, but no recovery was observed if less than one quarter of the leaf area remained intact.

After severe frost damage, productivity and growth of woody plants are reduced for years. As a consequence of the inadequate xylem growth of apple

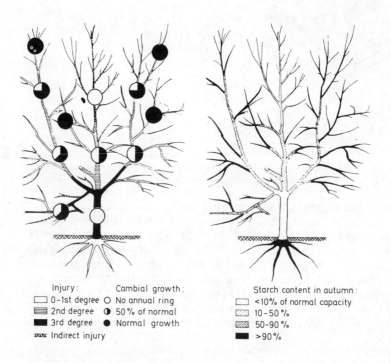

Injury:	Cambial growth:	Starch content in autumn:
☐ 0–1st degree	○ No annual ring	☐ <10% of normal capacity
▤ 2nd degree	◑ 50% of normal	▨ 10–50%
■ 3rd degree	● Normal growth	▩ 50–90%
∿ Indirect injury		■ >90%

Figure 14.8   *Extent of tissue injury, and amount of cambial growth and starch accumulation in apple trees one year after an early November frost. Injury degrees: 0 = no visible injuries; 1 = completely repairable slight injuries of cambium and xylem, frost rings visible; 2 = moderate to severe xylem and bark injuries, regeneration possible by intact cambium; 3 = irreparable injuries, no or little cambial activity; indirect injuries: secondary drying of twigs*

trees after being injured by a late autumn frost, in the following summers the critical level of water potential depression that induces stomatal closure is reached 1–2 hours earlier during daytime as compared to control trees (Körner and Schubert, personal communication). This results in a reduced CO_2 uptake and in shortening the daily assimilation period. A secondary effect of wood necrosis and narrow tree ring formation is the lack of storage capacity for reserve materials. In autumn much less starch and lipids can be deposited in frost-damaged trees (*Figure 14.8*). As a consequence, in the following spring carbohydrate mobilization will be insufficient and may become limiting for the new growth; the leaves remain smaller, less leaf area develops and the productivity of the tree is further diminished.

Conclusion

Low temperature stress restricts CO_2 uptake and productivity, endangers plant life, and delimits agricultural and horticultural utilization of land area. In most cases low dry matter production and even damage by frost are caused by exceeding the specific susceptibility limits by only a few degrees centigrade. For securing improved and more stable yields more detailed information is needed on the functional basis and the variability of resistance adaptation, and on responses of life functions to low temperatures and freezing. Until now only little interrelated research on the physiological background of cold-resistance and production processes has been done. Productivity research traditionally tends to emphasize highest yields at optimal conditions; however, preservation and improvement of plant productivity at suboptimal temperatures could be of similar importance for many crops (Stoner, 1978).

Acknowledgement

I am indebted to Dr H. Hilscher for the translation of the manuscript.

References

AMIN, J. V. (1969). *Physiol. Plant.*, **22**, 1184–1191

ASAHINA, E. (1956). *Contr. Inst. Low. Temp. Sapporo*, **10**, 83–126

ASAHINA, E. (1966). In *Cryobiology*, pp. 451–486. Ed. by H. T. Meryman. Academic Press, London

ASAHINA, E. (1978). In *Plant Cold Hardiness and Freezing Stresses*, pp. 17–36. Ed. by P. H. Li and A. Sakai. Academic Press, New York

AUSSENAC, G. (1973). *Ann. Sci. Forest.*, **30**, 141–155

BAUER, H., HUTER, M. and LARCHER, W. (1969). *Ber. Dtsch. bot. Ges.*, **82**, 65–70

BAUER, H., LARCHER, W. and WALKER, R. B. (1975). In *Photosynthesis and Productivity in Different Environments*, pp. 557–586. Ed. by J. P. Cooper. Cambridge University Press, Cambridge

BENDETTA, G. (1972). Dissertation. Innsbruck
BIEBL, R. (1962). *Protoplasmatologia*, Vol. XII, No. 1. Springer, Wien
BITTENBENDER, B. C. and HOWELL, G. S. (1976). *J. Am. Soc. hort. Sci.*, **101**, 135–139
BROOKING, I. R. (1976). *Austr. J. Pl. Physiol.*, **3**, 589–596
BUTTROSE, M. S. (1966). *Vitis*, **5**, 455–464
CASAS, I., REDSHAW, E. S. and IBÁÑEZ, M. L. (1965). *Nature*, **208**, 1348–1349
CHABOT, B. F. and BILLINGS, W. D. (1972). *Ecol. Monographs*, **42**, 163–199
CHANDLER, W. H. (1954). *Proc. Am. Soc. hort. Sci.*, **64**, 552–572
CHRISTIANSEN, M. N. (1971). *Beltwide Cotton Prod. Res. Conf.*
CRAIGHEAD, F. C. (1940). *J. Forest.*, **38**, 885–888
CREENCIA, R. P. and BRAMLAGE, W. J. (1971). *Pl. Physiol.*, **47**, 389–392
CROOKSTON, R. K., O'TOOLE, J., LEE, R., OZBUN, J. L. and WALLACE, D. H. (1974). *Crop Sci.*, **14**, 457–464
DILLEY, D. R., HEGGESTAD, H. E., POWERS, W.L. and WEISER, C. J. (1975). In *Crop Productivity Research Imperatives*, pp. 309–355. Ed. by A. W. A. Brown. Kettering Foundation, Yellow Springs, Ohio
DOWNTON, W. J. S. and HAWKER, J. S. (1975). In *Environmental and Biological Control of Photosynthesis*, pp. 81–88. Ed. by R. Marcelle. Junk, Den Haag
DRAKE, B. and RASCHKE, K. (1974). *Pl. Physiol.*, **53**, 808–812
DUKE, S. H., SCHRADER, L. E. and MILLER, M. G. (1977). *Pl. Physiol.*, **60**, 716–722
EAKS, I. L. (1960). *Pl. Physiol.*, **35**, 632–636
EAKS, I. L. (1976). *J. Am. Soc. hort. Sci.*, **101**, 538–540
FUKUSHIMA, T., YAMAZAKI, M. and TSUGIYAMA, T. (1979). *Sci. Horticult.*, **6**, 185–197
HÄRTEL, O. (1976). *Umschau*, **76**, 347–350
HEBER, U. W. (1967). *Pl. Physiol.*, **42**, 1343–1350
HEBER, U. W. and SANTARIUS, K. A. (1964). *Pl. Physiol.*, **39**, 712–719
HEBER, U. W. and SANTARIUS, K. A. (1973). In *Temperature and Life*, pp. 232–263. Ed. by H. Precht, J. Christophersen, H. Hensel and W. Larcher. Springer-Verlag, Berlin
HOWELL, G. S., STERGOIS, B. G. and STACKHOUSE, S. S. (1978). *Am. J. Ecol. Viticult.*, **29**, 187–191
ILKER, R., WARING, A. J., LYONS, J. M. and BREIDENBACH, R. W. (1976). *Protoplasma*, **90**, 229–252
IVORY, D. A. and WHITEMAN, P. C. (1978). *Austr. J. Pl. Physiol.*, **5**, 149–157
JAHNEL, H. (1969). *Flora*, **158**, 288–290
KANWISHER, J. W. (1966). In *Cryobiology*, pp. 487–494. Ed. by H. T. Meryman. Academic Press, London
KARNATZ, H. (1956a). *Der Züchter*, **26**, 178–187
KARNATZ, H. (1956b). *Der Züchter*, **26**, 307–315
KELLER, T. (1965). *Schweiz. Z. Forstw.*, **9**, 719–729
KEMMER, E. and SCHULZ, F. (1955). *Das Frostproblem im Obstbau*. Bayerischer Landwirtschaftsverlag, München
KIMBALL, ST. L. and SALISBURY, F. B. (1973). *Am. J. Bot.*, **60**, 1028–1033
KISHITANI, S. and TSUNODA, S. (1974). *Photosynth.*, **8**, 161–167
KISLYUK, I. M. (1964a). *Doklady Akad. Nauk*, **158**, 1434–1436
KISLYUK, I. M. (1964b). *Tsitolog. osnovy prisposobl. rast. k faktoram sredy*, pp. 167–184. Nauka, Moscow

KOH, S., KUMURA, A. and MURATA, Y. (1978a). *Jap. J. Crop. Sci.*, **47**, 69–74
KOH, S., KUMURA, A. and MURATA, Y. (1978b). *Jap. J. Crop. Sci.*, **47**, 75–81
KOHN, H. (1959). *Gartenbauwiss.*, **24**, 315–329
KOZLOWSKI, T. T. (1971). *Growth and Development of Trees*, Vol. II: *Cambial Growth, Root Growth, and Reproductive Growth*. Academic Press, New York
KOZLOWSKI, T. T. and KELLER, T. (1966). *Bot. Rev.*, **32**, 294–382
KOZUKUE, N. and OGATA, K. (1972). *J. Food Sci.*, **37**, 708–711
KRAMER, J. P. and WETMORE, T. H. (1943). *Am. J. Bot.*, **30**, 428–431
KRASAVTSEV, O. A. (1972). *Kalorimetriya rastenii pri temperaturakh nizhe nulya.* Nauka, Moscow
KWIATKOWSKA, M. (1970a). *Acta Soc. Bot. Polon.*, **39**, 347–360
KWIATKOWSKA, M. (1970b). *Acta Soc. Bot. Po.*, **39**, 361–371
LAPINS, K. (1965). *Can. J. Pl. Sci.*, **45**, 429–435
LARCHER, W. (1961). *Planta*, **56**, 575–606
LARCHER, W. (1969). *Photosynth.*, **3**, 167–198
LARCHER, W. (1970). *Oecol. Plant.*, **5**, 267–286
LARCHER, W. (1973). In *Temperature and Life*, pp. 194–231. Ed. by H. Precht, J. Christophersen, H. Hensel and W. Larcher. Springer-Verlag, Berlin
LARCHER, W. (1979). *Rheinisch-Westfälische Akad. Wiss.*, Vorträge N291
LARCHER, W. (1980). *Physiological Plant Ecology*, 2nd edn. Springer-Verlag, Berlin
LARCHER, W. and BAUER, H. (1982). In *Encyclopedia of Plant Physiology: Physiological Plant Ecology*, Vol. 13A. Springer-Verlag, Berlin (in press)
LEVITT, J. (1957). *Protoplasma*, **48**, 289–302
LEVITT, J. (1966). In *Cryobiology*, pp. 495–563. Ed. by H. T. Meryman. Academic Press, London
LEVITT, J. (1972). *Responses of Plants to Environmental Stresses*. Academic Press, New York
LEVITT, J. (1978). In *Plant Cold Hardiness and Freezing Stresses*, pp. 3–15. Ed. by P. H. Li and A. Sakai. Academic Press, New York
LI, P. H. and PALTA, J. P. (1978). In *Plant Cold Hardiness and Freezing Stresses*, pp. 49–71. Ed. by P. H. Li and A. Sakai. Academic Press, New York
LIN, S. S. and PETERSON, M. L. (1975). *Crop. Sci.*, **15**, 657–660
LUDLOW, M. M. and WILSON, G. L. (1971). *Austr. J. biol. Sci.*, **24**, 449–470
LUTZ, J. M. and HARDENBURG, R. E. (1968). *Agric. Handbook*, No. 66. Washington, US Government Printing Office
LYONS, J. M. (1973). *Ann. Rev. Pl. Physiol.*, **24**, 445–466
LYONS, J. M. and RAISON, J. K. (1970). *Pl. Physiol.*, **45**, 386–389
MÄCHLER, F. and NÖSBERGER, J. (1978). *Oecologia*, **35**, 267–276
MÄCHLER, F., NÖSBERGER, J. and ERISMANN, K. H. (1977). *Oecologia*, **31**, 79–84
MAGNOLER, A. and CAMBINI, A. (1968). *Mem. Staz. Sughero Tempio Pausania*, **25**, 1–23
MAGNOLER, A. and CAMBINI, A. (1970). *Forst Sci.*, **16**, 364–366
MARTIN, B., MÅRTENSSON, O. and ÖQUIST, G. (1978). *Physiol. Pl.*, **43**, 297–305
MAZUR, P. (1969). *Ann. Rev. Pl. Physiol.*, **20**, 419–448
MCWILLIAM, J. R. and FERRAR, P. J. (1974). in *Mechanisms of Regulation of Plant Growth*, pp. 467–476. Ed. by R. L. Bieleski, A. R. Ferguson and M. M. Cresswell. Royal Society of New Zealand, Wellington
MILLER, J. D. (1976). *Cold Tolerance in Sugar Cane Relatives*. Sugar y Azucar

MITSCHERLICH, G. and GADOW, K. (1968). *Allg. Forst- und Jagd-Ztg.*, **139**, 175–184

MITTELSTÄDT, H. (1969). *D.A.L. Tagungsber.*, **96**, 149–173

MOCK, J. J. and SKRDLA, W. H. (1978). *Euphytica*, **27**, 27–32

MODLIBOWSKA, I. (1956). *Rapp. gen. Congr. Pomologie Intern.*, pp. 83–111. Namur

MODLIBOWSKA, I. (1962). *16th Hortic. Congr.*, pp. 180–189

MODLIBOWSKA, I. and MONTGOMERY, H. B. S. (1947). *Ann. Rep. East Malling Res. Stat.*, pp. 165–168

MONANGE, Y. (1961). *Travaux Lab. Forest.*, **96**, 1–6

MORETTINI, A. (1961). *Acc. econ.-agr. georg.*, **8**, 1–40

MURATA, T. (1969). *Physiol. Plant.*, **22**, 401–411

NEGISI, K. (1966). *Bull. Tokyo Univ. For.*, **62**, 1–115

NEILSON, R. E., LUDLOW, M. M. and JARVIS, P. G. (1972). *J. appl. Ecol.*, **9**, 721–745

NOBEL, P. S. (1974). *Planta*, **115**, 369–372

OLIEN, C. R. (1973). *J. theor. Biol.*, **39**, 201–210

OLIEN, C. R. (1974). Res. Rep., 247; *Farm Sci.*, 1–20

OLIEN, C. R. (1978). In *Plant Cold Hardiness and Freezing Stresses*, pp. 37–48. Ed. by P. H. Li and A. Sakai. Academic Press, New York

OLIEN, C. R. and MARCHETTI, B. L. (1976). *Crop Sci.*, **16**, 201–204

O'NEIL, L. C. (1962). *Can. J. Bot.*, **40**, 273–280

ÖQUIST, G. (1980). This volume, pp. 53–80

OURECKY, D. K. and REICH, J. E. (1976). *Hortic. Sci.*, **11**, 413–414

PALTA, J. P. and LI, P. H. (1978). In *Plant Cold Hardiness and Freezing Stresses*, pp. 93–115. Ed. by P. H. Li and A. Sakai. Academic Press, New York

PANKRATOVA, N. M. (1956). *Bot. Zhurnal*, **41**, 263–266

PARKER, J. (1961). *Ecology*, **42**, 372–380

PARKER, J. (1963). *Bot. Rev.*, **29**, 123–201

PASTERNAK, D. and WILSON, G. L. (1972). *New Phytol.*, **71**, 682–689

PAVLETIČ, Z. and LIETH, H. (1958). *Ber. Dtsch. bot. Ges.*, **71**, 309–314

PEOPLES, T. R. and KOCH, D. W. (1978). *Crop Sci.*, **18**, 255–258

PHARIS, R. P., HELLMERS, H. and SCHUURMANS, E. (1970). *Photosynth.*, **4**, 273–279

PHILLIPS, P. J. and MCWILLIAM, J. R. (1971). In *Photosynthesis and Photorespiration*, pp. 97–104. Ed. by M. D. Hatch, C. B. Osmond and R. O. Slatyer. Wiley, New York

PISEK, A. (1958). *Gartenbauwiss.*, **23**, 54–74

PISEK, A. (1973). In *Temperature and Life*, pp. 102–133. Ed. by H. Precht, J. Christophersen, H. Hensel and W. Larcher. Springer-Verlag, Berlin

PISEK, A. and EGGARTER, H. (1959). *Gartenbauwiss.*, **24**, 446–456

PISEK, A. and KEMNITZER, R. (1968). *Flora B*, **157**, 314–326

PISEK, A., LARCHER, W. and UNTERHOLZNER, R. (1967). *Flora B*, **157**, 239–264

PISEK, A. and WINKLER, E. (1958). *Planta*, **51**, 518–543

POLSTER, H. and FUCHS, S. (1963). *Archiv Forstwesen*, **12**, 1011–1024

POMEROY, M. K. (1977). *Pl. Physiol.*, **59**, 250–255

PRECHT, H. (1973). In *Temperature and Life*, pp. 325–352. Ed. by H. Precht, J. Christophersen, H. Hensel and W. Larcher. Springer-Verlag, Berlin

PROEBSTING, E. L. and MILLS, H. H. (1978). *J. Am. Soc. hortic. Sci.*, **103**, 191–198

QUAMME, H. A. (1976). *Can. J. Pl. Sci.*, **56**, 493–500

RAKITINA, Z. G. (1960). *Trudy konf. fiziol. ustoichivost rast.*, pp. 278–284. Nauka, Moscow

ROGERS, W. S. (1952). Report 13. *Int. Hortic. Congress*, pp. 1–6

ROWLEY, J. A., TUNNICLIFFE, C. G. and TAYLOR, A. O. (1975). *Austr. J. Pl. Physiol.*, **2**, 447–451

SAKAI, A. (1968). *Contr. Low Temp. Sci. B*, **15**, 1–14

SALISBURY, F. B., KIMBALL, ST. L., BENNETT, B., ROSEN, P. and WEIDNER, M. (1973). *Space Life Sci.*, **4**, 124–138

SAMYGIN, G. A. (1974). *Pritchiny vymerzaniya rastenii*. Nauka, Moscow

SCHNELLE, F. (1963). In *Frostschutz im Pflanzenbau*, Vol. 1, pp. 391–407. Ed. by F. Schnelle. Bayerischer Landwirtschaftsverlag. München

SEELEY, E. J. and KAMMERECK, R. (1977). *J. Am. Soc. hortic. Sci.*, **102**, 282–286

SEMIKHATOVA, O. A. (1953). *Exp. Bot.*, **9**, 132–155

SEMIKHATOVA, O. A. (1962). *Bot. Zh.*, **47**, 636–644

SENSER, M. and BECK, E. (1977). *Planta*, **137**, 195–201

SHIRAHASHI, K., HAYAKAWA, S. and SUGIYAMA, T. (1978). *Pl. Physiol.*, **62**, 826–830

SHNEYOUR, A., RAISON, J. K. and SMILLIE, R. M. (1973). *Biochem. Biophys. Acta*, **292**, 152–161

SINGH, J., DE LA ROCHE, I. and SIMINOVITCH, D. (1977). *Pl. Physiol.*, **60**, 713–715

SMILLIE, R. M., CRITCHLEY, ch., BAIN, J. M. and NOTT, R. (1978). *Pl. Physiol.*, **62**, 191–196

SOCHANOWICZ, B. and KANIUGA, Z. (1979). *Planta*, **144**, 153–159

SOSINSKA, A., MALESZEWSKI, S. and KACPERSKA-PALACZ, A. (1977). *Z. Pfl.-physiol.*, **83**, 285–292

STEWART, W. S. and BANNISTER, P. (1974). *Flora*, **163**, 415–421

STONER, A. (1978). *Hortic. Sci.*, **13**, 684–686

STRAIN, B. R. (1966). *Forest Sci.*, **12**, 334–337

SZABOLCS, V. (1967). *Bull. Agr. Coll. Keszthely*, **9**, 1–83

TAYLOR, A. O. and CRAIG, A. S. (1971). *Pl. Physiol.*, **47**, 719–725

TAYLOR, A. O., JEPSEN, N. M. and CHRISTELLER, J. T. (1972). *Pl. Physiol.*, **49**, 798–802

TAYLOR, A. O. and ROWLEY, J. A. (1971). *Pl. Physiol.*, **47**, 713–718

TAYLOR, A. O., SLACK, C. R. and MCPHERSON, H. G. (1974). *Pl. Physiol.*, **54**, 696–701

TILL, O. (1956). *Flora*, **143**, 499–542

TIMMIS, R. (1977). *Can. J. For. Res.*, **7**, 19–22

TORIYAMA, K. (1976). In *Climatic Change and Food Production*, pp. 237–243. Ed. by K. Takahashi and M. M. Yoshino. University of Tokyo Press, Tokyo

TRANQUILLINI, W. (1957). *Planta*, **40**, 612–661

TSCHÄPE, M. (1972). *Biochem. Phys. Pfl.*, **163**, 81–92

VAN HASSELT, ph. R. (1972). *Acta Bot. Neerl.*, **21**, 539–548

VAN HASSELT, ph. R. and STRIKWERDA, J. T. (1976). *Physiol. Plant.*, **37**, 253–257

VOINIKOV, V. K. (1978). *Fiziol. rast.*, **25**, 761–766

WEISE, G. (1961). *Biol. Zentralbl.*, **80**, 137–166

WEISE, G. and POLSTER, H. (1962). *Biol. Zentralbl.*, **81**, 129–143

WILSON, J. M. (1978). *New Phytol.*, **80**, 325–334

YELENOSKY, G. and YOUNG, R. (1978). *Citrus Ind.*, **59**, 5–14

ZELAWSKI, W. and KUCHARSKA, J. (1967). *Photosynth.*, **1**, 207–213

15

SALT TOLERANCE

R. G. WYN JONES
Department of Biochemistry and Soil Science, University College of North Wales, Bangor, UK

Introduction

Substantial areas of the earth's potentially productive lands are affected by soil salinity and alkalinity. Ponnamperuma (1977) estimated that there are 381 million hectares of saline soils and the problem is increasing because of inadequate irrigation and drainage practices. In Pakistan, some 10 out of the 15 million hectares of canal-irrigated land are becoming saline or water-logged and thousands of hectares are going out of production annually (Mohammad, 1978). Excess salts in the soil profile and the associated problems of alkalinity and water logging are therefore major limitations on crop yield in many arid countries, although poor cultivation practices and inadequate fertilization may also contribute to the low yields.

The agronomic problem of salinity is compounded by the relatively low salt tolerance of many of the major crop plants (Maas and Hoffman, 1977). However, salt tolerance itself is not an uncommon phenomenon in higher plants. Perhaps the most extreme examples are afforded by mangrove species, e.g. Avicennia, Aegilitis, Rhizophora and by the submerged sea 'grasses', e.g. Halophila, Posidonia and Zostera (Waisel, 1972). Members of many other families, notably the Chenopodiaceae, are able to grow in inland and estuarine saline habitats. Flowers, Trobe and Yeo (1977) have prepared a dendrogram to illustrate the broad distribution of the halophytic characteristic in the orders of flowering plants.

Since this characteristic is so widespread, the development of tolerant agronomic species either by selective breeding from existing crops or by developing completely new crops from native halophytes appears feasible. This assessment is further encouraged by the marked varietal differences in salt sensitivity found in the Gramineae and the Leguminoseae (Maas and Hoffman, 1977). This potential has been dramatically highlighted by Epstein and Norlyn (1977), who selected a highly tolerant barley cultivar capable of growing in sea water. There can be no doubt that the breeding of high-yielding, tolerant cultivars of the major food crops will be of immense significance to many Third World countries, particularly as the alternative course of extensive land reclamation and drainage schemes is so costly. Thus, both from a practical viewpoint and because of the insight it may provide into membrane transport phenomena in higher plants, an understanding of the physiological and biochemical bases of salt tolerance is potentially extremely rewarding.

Factors Involved in Salt Stress

Three major potential limitations on the growth of plants in saline habitats have been postulated (Bernstein and Hayward, 1958; Strogonov, 1964; Gauch, 1972; Levitt, 1972; Greenway, 1973; Maas and Nieman, 1978). These may be broadly characterized as:

(1) water stress;
(2) specific ion stress or toxicity;
(3) ion imbalance stress or induced nutrient deficiency.

There is no concensus as to the relative importance of these potentially detrimental stresses, and indeed various authors differ widely in their assessments. Although these stresses may be formally recognized as separate entities (Levitt, 1972), in physiological and biochemical terms they may not be so clearly delineated and may indeed be closely interrelated.

Before considering these factors in greater detail, a number of general points must be made. No exclusive classification of higher plants into halophytes and glycophytes can be sustained; rather, a continuum ranging from highly tolerant to extremely sensitive species is observed. The position of an individual species or cultivar in this continuum is not constant but may change according to factors such as plant age or atmospheric relative humidity (Maas and Hoffman, 1977; Hoffman and Jobes, 1978). Species also differ in their tolerance of secondary associated stresses such as alkalinity, high or low temperature or anaerobiosis, all of which are factors that may be important in determining field performance.

Although salinity decreases plant growth, Maas and Nieman (1978) emphasize the absence in many cases of gross toxicity symptoms and the continuation of an orderly ontogeny with deoxyribonucleic acid (DNA), ribonucleic acid (RNA) and protein levels per unit organic matter remaining fairly constant. However, apparently contradicting this observation, there are also well-known examples of necrosis in response to specific ion toxicities, e.g. Cl^- toxicity in avocado (Bingham, Fenn and Oertli, 1968). It is quite characteristic of this field that unqualified generalizations can rarely be made and that many apparently contradictory results have been reported.

WATER STRESS

Since the water potential (Ψ_E) of a saline growth medium or soil is lowered by the osmotic potential of the dissolved solutes (e.g. sea water, $\Psi_{sw} \equiv -2.4$ MPa $\equiv 1000$ osmol m^{-3}), the intracellular water potential must decrease either by increasing the number of solute species (increasing π_{sap}) or by increasing the water binding capacity (Ψ_{matrix}). In the absence of such responses the tissue will loose turgor pressure and dehydrate. Since cell elongation is turgor-dependent and contributes massively to vegetative growth (Hsiao, 1973), growth will quickly be inhibited.

The major evidence that water stress is a significant factor in the inhibition of growth by external salts comes from early work which shows that iso-osmotic concentrations of different salt solutions produce a similar reduction in the growth of sensitive glycophytes (*Figure 15.1*; Gauch and Wadleigh,

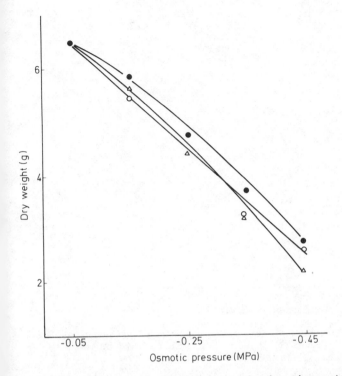

Figure 15.1   *The dry weight yield of dwarf red kidney bean plants as influenced by the osmotic pressure of the growth medium produced by the addition of NaCl (•); CaCl₂ (o) and Na₂SO₄ (△). (Redrawn from Gauch and Wadleigh, 1945)*

1945). Thus the osmotic potential and not the ionic composition of the medium was taken to be crucial, but these data have not been universally reproduced (*see* later). Various observations (Hoffman and Rawlins, 1971; O'Leary, 1975) that glycophytic plants are more resistant to salt stress in high rather than low relative humidities also implicate water stress. However, since net salt uptake, especially the transport of potentially toxic ions, Na^+ and Cl^-, may vary with water flow rates and consequently with relative humidity (Pitman, 1975; Pitman and Cram, 1977), this observation is also equivocal. For example, Storey and Pitman (unpublished data) have recently observed that in the halophyte *Atriplex spongiosa* after four days' growth in 300 mol m^{-3} NaCl at 95–100 per cent relative humidity (RH), the shoot Cl^- content scarcely changed but at 40–45 per cent RH the Cl^- level increased from 7 to 93 mol m^{-3}. Similarly, Pitman (1965) found that in barley growing at a high transpiration rate, the K^+/Na^+ ratio decreased, presumably reflecting a high mass flow of cations to the root and subsequent $K^+ : Na^+$ exchange associated with the extra salt load. Greenway (1965) also reported an increased Cl^- uptake into barley shoots at low per cent of RH but the effect was far less dramatic than that observed in *A. spongiosa*.

Crassulacean acid metabolism (CAM) is probably a metabolic adaptation to semi-arid habitats (Osmond, 1978). In *Mesembryanthenum crystallinum* a shift from conventional C_3 metabolism to CAM is induced by NaCl stress

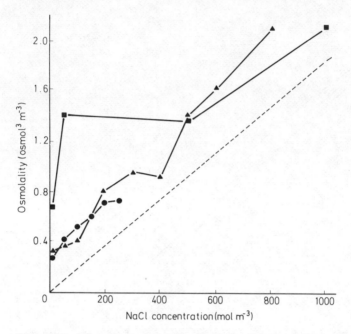

Figure 15.2   *Effect of growth at increasing NaCl salinity on the leaf sap osmolality of* Suaeda monoica (■), Atriplex spongiosa (▲), Hordeum vulgare *c.v. California Mariout* (●). – – – = *Osmolality of growth medium. (Redrawn from Storey and Wyn Jones, 1978a, 1979)*

(Winter and Von Willert, 1972) and by other treatments likely to promote water stress, which again implies that water stress is a major factor in salt toxicity. However, Bloom (1979) failed to induce CAM in *M. crystallinum* by a water stress treatment (polyethylene glycol 6000) and suggested that salt accumulation in the tissue is a requirement for the induction of CAM. Once again, the relationship between water and specific ion stresses is equivocal.

Most plants are able to increase their sap osmolalities in response to a hypersaline stress (Bernstein, 1961; Waisel, 1972; Flowers, 1975). (Changes in Ψ_{matrix} are more problematical and will not be discussed in this chapter.) Osmotic adaptation in the shoots of three contrasting species is illustrated in *Figure 15.2*. In barley and *A. spongiosa* an approximately constant $\Delta\pi$ between leaf sap and external medium is maintained; that is, they apparently behave as *osmoconformers*. *Suaeda monoica* appears to operate preferentially at a high but constant π_{sap} over a wide range of salinity—apparently an *osmoregulator*. In the two chenopods, net salt accumulation (mainly NaCl) accounts for the initial changes in π_{sap} but in barley the situation is complex. $K^+ : Na^+$ exchange takes place in response to an increasing external NaCl concentration but the $[K^+ + Na^+]$ content per unit dry weight remains constant up to a threshold, indeed an osmotic pressure plateau is observed under certain conditions (Pitman, 1965; Storey and Wyn Jones, 1978a). Nevertheless, some osmotic adjustment is achieved by a decrease in the tissue water content. This drop in tissue hydration is rather typical of many Gramineae including halo-tolerant species and contrasts with the succulence induced in some dicotyledenous species, particularly chenopods, by moderate salinities (Jennings, 1976; Storey and Wyn Jones, 1979).

A further influence of relative humidity on salt tolerance has been observed in *Atriplex halimus* (Gale, Naaman and Poljakoff-Mayber, 1970) in which, at low relative humidities but not at high per cent of RH, moderate salinity (NaCl) promoted growth (*Figure 15.3*). This suggests that high tissue salt levels and therefore high π_{sap} are required for growth under conditions of high transpirational water loss.

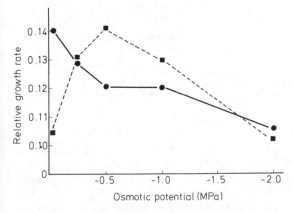

Figure 15.3   *Influence of atmospheric relative humidity (• = 63 per cent RH; ▪ = 27 per cent RH) on the salinity (NaCl) dependence of the relative growth rate of* Atriplex halimus. *(Data from Gale, Naaman and Poljakoff-Mayber, 1970)*

In some halophytes, as will be discussed later, organic solutes make a major contribution to π_{sap}. However, although this and the other physiological differences already discussed are important, there is no doubt that some measure of osmotic compensation is the normal response of plants to salinity. This may imply some degree of turgor homeostasis (Cram, 1976) and a constant turgor has been recorded in a few instances (Hoffman and Rawlins, 1971). However, growth has been widely found to be inhibited despite the maintenance of a constant $\Delta\pi$ between leaf sap and external solution at high salinity (*see Figure 15.7*). Many factors could give rise to this inhibition and some of them are discussed in a later section.

SPECIFIC ION STRESSES

There are many examples of glycophytes and even halophytes in which their tolerance to NaCl has been correlated with their ability to exclude Cl^- and/or Na^+ from the shoots (*Table 15.1*). Various physicochemical mechanisms are involved in the exclusion of these ions from the shoots (Jacoby, 1964), including active Na^+ efflux from roots (Jeschke, 1977) and absorption by specialized xylem parenchyma cells (Yeo *et al.*, 1977). Such observations strongly suggest that specific ion toxicity in the shoot is a crucial factor in salt stress.

Table 15.1 EXAMPLES OF SPECIES WHERE SODIUM CHLORIDE
TOLERANCE HAS BEEN RELATED TO ION EXCLUSION

Family	Species	Excluded ion	Reference
Cultivars			
Gramineae	Barley	Cl^-, (Na^+)	Greenway (1973)
			Storey and Wyn Jones (1978)
	Festuca rubra	Cl^-, Na^+	Rozema *et al.* (1978)
	Triticum aestivum	Cl^-, Na^+	Torres and Bingham (1973)
	Agropyron elongatum	Cl^-, Na^+	Shannon (1978)
Leguminoseae	Soya bean	Cl^-	Abel (1969)
			Läuchli and Wieneke (1979)
Rootstocks	Acacado	Cl^-	Downton (1978)
	Vitris vinerifera	Cl^-	Ehlig (1960)
			Downton (1977)
	Citrus crops	Cl^-, Na^+	Maas and Hoffman (1977)
	Stone-fruit crops	Cl^-, Na^+	Maas and Hoffman (1977)

Paradoxically, when the ion relations of halophytes and glycophytes are compared in a wider context, many halophytes are found to accumulate Na^+ and Cl^- in order to adjust osmotically. It seems that halophytes have higher Na^+ and lower K^+ affinities than glycophytes (Collander, 1941; Storey and Wyn Jones, 1979). Many examples have been recorded of increased growth and greater succulence in halophytes subject to Na^+ and/or Cl^- in the range 50–200 mol m^{-3} (Jennings, 1976; Flowers, Trobe and Yeo, 1977). High NaCl concentrations are of course inhibitory but, curiously, in several instances K^+ salts have been found to be more inhibitory than Na^+ salts (e.g. *Atriplex inflata* and *nummularia*: Ashby and Beadle, 1957; *Puccinellia peisonis*: Stelzer and Läuchli, 1977a). The ability of Na^+ apparently to replace, to a certain extent, K^+ in a range of halophytes has also been widely discussed (Marschner, 1971). Nevertheless, the apparently lower affinity of halophytes for K^+ (using the $S_{K:Na}$ as a criterion) is probably misleading and reflects, not a low absolute affinity for K^+, but a relatively high rate of Na^+ uptake. It has been noted (Wyn Jones and Storey, 1978) that in some plants, a constant or even increasing K^+ selectivity is maintained as the external salt concentration increases. There are also several reports of NaCl promoting increased K^+ uptake into the roots of halophytes (*Triglochin maritima*: Parham, 1971; Jefferies, 1973; *Suaeda monoica*: Storey and Wyn Jones, 1979).

Although the leaves of many halophytes contain high electrolyte concentrations, these plants are not uncontrolled ion accumulators. Many halophytes have salt glands or bladders (Lüttge, 1975) which selectively export Na^+ and Cl^- from the leaves (*Figure 15.4*). Scholander *et al.* (1962) observed that in mangroves the salt content of the xylem sap of species with salt glands was higher than that of species lacking these organs, which suggests that the salt content of the leaves could be controlled by alternative mechanisms in the leaves or the roots. Furthermore, a simple calculation based on the transpiration rates of halophytes (2–5 mg gFW^{-1} min^{-1}; although these are lower on average than those of glycophytes; Waisel, 1972)

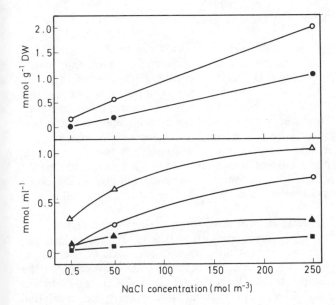

Figure 15.4   *Distribution of Na$^+$ ($\triangle$, $\blacktriangle$) and Cl$^-$ ($\circ$, $\bullet$, $\square$, $\blacksquare$) between salt glands (open symbols) and lamina (closed symbols) of* Atriplex spongiosa *grown under increasing salinity. Data in upper graph on dry weight basis and those in lower graph are expressed as concentrations in tissue water. (Data from Osmond* et al., *1969)*

and assuming a root solution salinity of only 100 mol m^{-3} NaCl, shows that the sap salt concentration would increase by about 10–20 mol m^{-3} per hour of transpiration. Clearly, excessively high sap salinities would build up very quickly unless uptake is tightly controlled.

Transpiration-dependent mass flow of ions to the root surface will increase the salt stress and place an extra strain on the exclusion mechanisms, particularly for Na$^+$ and Cl$^-$. The evolution of a second endodermis in *P. peisonis* (Stelzer and Läuchli, 1977b) and the decrease in the number of cortical cells and the increase in the width of the caparian strip found in a number of halophytes (Poljakoff-Mayber, 1975) may be seen as morphological adaptations to accommodate this problem. It would appear that both the apoplasmic and the symplasmic pathways of ion movement must be regulated. Control of apoplasmic flow by adcrustation in the cell walls is also a feature of the morphology of salt glands in halophytes (Lüttge, 1975).

Both glycophytes and halophytes therefore regulate the accumulation rates of specific ions and differ quantitatively not qualitatively in their ability to do this. Nevertheless it is a feature of certain halophytes that they can utilize electrolytes (including Na$^+$ and Cl$^-$) as major leaf osmotica without deleterious consequences. This leads to the well-recognized paradox that the cytoplasmic enzymes and organelles from halophytes and glycophytes alike are inhibited by the NaCl or KCl concentrations observed in the saps of these halophytes (Flowers, 1975; Osmond, 1976; Flowers, Trobe and Yeo, 1977; Wyn Jones *et al.*, 1977; Wyn Jones, Brady and Speirs, 1979).

ION IMBALANCE STRESS

It is possible that high concentration of Na^+, Cl^- or other ions found in saline soils, Mg^{2+}, SO_4^{2-} etc. may induce deficiencies in other essential nutrient ions such as K^+, or HPO_4^{2-} (HPO_4^-) or NO_3^- (Gauch, 1972; Levitt, 1972). The exchange of shoot K^+ by Na^+ has been noted earlier and this sparing of K^+ by Na^+ has been observed to be closely related to salt tolerance (Marschner, 1971). However, since K^+ has essential biochemical functions (*see* later) a decline in tissue K^+ beyond a specific level could produce deficiency. In this it may be necessary to distinguish between the initial exchange of K^+ for Na^+ at low salinities observed in tolerant species and the steady decline in shoot K^+ levels found in some glycophytes as the NaCl stress increases (Wyn Jones and Storey, 1978); the latter could well lead to K^+ deficiency.

Weigel (1968, 1970) found an inverse relationship between external NO_3^- and internal Cl^- (*see also* Torres and Bingham, 1973) but, in general, increased nitrogen fertilization will not diminish salt toxicity (Bernstein, Francois and Clarke, 1974). Sodium and calcium chlorides decreased the phosphorus concentration in corn leaves (Nieman and Clarke, 1976) but Bernstein, Francois and Clarke (1974) found that salinity induced lethal phosphate levels in some plants. Maas and Nieman (1978) suggested that salinity damages the mechanisms that control the intracellular phosphate concentration and the cell's energy charge.

Another potent source of nutrient imbalance is that increased Na^+ levels will lead to a displacement of Ca^{2+} from exchange sites on membranes and cell walls; these problems will be discussed later.

STRESS INTERACTIONS

The data discussed so far have not led to a clear definition of the primary lesion in salt stress. Indeed, further problems and paradoxes have been raised. While Cl^- and Na^+ exclusion from the shoots is the major tolerance mechanism in many glycophytes, some halophytes can accommodate high shoot Na^+ and Cl^- levels. Nevertheless, even in these tissues there is little evidence of a cytoplasmic biochemical tolerance of high inorganic electrolyte concentrations. This paradox has led to the hypotheses of electrolyte compartmentation and the accumulation of specific cytoplasmic compatible solubles; these, for simplicity, will be referred to hereafter as 'cytosolutes'. However, these hypotheses do not entirely resolve the specific ion toxicity-water stress dilemma.

Oertli (1968) pointed out that the rapid accumulation of salts in the cell walls of leaves could lead to the localized dehydration of tissue if the rate of xylem transport was not matched by the rate of cellular salt uptake. The transient accumulation of proline, a possible internal marker for salt stress (Hanson, Nelson and Everson, 1977; Wyn Jones and Storey, 1978), in leaves following salt stress (Ahmad and Wyn Jones, 1979) may be interpreted to agree with Oertli's concept. However, other types of specific ion/water stress interactions can be envisaged. As Ca^{2+} is important in maintaining the selective permeability of membranes and Na^+ frequently enhances membrane leakage rates (Wyn Jones and Lunt, 1967; Epstein, 1972), the displacement of Ca^{2+} by accumulated Na^+ in the cell walls might well influence the cations bound to the plasmalemma and increase the leakage across that membrane (Marschner and Mix, 1973; Poovaiah and Leopold, 1976). Since turgor maintenance and osmotic adaptation require solute accumulation, an increase in cellular permeability would impair this homeostatic process and could increase the energy and/or carbon demand for osmoregulation so much as to resemble a 'futile cycle'. Similarly, a decrease in root membrane selective permeability would lead to a catastrophic uptake of the very ions likely to increase toxicity. It is very relevant that differences in membrane lipid composition in tolerant and non-tolerant crops have been observed (Stuiver, Kuiper and Marschner, 1978) while Kuiper (1968) also found that the phosphatidyl choline content of grape roots was inversely related to Cl^- transport.

Further interactions between specific ion and water stresses can be envisaged. If the elastic modulus of the cell wall were increased by the cell wall salt-load, this might decrease turgor-dependent growth (Zimmerman, 1978), conversely, a drop in the elastic modulus would decrease the solute accumulation required to promote elongation. These stresses could also interact via salt-induced changes in the proton-buffering capacity of the cell wall (Demarty, Morvan and Thellier, 1978) or in the acid-lability of cell wall bonds. It may be relevant that, in the few reported data, the cation exchange capacities of the cell walls of halo-tolerant species are higher than those of glycophytes (barley leaf stress, $3.5\,\mu mol\,gFW^{-1}$, roots $2.0\,\mu mol\,gFW^{-1}$; beetroot slices $12\,\mu mol\,gFW^{-1}$; Atriplex leaf slices, $16\,\mu mol\,gFW^{-1}$: Walker and Pitman, 1976). A salt-induced change in root hydraulic conductivity would also affect water relationships but, although this was reported by O'Leary (1969), Shalhevet *et al.* (1976) failed to observe such an effect.

A close relationship between water stress and ion toxicity can therefore be readily visualized even without recourse to possible hormonal interactions, which lie outside the scope of this chapter. It seems that at a molecular or biochemical level, the rigorous distinction between specific ion and water stresses would appear to have little meaning. At present no discrete universal lesion can be recognized and indeed the scope for interaction and the complexity of the mechanisms involved in tolerance are so great, that there is no *prima facie* case for expecting there to be a single universal lesion.

The Role of Compartmentation in Tolerance

Solute compartmentation between cytosol and vacuole is probably an important facet of salt tolerance and, indeed, of many other aspects of the physiological biochemistry of plants. Flowers (1975) and Osmond (1976) surveyed the indirect evidence for compartmentation arising from the solute sensitivity of enzymes. Recently, Jeschke (1979) has reviewed the physiological evidence that K^+ is selectively retained in the cytoplasm while Na^+ is occluded in the vacuoles of cereals. *Figure 15.5* shows an example of these data in which Na^+ is very low in root tip cells of low vacuolation but is high in vacuolated tissue. In a complementary paper Wyn Jones, Brady and

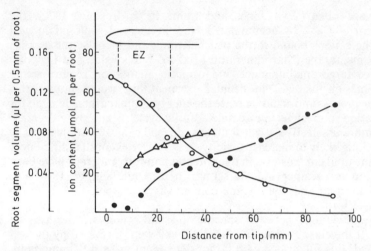

Figure 15.5   *The longitudinal profiles of the K^+ ($\circ$) and Na^+ ($\bullet$) content ($\mu mol\ ml$ per root) and the segment volume ($\triangle$) of aerated barley roots grown after germination for four days in $1\ mol\ m^{-3}\ Na^+$ solution. EZ = Zone of cell enlargement. (Redrawn from Jeschke and Stelter, 1976)*

Speirs (1979) discussed the extensive comparative biochemical evidence for a rather consistent inorganic electrolyte composition in the cytoplasms (cytosols) of eukaryotic cells and explored a possible biochemical basis for this pattern, namely protein synthesis. The consensus of these articles points to a cytoplasm which contains K^+ in the range 80–100 mol m^{-3} and which largely excludes free Na^+ and Cl^- ion (possibly as low as 30 or less mol m^{-3}): these latter ions and organic acids are preferentially concentrated in the vacuole (*Figure 15.6*).

I will not in this chapter reconsider the extensive data which gave rise to this model and which also lead to two further propositions: (a) that non-toxic cytosolutes are accumulated to maintain a water potential equilibrium across the tonoplast, and (b) that the tonoplast of some halophytes is better able to sustain ionic gradients and solute compartmentation than that of glycophytes. However, it should be emphasized that this model will account for several of the paradoxes discussed earlier: the salt sensitivity of enzymes and organelles from halophytes; the correlation of tolerance with Cl^- and Na^+ exclusion in many glycophytes while some halophytes to an extent accumulate these ions; the greater sensitivity of some halophytes to K^+ salts rather than Na^+ salts since the former may be more difficult to occlude in the vacuole (and possibly to exclude from the cell); the promotion of succulence (vacuole development) by Na^+ and/or Cl^- in some halophytes.

The accumulation of organic cytosolutes instead of inorganic electrolytes is clearly established in lower plants of low vacuolation (Hellebust, 1976; Kauss, 1977) and in the tissues of invertebrate animals (Prosser, 1973). The situations in vacuolated macro-marine algae is rather obscure but cytoplasmic osmotic adaptation in them may well resemble that in the micro-algae (Dickson, Wyn Jones and Pitman, unpublished data). In agreement with this concept, Wiencke and Läuchli (1978) observed the induction of vacuolation

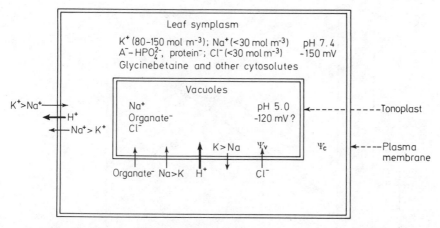

Figure 15.6 *Idealized model of solute compartmentation in higher plant cells accumulating sodium salts and glycinebetaine. The values quoted are intended to indicate a hypothetical 'norm' and not absolute unchangeable values. Thick arrow = active electrogenic transport. Thin arrow = selective transport (no energetic deduction).*

$$\Psi_v = \Psi_c$$
$$\Psi_v = \Psi^\pi_{Na\ salts}$$
$$\Psi_c = \Psi^\pi_{K\ salts} + \Psi^\pi_{cytosolutes} + \Psi^\pi_{other\ salts} + \Psi_{matrix}$$

(Based on data from Raven and Smith, 1977; Jeschke, 1979; Wyn Jones, Brady and Speirs, 1979)

by hypersaline stress in Porphyra and suggested that it occurs to accommodate the extra salts entering the cell under these conditions.

In higher plants, a range of organic cytosolutes have been proposed including glycinebetaine and possibly other betaines (Wyn Jones and Storey, 1981). An illustration of the physicochemical data which support this hypothesis for glycinebetaine acting as cytosolute is given in *Figure 15.7*, which shows the close correlation between osmotic adjustment and glycinebetaine accumulation in *Atriplex spongiosa*. Glycinebetaine at fairly high concentrations does not impair enzyme or organelle function and may have some protective capability (Shkedy-Vinkler and Avi-Dor, 1975; Larkum and Wyn Jones, 1979; Pollard and Wyn Jones, 1979). Since translation during protein synthesis may be a primary determinant of the cytosolic free ion concentrations, it has been important to determine the influence of glycinebetaine on this *in vitro* system (*Figure 15.8*). It will be seen that glycinebetaine is again relatively non-toxic up to 300–400 mol m^{-3}. Further evidence is available that cytosolic accumulation of glycinebetaine influences tonoplast Na$^+$ fluxes so as to increase the vacuolar Na$^+$ content (Ahmad, 1978; Ahmad and Wyn Jones, 1978). Thus an integrated regulation of cytosolic and vacuolar events may be tentatively suggested which allows both the maintenance of $\Delta\Psi = 0$ across the tonoplast and the steeper solute gradients in response to increased salt stress.

Proline has also been advocated as a cytosolute in halophytes (Stewart and Lee, 1974; Treichel, 1975). However, the accumulation of this compound is a near-universal response to water stress (Hsiao, 1973) and may indeed be a measure of internal water stress (Hanson, Nelson and Everson, 1977) so

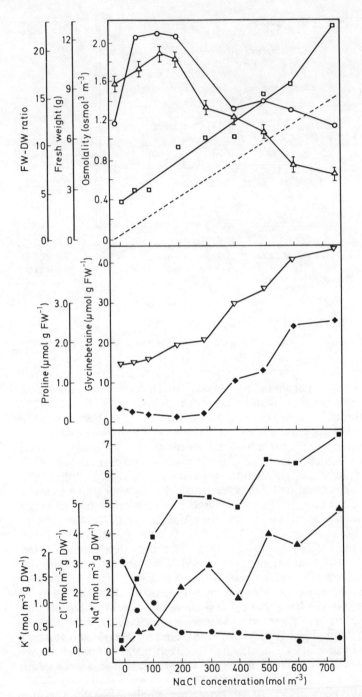

Figure 15.7 *Influence of growth at increasing NaCl salinity on growth and various other parameters of* Atriplex spongiosa. *△ = Fresh weight yield. ○ = Fresh weight/dry weight ratio. □ = Shoot sap osmolality. ▽ = Glycinebetaine. ♦ = Proline. ● = K⁺. ■ = Na⁺. ▲ = Cl⁻. - - - = Osmolality of the growth medium. (Data from Storey and Wyn Jones, 1979)*

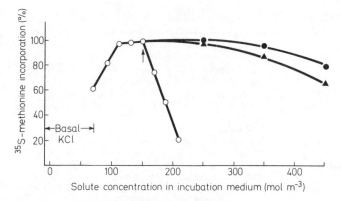

Figure 15.8 *Effect of potassium acetate, proline and glycinebetaine on* in vitro *protein synthesis in a wheat germ system primed with wheat leaf mRNA (Wyn Jones, Brady and Speirs, 1979). In addition to the basal KCl concentration (70 mol m⁻³), potassium acetate was added (○). At the concentration of potassium salts which gave the maximum rate of incorporation, glycinebetaine (●) or proline (▲) was added to give the final solute concentration. (Unpublished data of Speirs, Brady and Wyn Jones)*

interpretation of these data is complex. Nevertheless, proline may be the major cytosolute in the salt adaptation of some species, e.g. *Triglochin maritima* (Stewart and Lee, 1974) and *Puccinellia maritima* (Stewart *et al.*, 1979). Rather better-documented examples of proline accumulation exist in the micro-algae and bacteria (Hellebust, 1976). It may be noted that in *A. spongiosa (Figure 15.7)* proline is only accumulated at concentrations of NaCl which inhibit growth and which decrease the tissue succulence.

Schobert (1977) has proposed an alternative hypothesis for the function of proline in plants, in which she proposes that it protects the hydration of proteins by hydrophilic binding of water rather than acting as an osmotic solute. She has found significant proline-protein interaction (Schobert and Tschesche, 1978) but only at very high proline concentrations (4–6 mol³ m⁻³), which have not been reported in higher plants and would probably cause osmotic disequilibrium under normal conditions. The value of her hypothesis therefore seems restricted to very extreme conditions.

The list of possible cytosolutes in higher plants is increasing (*Table 15.2*) and includes sugar alcohols as well as dipolar amino acids and their derivatives. An interesting possible candidate for this role is dimethyl-sulphoniopropionate, which occurs widely in marine algae (Challenger, 1959; Dickson and Wyn Jones, unpublished data). Sulphonic compounds have been found in higher plants: Spartina species (Larher, Hamelin and Stewart, 1977; Storey and Wyn Jones, 1978b), Posidonia species (Wyn Jones and Hughes, unpublished data) and *Wedelia biflora* (Gorham and Storey, unpublished data). In the latter species and in *Ulva lactuca* the concentration changes in responses to salinity but this was not observed in Spartina (Storey, 1976; Stewart *et al.*, 1979). The particular solute accumulated in plants appears to have a strong taxonomic basis (Wyn Jones and Storey, 1981), for example, glycinebetaine (Chenopodiaceae), beta-alaninebetaine (Plumba-ginoceae), sorbitol (Plantaginaceae), prolinebetaine (Capparaceae). No

Table 15.2 EXAMPLES OF PUTATIVE CYTOSOLUTES IN VARIOUS SPECIES

Solutes	Family	Species	Reference
Glycinebetaine	Chenopodiaceae	*Suaeda monoica*	Storey and Wyn Jones (1979)
		Suaeda maritima	Flowers and Hall (1978)
		Atriplex spongiosa	Storey and Wyn Jones (1979)
		Spinaceae oleracea	Coughlan and Wyn Jones (1980)
		Beta vulgaris	Wyn Jones *et al.* (1977)
	Gramineae	*Spartina* × *townsendii*	Storey and Wyn Jones (1978b)
		Diplachne fusa	Sandhu *et al.* (1981)
Proline	Gramineae	*Puccinellia maritima*	Stewart *et al.* (1979)
		Triglochin maritima	Stewart and Lee (1974)
Sorbitol	Plantaginaceae	*Plantago maritima*	Ahmad, Larher and Stewart (1978)
		Plantago capensis	Gorham and Wyn Jones (unpublished data)
Prolinebetaine	Leguminoseae	*Medicago sativa*	Wyn Jones and Owen (1981)
Beta-dimethyl-sulphonio-propionate	Compositae	*Wedelia biflora*	Gorham and Storey (unpublished data)

doubt there will be further developments in this field in the next few years. The major problem in the study of potential cytosolutes in higher plants and vacuolated algae remains the absence of definitive data to prove the cellular localization of these water-soluble compounds.

While it is common to find high electrolyte concentrations in the shoots of halophytes (Waisel, 1972), some have relatively low concentrations of Na^+ and Cl^-. In these, very high levels of soluble sugars have been observed (*Table 15.3*; Albert and Popp, 1978). It is important to emphasize that these sugar levels contribute significantly to the osmotic pressure of the *total* leaf sap, unlike the putative cytosolutes, glycinebetaine or sorbitol, whose concentrations are such that they would only make a major contribution to osmotic adjustment if localized in the cytoplasm (cytosol). These sugar-accumulating, NaCl-excluding species are monocots predominantly from the Juncaceae and Cyperaceae and also some Gramineae. They also have leaf K^+ levels rather higher than those observed in high salt euhalophytes, typically Chenopodiaceae (*Table 15.4*). This pattern of solute accumulation closely resembles that found in many xerophytic species such as Eucalyptus species, in which sugars contribute massively to the sap osmotic potential (*Table 15.3*; Storey and Pitman, unpublished data). In such dry habitats there may frequently be a lack of available inorganic ions which could be used to contribute to the osmotic potential. Crawford (1978) emphasized the xeromorphic character of many marsh plants such as Juncus and Carex species, although adapted to anaerobic waterlogged soil and with no obvious water deficiency problems. He suggests that this reflects a need to reduce the mass flow to the roots of the phytotoxic chemicals—nitrite, Mn^{II}, Fe^{II} and sulphite ions—found in anaerobic soils. Reeds, rushes and sedges constitute the bulk of the vegetation mass of non-saline wet lands (Crawford, 1978) and it may be that the salt-tolerant species of these families have followed a different evolutionary path and retained a low capacity for ion accumulation

285

Table 15.3 EXAMPLES OF THE CONTRIBUTION OF SOLUTES TO OSMOTIC POTENTIAL OF HALOPHYTIC AND XEROPHYTIC PLANTS

Species	π (MPa)	K^+ (mol m^{-3})	Na^+ (mol m^{-3})	Mg^{2+} (mol m^{-3})	Ca^{2+} (mol m^{-3})	Sugars	Glycine-betaine	Amino acids
Salt marshes*								
Salicornia europ.	−2.0	56	458	—	—	12	34	9 (1 Pro)†
Spergularia spp.	−2.3	62	350	—	—	25	6	31 (7 Pro)
Triglochin maritima	−2.2	168	142	48	16	33	<4	— (81 Pro)
Juncus maritimus	−1.7	193	132	—	—	135	<4	33 (1–2 Pro)
Scirpus maritimus	−2.1	147	152	—	—	320	<4	45 (1 Pro)
Arid sandy soils**								
Acacia hakeoides	−2.0	200	3	75	30	260	<4	35
Eucalyptus socialis	−3.0	265	0.5	70	20	375	<4	10
Pittosporum phillyreoides	−2.2	355	35	100	50	365	<4	15

* Data from Gorham, Hughes and Wyn Jones (1980). ** Data from Storey and Pitman (unpublished data).
† Pro = Proline.

Table 15.4 SOLUTE CONCENTRATIONS AND π OF PHLOEM SAP FROM *ASTER TRIPODIUM* GROWING IN FRESH AND SEA WATER. (DATA FROM DOWNING, 1979)

	Phloem sap	
	Fresh water	Sea water
π (osmol kg^{-1})	599 ± 20	1544 ± 50
K$^+$ (mol m^{-3})	86	106
Na$^+$ (mol m^{-3})	0.35	31
Cl$^-$ (mol m^{-3})	5.4	28.3
Sucrose equivalents (mol m^{-3})	302	672
Approx. π due to recorded solutes (osmol kg^{-1})	460	920

in contrast to the euhalophytic types, which are often succulent meso- or xeromorphs (Waisel, 1972). Even in these latter species the regulation of salt uptake is essential, as has been discussed, and very high transpiration rates would only increase the salt load in the rhizosphere. It is therefore not surprising that so many halophytes are also C$_4$ plants.

It is becoming increasingly clear that various integrated strategies have evolved to produce salt tolerance, which involve biochemical, physiological and morphological features. It may be suggested that the Chenopodiaceae afford an example of one such strategy, although of course with variants. These plants exploit high concentrations of NaCl or organate in order to osmoregulate in their vacuoles (and possibly to induce succulence) while K$^+$ is concentrated in the cytoplasm together with the cytosolute glycinebetaine, which itself modifies the tonoplast ionic gradients. Such species may need to discriminate *less* strongly in their roots against Na$^+$ (and Cl$^-$) than high K$^+$ species, e.g. *Puccinellia peisonis* (Stelzer and Läuchli, 1978) or glycophytes such as barley (Jeschke and Stelter, 1976). In Puccinellia the morphological adaptations might minimize transpiration-dependent Na$^+$ and Cl$^-$ uptake but in *Atriplex spongiosa* this was permitted (Storey and Pitman, unpublished data). In *A. halimus* the promotion of growth by NaCl at low per cent RH was observed (Gale, Naaman and Poljakoff-Mayber, 1970). This was ascribed to improved water relationships due to a decrease in transpiration, partly due to an increase in stomatal resistance but more to an increase in mesophyll resistance under conditions likely to increase the NaCl content of the leaves. This took place without any marked effect on photosynthesis (Kaplan and Gale, 1972). Another factor which may distinguish the adaptational strategy of *P. peisonis* from members of the Chenopodiaceae is that it accumulates proline and not glycinebetaine. Accumulators of the latter are characteristically high in Na$^+$ (Storey, Ahmad and Wyn Jones, 1977). However, *Plantago maritima* (Ahmad, Larher and Stewart, 1978) affords an example of a high Na$^+$/K$^+$ plant which accumulates not glycinebetaine but sorbitol to discourage overenthusiastic generalizations. By analogy with mangroves species (Scholander *et al.*, 1962), the degree of ion discrimination and exclusion by the roots may also be related to the presence or absence of salt glands or other Na$^+$- and Cl$^-$-specific export systems in the shoots.

A very wide range of morphological, physiological and biochemical adaptations related to salt tolerance may therefore be described in different species. Nevertheless, some unifying concepts may be tentatively proposed. All the tissues of the plant must maintain a balance between the demands, potentially conflicting demands, of water balance, turgor homeostasis and selective ion accumulation. At a cellular level, it may be proposed that all the strategies lead to turgid leaf cells whose cytoplasms have a high K^+/Na^+ free ion ratio and a moderate total inorganic ion content. Within this framework some cells or even whole leaves may develop excess salt loads to protect the growing apices from salt, mainly NaCl, toxicity. Greenway (1973) has emphasized the importance of growth itself as a mechanism to avoid salt toxicity. It is also probable that Cl^- is partially excluded from the cytoplasms and indeed Cl^- may be as cytotoxic as Na^+ (Wyn Jones, Brady and Speirs, 1979). Some cytochemical studies (e.g. Harvey, Flowers and Hall, 1976) have suggested that Cl^- is mainly accumulated in the vacuole of halophytes. However other work (e.g. Stelzer and Läuchli, 1978) has shown Cl^- in the lumen of the endoplasmic reticulum and the plasmadesmata. Thus its subcellular distribution is far from clear and there may be compartmentation at a more sophisticated level than simply cytoplasm versus vacuole. It is not unlikely that a rather similar situation appertains in root cells, which of course generate a lower osmotic pressure than leaf cells, but a critical examination of these data must be excluded from this chapter for lack of space.

Some Implications of High and Low Salt Strategies

Since Na^+ and Cl^- and other ions are common in saline environments, they can be regarded as potentially 'cheap' although dangerously toxic osmotica for the plants. If the Na^+ concentration in the external solution exceeds that in the cell, Na^+ will enter a cell of conventional membrane potential (approximately -100 mV) passively. There are indeed many examples in plants (Pitman, 1976; Raven, 1976) where energy is expended to pump Na^+ out of cells. Anion accumulation will require an expenditure of energy either to take in Cl^- or another anion against its electrochemical potential gradient, or to synthesize malate or another organic anion from fixed CO_2. Since the osmotic coefficient of NaCl is about 0.92 (Robinson and Stokes, 1955), it is possible to compute the approximate energy cost of the accumulation of an osmole of NaCl assuming the ions move independently and that 1 mole of adenosine triphosphate (ATP) is required to move 1 mole of Cl^-. This would suggest a maximum of 0.54 moles of ATP required per osmole of NaCl in contrast to the approximately 13 moles of ATP required for 1 osmole of K_2^+ malate^{2-} (assuming an osmotic coefficient of 0.8), if malate is synthesized from gaseous CO_2 via the Calvin cycle and phosphoenol pyruvate (PEP) carboxylase.

In contrast, sugar accumulation is much more energy- and carbon-demanding. The synthesis of 1 mole (1 osmole) of glucose from CO_2 requires 54 moles of ATP (Mahler and Cordes, 1969). Furthermore, the presence of 300 mol m^{-3} of C_6 sugars in the cell sap could account for 20–30 per cent of the dry weight of the tissue. This analysis implies that osmotic adjustment by salt accumulation is less energy- and carbon-demanding and in these terms more compatible with high yields. A further factor may be the ability of certain species such as chenopods (Storey and Wyn Jones, 1979) to grow at lower tissue K^+ levels; this could be interpreted as an ability to compartment K^+ in the cytoplasm and Na^+ in the vacuole and is related to the concept of Na^+ sparing of K^+, noted previously (Marschner, 1971; Mengel and Kirkby, 1978). However, the palatability of high Na/K fodder to animals may lead to another problem in the agronomic exploitation of such species.

Many of these high electrolyte species are thought to utilize nitrogen dipoles as cytosolutes, and this may itself create an additional nutritional demand in low nitrogen habitats (Storey, Ahmad and Wyn Jones, 1977; Stewart *et al.*, 1979). It is interesting to note that nitrogen fixation has been reported to be associated with a number of estuarine grasses (Jones, 1974; Portiguin, 1978) but not, as far as I am aware, with members of the Chenopodiaceae. Jones (1974) suggests that fixation by blue-green algae may contribute to the nitrogen economy of the lower *Salicornia* marsh. However, from a study of the effects of added inorganic nutrients on a salt marsh, Jefferies and Perkins (1977) concluded that primary productivity on the marsh was not limited by nitrogen shortage *per se* but by water and specific ion stresses.

In the tolerant glycophyte barley, the grain salt levels are lower than those of the vegetative tissue (Greenway *et al.*, 1965). The developing grain is fed by the phloem, and Downing (1979) has demonstrated that even in the halophyte *Aster tripolium* both Na^+ and Cl^- are largely excluded from the phloem (*Table 15.4*). These data are of outstanding interest in that Raven (1977) has concluded that the phloem contents are modified cytosol and the data obtained with *Aster tripolium* on K^+, Na^+ and Cl^- concentrations are in excellent agreement with the compendium of data from the cytoplasms of eukaryotic cells previously considered. This evidence strongly supports the hypothesis that these ions are partially excluded from the cytoplasms of halophytes, advanced in this chapter (*see also* Wyn Jones, Brady and Speirs, 1979). Further, they lead to the supposition that phloem-fed tissues should be relatively low in Na^+ and Cl^- but have a very large requirement for other cytosolutes in order to maintain their water balance. Thus it may be possible to obtain relatively salt-free grain from euhalophytic crops. However, since the growing apical meristems are also phloem fed, the problems of turgor regulation in these cells may be more difficult and may place greater demands on the carbon and nitrogen economies than that in more mature cells.

Acknowledgements

I am very grateful to Dr R. Storey and Professor M. G. Pitman for permission to quote their unpublished data and for invaluable discussions on this topic over recent years. I would also like to acknowledge the receipt of an NERC award which is supporting the work quoted as our own unpublished data, and my debt to Dr John Gorham for his ready cooperation in preparing this chapter.

References

ABEL, G. H. (1969). *Crop Sci.*, **9**, 697–698
AHMAD, I., LARHER, F. and STEWART, G. R. (1978). *New Phytol.*, **82**, 671–678
AHMAD, N. (1978). Ph D Thesis, University of Wales, Cardiff
AHMAD, N. and WYN JONES, R. G. (1978). *Proc. Fed. Eur. Soc. Pl. Physiol.*, *Edinburgh*, p. 190
AHMAD, N. and WYN JONES, R. G. (1979). *Pl. Sci. Lett.*, **15**, 231–237
ALBERT, R. and POPP, M. (1978). *Ecol. Plant.*, **13**, 27–42
ASHBY, W. C. and BEADLE, N. C. W. (1957). *Ecology*, **38**, 344–352
BERNSTEIN, L. (1961). *Am. J. Bot.*, **48**, 909–918
BERNSTEIN, L., FRANCOIS, L. C. and CLARKE, R. A. (1974). *Agron. J.*, **66**, 412–421
BERNSTEIN, L. and HAYWARD, H. E. (1958). *Ann. Rev. Pl. Physiol.*, **9**, 25–46
BINGHAM, F. T., FENN, L. B. and OERTLI, J. J. (1968). *Soil Sci. Soc. Am., Proc.*, **32**, 249–252
BLOOM, A. J. (1979). *Pl. Physiol.*, **63**, 749–753
CHALLENGER, F. (1959). *Aspects of the Organic Chemistry of Sulphur*, pp. 32–72. Butterworths, London
COLLANDER, R. (1941). *Pl. Physiol.*, **16**, 691–720
COUGHTON, S. J. and JORIES, R. G. W. (1980). *J. exp. Bot.*, **31**, 883–893
CRAM, W. J. (1976). In *Encyclopaedia of Plant Physiology NS. 2 Transport*, Vol. B, pp. 284–316. Ed. by U. Lüttge and M. G. Pitman. Springer-Verlag, Berlin
CRAWFORD, R. M. M. (1978). *Naturwiss.*, **65**, 194–201
DEMARTY, M., MORVAN, C. and THELLIER, M. (1978). *Pl. Physiol.*, **62**, 477–481
DOWNING, N. (1979). PhD Thesis, Cambridge University
DOWNTON, W. J. S. (1977). *Sci. Hortic.*, **7**, 249–253
DOWNTON, W. J. S. (1978). *Austr. J. agric. Res.*, **29**, 523–534
EHLIG, C. F. (1960). *Prod. Am. Soc. hortic. Sci.*, **76**, 323–331
EPSTEIN, E. (1972). *Mineral Nutrition of Plants: Principles and Perspectives.* Wiley, New York
EPSTEIN, E. and NORLYN, J. D. (1977). *Science*, **197**, 249–251
FLOWERS, T. J. (1975). In *Ion Transport in Plant Cells and Tissues*, pp. 309–334. Ed. by D. A. Baker and J. L. Hall. North-Holland, Amsterdam
FLOWERS, T. J. and HALL, J. L. (1978). *Ann. Bot.*, **42**, 1057–1063
FLOWERS, T. J., TROBE, P. F. and YEO, A. R. (1977). *Ann. Rev. Pl. Physiol.*, **28**, 89–121
GALE, J., NAAMAN, R. and POLJAKOFF-MAYBER, A. (1970). *Austr. J. biol. Sci.*, **23**, 947–952
GAUCH, H. G. (1972). *Inorganic Plant Nutrition*, pp. 395–426. Dowden Hutchinson and Ross, Stroudsburg, PA
GAUCH, H. G. and WADLEIGH, C. H. (1945). *Soil Sci.*, **59**, 139–153
GORHAM, J., HUGHES, U. and WYN JONES, R. E. (1980). *Pl. Cell Env.*, **3**, 309–318
GREENWAY, H. (1965). *Austr. J. biol. Sci.*, **18**, 249–268
GREENWAY, H. (1973). *J. Austr. Inst. agric. Sci.*, **39**, 24–34
GREENWAY, H., GUNN, A., PITMAN, M. G. and THOMAS, D. A. (1965). *Austr. J. biol. Sci.*, **18**, 525–540
HANSON, A. D., NELSON, C. E. and EVERSON, E. H. (1977). *Crop Sci.*, **17**, 720–726

HARVEY, D. M. R., FLOWERS, T. J. and HALL, J. L. (1976). *New Phytol.*, **77**, 319–323

HELLEBUST, J. A. (1976). *Ann. Rev. Pl. Physiol.*, **27**, 485–505

HOFFMAN, G. J. and JOBES, J. A. (1978). *Agron. J.*, **70**, 765–769

HOFFMAN, G. J. and RAWLINS, S. L. (1971). *Agron. J.*, **63**, 877–880

HSIAO, T. C. (1973). *Ann. Rev. Pl. Physiol.*, **24**, 519–570

JACOBY, B. (1964). *Pl. Physiol.*, **39**, 445–449

JEFFERIES, R. L. (1973). In *Ion Transport in Plants*, pp. 297–321. Ed. by W. P. Anderson. Academic Press, London

JEFFERIES, R. L. and PERKINS, N. (1977). *J. Ecol.*, **65**, 867–882

JENNINGS, D. H. (1976). *Biol. Rev.*, **51**, 543–586

JESCHKE, W. D. (1977). In *Regulation of Cell Membrane Activities in Plants*, pp. 63–78. Ed. by E. Marrè and O. Cifferi. North-Holland, Amsterdam

JESCHKE, W. D. (1979). In *Recent Advances in the Biochemistry of Cereals*, pp. 37–62. Ed. by D. L. Laidman and R. G. Wyn Jones. Academic Press, London

JESCHKE, W. D. and STELTER, W. (1976). *Planta*, **128**, 107–112

JONES, K. (1974). *J. Ecol.*, **62**, 553–565

KAPLAN, A. and GALE, J. (1972). *Austr. J. biol. Sci.*, **25**, 895–903

KAUSS, H. (1977). In *International Review of Biochemistry*, Vol. 13: *Plant Biochemistry* II, pp. 199–140. Ed. by D. H. Northcote. University Park Press, Baltimore

KUIPER, P. J. C. (1968). *Pl. Physiol.*, **43**, 1367–1371

LARHER, F., HAMELIN, J. and STEWART, G. P. (1977). *Phytochem.*, **16**, 2019

LARKUM, A. W. D. and WYN JONES, R. G. (1979). *Planta*, **145**, 393–394

LÄUCHLI, A. and WIENEKE, J. (1979). *Z. Pfl.-ernähr. Bodenk.*, **142**, 3–13

LEVITT, J. (1972). *Responses of Plants to Environmental Stresses*. Academic Press, New York

LÜTTGE, U. (1975). In *Ion Transport in Plant Cells and Tissues*, pp. 335–375. Ed. by D. A. Baker and J. L. Hall. North-Holland, Amsterdam

MAAS, E. U. and HOFFMAN, G. J. (1977). *J. Irr. Drain. Div.*, **103**, 115–134

MAAS, E. U. and NIEMAN, R. H. (1978). In *Crop Tolerance to Sub-optimal Land Conditions*, pp. 277–298. American Society of Agronomy

MAHLER, H. R. and CORDES, E. H. (1969). *Biological Chemistry*, pp. 498–502. Harper & Row, New York

MARSCHNER, H. (1971). *Potassium Biochem. Physiol. Coll. of the Int. Potash Inst.*, **8**, 50–63

MARSCHNER, H. and MIX, G. (1973). *Z. Pfl.-ernähr. Bodenk.*, **136**, 203–219

MENGEL, K. and KIRKBY, E. A. (1978). *Principles of Plant Nutrition*. International Potash Institute, Berne

MOHAMMAD, S. (1978). Workshop on Membrane Biophysics and Development of Salt Tolerance in Plants. University of Agriculture, Faisalabad, Pakistan

NIEMAN, R. H. and CLARKE, R. A. (1976). *Pl. Physiol.*, **57**, 157–161

OERTLI, J. J. (1968). *Soil Sci.*, **105**, 216–222

O'LEARY, J. W. (1969). *Israel J. Bot.*, **18**, 1–9

O'LEARY, J. W. (1975). *Pl. Soil*, **42**, 717–721

OSMOND, C. B. (1976). In *Encyclopaedia of Plant Physiology*, N.S. Vol. II: *Transport in Plants*, Part A: *Cells*, pp. 347–372. Ed. by U. Lüttge and M. G. Pitman. Springer-Verlag, Berlin

OSMOND, C. B. (1978). *Ann. Rev. Pl. Physiol.*, **29**, 379–414

OSMOND, C. B., LÜTTGE, U., WEST, K. R., PALLAGHY, C. K. and SHACHAR-HILL, B. (1969). *Austr. J. biol. Sci.*, **22**, 797–814

PARHAM, M. R. (1971). PhD Thesis, University of East Anglia (quoted in Flowers, Trobe and Yeo, 1977)

PITMAN, M. G. (1965). *Austr. J. biol. Sci.*, **18**, 10–24

PITMAN, M. G. (1975). In *Ion Transport in Plant Cells and Tissues*, pp. 267–308. Ed. by D. A. Baker and J. L. Hall. Elsevier, Amsterdam

PITMAN, M. G. (1976). In *Encyclopaedia of Plant Physiology*, N.S. Vol. II: *Transport in Plants*, Part B: *Tissues and Organs*, pp. 95–127. Ed. by U. Lüttge and M. G. Pitman. Springer-Verlag, Berlin

PITMAN, M. G. and CRAM, W. J. (1977). In *Integration of Activity in the Higher Plant*, pp. 391–424. Ed. by D. H. Jennings. Cambridge University Press, Cambridge

POLJAKOFF-MAYBER, A. (1975). In *Plants in Saline Environments*, pp. 97–117. Ed. by A. Poljakoff-Mayber and J. Gale. Springer-Verlag, Berlin

POLLARD, A. and WYN JONES, R. G. (1979). *Planta*, **144**, 291–298

PONNAMPERUMA, F. N. (1977). In *Plant Responses to Salinity and Water Stress*, p. 32. Ed. by W. J. S. Downton and M. G. Pitman. Association for Sciences Co-operation in Asia, Mildura, Australia

POOVAIAH, D. W. and LEOPOLD, A. C. (1976). *Pl. Physiol.*, **58**, 182–185

PORTIGUIN, D. G. (1978). *Aqu. Bot.*, **4**, 193–210

PROSSER, C. L. (1973). *Comparative Animal Physiology*. W. B. Saunders, Philadelphia

RAVEN, J. A. (1976). In *Encyclopaedia of Plant Physiology*, N.S. Vol. II: *Transport in Plants*, Part A: *Cells*, pp. 129–188. Ed. by U. Lüttge and M. G. Pitman. Springer-Verlag, Berlin

RAVEN, J. A. (1977). *New Phytol.*, **79**, 465–480

RAVEN, J. A. and SMITH, F. A. (1977). In *Regulation of Cell Membrane Activities in Plants*, pp. 25–40. Ed. by E. Marrè and O. Ciferri. North-Holland, Amsterdam

ROBINSON, R. A. and STOKES, R. H. (1955). *Electrolyte Solutions*. Butterworth, London

ROZEMA, J., ROSEMA-DIJST, E., FREIJSEN, A. H. J. and HUBER, J. J. L. (1978). *Oecologia*, **34**, 329–341

SANDHU, G. R., ASLAM, Z., SALIM, M., SATTAR, A., QURESHI, R. H. and WYN JONES, R. G. (1981). *Pl. Cell Env.* (submitted)

SCHOBERT, B. (1977). *J. theor. Biol.*, **68**, 17–26

SCHOBERT, B. and TSCHESCHE, H. (1978). *Biochim. Biophys. Acta*, **541**, 270–277

SCHOLANDER, P. F., HAMMEL, H. T., HEMMINGSEN, E. and GAREY, W. (1962). *Pl. Physiol.*, **37**, 722–729

SHALHEVET, J., MAAS, E. V., HOFFMAN, G. J. and OGATA, J. (1976). *Physiol. Plant.*, **38**, 224–232

SHANNON, M. C. (1978). *Agron. J.*, **70**, 719–722

SHKEDY-VINKLER, C. and AVI-DOR, Y. (1975). *Biochem. J.*, **150**, 219–226

STELZER, R. and LÄUCHLI, A. (1977a). *Z. Pfl.-physiol.*, **83**, 35–42

STELZER, R. and LÄUCHLI, A. (1977b). *Z. Pfl.-physiol.*, **84**, 95–108

STELZER, R. and LÄUCHLI, A. (1978). *Z. Pfl.-physiol.*, **88**, 437–448

STEWART, G. R., LARHER, F., AHMAD, I. A. and LEE, J. A. (1979). In *Ecological Processes in Coastal Environments*, pp. 211–219. Ed. by R. L. Jefferies and A. J. Davy. Blackwell, Oxford

STEWART, G. R. and LEE, J. A. (1974). *Planta*, **120**, 279–289
STOREY, R. (1976). PhD Thesis, University of Wales, Cardiff
STOREY, R., AHMAD, N. and WYN JONES, R. G. (1977). *Oecologia*, **27**, 319–332
STOREY, R. and WYN JONES, R. G. (1978a). *Austr. J. Pl. Physiol.*, **5**, 801–816
STOREY, R. and WYN JONES, R. G. (1978b). *Austr. J. Pl. Physiol.*, **5**, 831–838
STOREY, R. and WYN JONES, R. G. (1979). *Pl. Physiol.*, **63**, 156–162
STROGONOV, B. P. (1964). *Physiological Basis of Salt Tolerance of Plants (as Affected by Various Types of Salinity)*. Akad. Nauk. SSSR. Translated from Russian, Israel Progressive Science Translation, Jerusalem
STUIVER, C. E. E., KUIPER, P. J. C. and MARSCHNER, H. (1978). *Physiol. Plant.*, **42**, 124–128
TORRES, B. C. and BINGHAM, F. T. (1973). *Soil Sci. Soc. Am., Proc.*, **37**, 711–715
TREICHEL, S. (1975). *Z. Pfl.-physiol.*, **76**, 56–68
WAISEL, Y. (1972). *Biology of Halophytes*. Academic Press, New York
WALKER, N. A. and PITMAN, M. G. (1976). In *Encyclopaedia of Plant Physiology*, N.S. Vol. II: *Transport in Plants*, Part A: *Cells*, pp. 93–128. Ed. by U. Lüttge and M. G. Pitman. Springer-Verlag, Berlin
WEIGEL, R. C. Jr (1968). MS Thesis, University Maryland (cited by Gauch, 1972)
WEIGEL, R. C. Jr (1970). PhD Thesis, University Maryland (cited by Gauch, 1972)
WIENCKE, C. and LÄUCHLI, A. (1978). *Ninth Int. Congr. Electr. Micros., Toronto*, **2**, 412–413
WINTER, K. and VON WILLERT, D. J. (1972). *Z. Pfl.-physiol.*, **267**, 166–170
WYN JONES, R. G., BRADY, C. J. and SPEIRS, J. (1979). In *Recent Advances in the Biochemistry of Cereals*, pp. 63–104. Ed. by D. L. Laidman and R. G. Wyn Jones. Academic Press, London
WYN JONES, R. G. and LUNT, O. R. (1967). *Bot. Rev.*, **33**, 407–425
WYN JONES, R. G. and OWEN, E. D. (1981). *Medicago. sativa.* in preparation
WYN JONES, R. G. and STOREY, R. (1978). *Austr. J. Pl. Physiol.*, **5**, 817–829
WYN JONES, R. G. and STOREY, R. (1981). In *Physiology and Biochemistry of Drought Resistance in Plants*. Ed. by L. G. Paleg and D. Aspinall. Academic Press, London
WYN JONES, R. G., STOREY, R., LEIGH, R. A., AHMAD, N. and POLLARD, A. (1977). In *Regulation of Cell Membrane Activities in Plants*, pp. 121–136. Ed. by E. Marrè and O. Ciferri. Elsevier, Amsterdam
YEO, A. R., KRAMER, D., LÄUCHLI, A. and GULLASCH, I. (1977). *J. exp. Bot.*, **28**, 17–30
ZIMMERMAN, U. (1978). *Ann. Rev. Pl. Physiol.*, **29**, 121–148

16

AIR POLLUTION AND PLANT PRODUCTIVITY

M. H. UNSWORTH
*Department of Physiology and Environmental Studies, University of
Nottingham, UK*

Introduction

It is appropriate that this topic has been included in a session concerned with
the influences of environmental stresses on plant productivity. In the past
there has been a tendency for effects of air pollution to be considered as a
branch of plant pathology, and some physiologists seem to have implied that
the subject was of little interest from either a fundamental or an applied
basis. In fact by neglecting the influence of air pollution in physiological
experiments, plant scientists have sometimes reached some surprising
conclusions. For example, Brown and Escombe (1902) investigated the
influences of atmospheric CO_2 concentrations on the growth of a range of
plant species and found that plants in raised CO_2 concentrations lost their
leaves, aborted flowers and generally developed abnormally. They concluded
that all plants in the experiment:

> 'appear to be accurately "tuned" to an atmospheric environment of 300
> parts of CO_2 per million, and that the response which they make to slight
> increases in this amount are in a direction altogether unfavourable to their
> growth and reproduction. It is not too much to say that a comparatively
> sudden increase of CO_2 in the air to an extent of but two or three times the
> present amount, would result in the speedy destruction of nearly all our
> flowering plants.'

The additional CO_2 in Brown and Escombe's experiments was generated
by the action of hydrochloric acid on marble chips, so it is probable that
gaseous pollutants were also produced and caused the effects which may have
set back the introduction of CO_2 enrichment in glasshouse production by
several decades! Pollution in glasshouses will figure again, later in this paper.
Visible damage to vegetation by air pollution has been recorded for at least
100 years, and it was spectacular examples of such effects around smelters in
North America and Europe, where entire ecosystems were destroyed over
thousands of hectares, that stimulated much of the early research on responses
to pollutants. Such work concentrated on extreme responses, in particular
acute injury—the collapse of cells and destruction of leaf area, and chronic
injury—chlorosis and early senescence. The existence of a third form of
injury, invisible injury, has been hotly disputed since it was first proposed by
Wislicenus (1914) and Stoklasa (1923). Reviewing the subject in 1951,

Thomas emphasized what was probably the majority view on invisible injury at the time: 'No experimental support for this theory has appeared, but on the contrary, a large amount of experimental work has been done which demonstrates that this theory is entirely without foundation.'

In the frequent reviews since then (Heck and Brandt, 1977; Hallgren, 1978; Ormrod, 1978) it has become increasingly clear that the situation is less clear-cut than Thomas suggested. The aim of this paper is to summarize how recent work with a number of air pollutants is revealing responses with important implications for agricultural production.

Air Pollutants and their Concentrations

To set the scene, *Table 16.1* lists some of the most important air pollutants in the UK and indicates typical concentrations at unpolluted sites and in urban and rural areas.

Sulphur dioxide (SO_2) arises predominantly from the combustion of fossil fuels and from smelting ores. Detailed records of SO_2 in the UK are published regularly by the Department of the Environment. Oxides of nitrogen, NO_x (NO_2 and NO), are produced in the combustion of hydrocarbons and so are important constituents of vehicle exhaust gases (Derwent and Stewart, 1973). They are also produced when hydrocarbons are burnt to provide CO_2 enrichment in glasshouses (Capron and Mansfield, 1975). Ozone is generated as a result of complex photochemical reactions in the atmosphere when oxides of nitrogen and hydrocarbons are exposed to bright sunlight (Cox *et al.*, 1975). The daily mean concentrations in *Table 16.1* should be regarded only as order of magnitude indications of ambient concentrations. Actual concentrations observed will depend critically on atmospheric mixing (a function of wind speed and temperature), wind direction and topography, so that short period fluctuations, over periods of about three minutes, may exceed the mean by a factor of 5 or more (Martin and Barber, 1973).

Table 16.1   TYPICAL DAILY MEAN CONCENTRATIONS OF AIR POLLUTANTS

Pollutant	Concentration $ppb\ (=nl/l=10^{-3}\ ppm)$	
Sulphur dioxide		
Background	<5	
Urban	50–100	
Rural	5–50	
Ozone		
Background	20–40	
Photochemical smog	100–200	
Oxides of nitrogen	NO_2	*NO*
Background	1–4	~2
Urban	10–50	10–50
Glasshouse	~100	300–400

SULPHUR DIOXIDE AND PLANT GROWTH

Before the late 1960s, most studies in which plants were exposed to SO_2 used very high SO_2 concentrations (> 1 ppm), relevant to sites close to point sources with inefficient chimneys, but irrelevant to modern sources or to the spatially distributed pollution common in industrialized regions of Europe and North America. In addition, the conditions of exposure of laboratory plants often allowed little air movement so that rates of uptake of SO_2 were severely restricted (Mansfield, 1973). Some of the first experiments in which effects of the urban atmosphere on plant growth were studied were designed by Bleasdale (1973) in Manchester, 1952. He grew S23 ryegrass in glasshouses, one of which received ambient air while the other received air scrubbed clean of pollutants. In none of his experiments was there significant visible injury, but there were large reductions in yield in the polluted air. *Table 16.2* summarizes one set of his results obtained over 120 days in winter when the daily mean SO_2 concentration was always below 90 ppb (and in fact exceeded 60 ppb on only four days). Substantial yield reductions were found in plants growing in both high and low fertility soils.

Table 16.2 GROWTH OF S23 RYEGRASS IN AMBIENT AND FILTERED AIR (FROM BLEASDALE, 1973)

	Dry weight (mg shoots per plant)	
	High fertility soil	Low fertility soil
Filtered air	99	73
Ambient air	72	59
% reduction	27	19

Plants were grown in glasshouses in Manchester, over 120 days in winter. Daily mean SO_2 concentration did not exceed 90 ppb, and exceeded 60 ppb on only four days.

Although Bleasdale's experiments suggested that 'invisible injury' had occurred, any one of several pollutants may have been responsible, or interactions between pollutants may have taken place. Experiments with a wide range of species in the German Ruhr, reviewed by Guderian (1977), used either 'filtered' versus 'ambient' air cabinets, or natural gradients of pollutant concentration to demonstrate significant yield reductions in polluted air. Later, more controlled experiments were designed (Bell and Clough, 1973) in which all plants were grown in filtered air, and SO_2 was added to one set of 'treated' plants. *Table 16.3* summarizes some of the results of Bell and Clough for S23 ryegrass grown for 180 days over winter in unheated cabinets out of doors. The mean SO_2 concentration was 67 ppb in the treatment and 3 ppb in the control. As well as large reductions in productivity there were striking decreases in the leaf area duration of the treated plants.

Since this important demonstration that low concentrations of SO_2 alone could reduce growth of S23 ryegrass there have been numerous similar studies, some of which are summarized in *Table 16.4*. The table shows that, even with this cultivar, there is still no clear picture of the response of yield

Table 16.3 FINAL YIELDS OF S23 RYEGRASS GROWN FOR 180 DAYS OVER WINTER IN AIR CONTAINING EITHER 67 ppb OR 3 ppb SO$_2$. (FROM BELL AND CLOUGH, 1973)

	*% change**				
	Tillers	*Living leaves*	*Dead leaves*	*Stubble*	*Leaf area*
Numbers	−41	−45	+88	−	−51
Dry weight	−	−52	+78	−55	−

*Tabulated values are percentage change ((SO$_2$ − control)/control), i.e. negative values indicate reduction in SO$_2$ treatment.

Table 16.4 GROWTH REDUCTIONS IN S23 PERENNIAL RYEGRASS EXPOSED TO SO$_2$ COMPARED WITH CONTROLS IN CLEAN AIR. (AFTER HORSMAN, ROBERTS AND BRADSHAW, 1979)

SO$_2$ concentration (ppb)	*Duration of experiment (days)*	*% reduction in dry weight of living leaves*	*Reference*
250	56	38	Horsman, Roberts and Bradshaw (1979)
230	35	0	Horsman, Roberts and Bradshaw (1979)
140	77	24	Lockyer, Cowling and Jones (1976)
120	28	22	Ashenden and Mansfield (1977)
120	28	0	Ashenden and Mansfield (1977)
120	63	46	Bell and Clough (1973)
70	77	0	Lockyer, Cowling and Jones (1976)
67	180	52	Bell and Clough (1973)
46	59	0	Lockyer, Cowling and Jones (1976)
35	77	0	Lockyer, Cowling and Jones (1976)
19	85	0	Cowling and Lockyer (1978)
15	175	68	Bell (1976)
15	175	0	Bell (1976)

to SO$_2$. Probably a number of features of experimental design accounts for the variations between experiments. In particular, rates of growth in the different experiments varied considerably, as a result of various arrangements of lighting and temperature. There is evidence that rapidly growing plants are able to utilize absorbed SO$_2$ by oxidizing it to sulphate, whereas when growth is slow there is a build-up of toxic concentrations of sulphite. A second feature is that the various exposure systems had different rates of air movement and different sward structures, so that rates of uptake of SO$_2$ may have been quite different even when SO$_2$ concentrations were similar. Ashenden and Mansfield (1977) demonstrated the importance of air movement in controlling leaf boundary layer resistances to gas uptake. They exposed S23 ryegrass in wind tunnels at two wind speeds. *Table 16.5* shows that total shoot dry weight and leaf area were significantly suppressed by 110 ppb SO$_2$ when the wind speed was 0.42 m s^{-1}, but at 0.17 m s^{-1} these measurements did not differ significantly between treatment and control. They concluded that at the lower wind speed the large boundary layer resistance (typically about 800 s m^{-1} within the sward) limited rates of uptake of SO$_2$; at the higher wind speed, boundary layer resistance was an

Table 16.5 YIELDS OF S23 RYEGRASS AFTER BEING GROWN FOR FOUR WEEKS IN WIND TUNNELS IN ATMOSPHERES CONTAINING 110 ppb SO_2 OR NO DETECTABLE SO_2. SEPARATE EXPERIMENTS WERE UNDERTAKEN, WITH WIND SPEEDS OF 0.42 m s^{-1} AND 0.17 m s^{-1}. (FROM ASHENDEN AND MANSFIELD, 1977)

Experiment No.	SO_2 concentration (ppb)	Wind speed (m s^{-1})	Total shoot dry weight (g)	Leaf area (cm^2)	No. of fully expanded leaves
1	0	0.42	1.006	102.95	30.45
	110	0.42	0.829	88.98	31.10
Significance			<0.01	<0.01	NS
2	0	0.17	0.984	159.44	40.25
	110	0.17	1.119	173.41	47.40
Significance			NS	NS	<0.05

NS = Not significant.

order of magnitude smaller. Interestingly, the number of fully expanded leaves was significantly larger at the low wind speed for the plants growing in polluted air, and the authors suggested that fumigation conditions which lead to a low uptake rate of SO_2 may perhaps induce non-deleterious responses in some plants. To avoid unknown influences of experimental design on responses it would be better to relate responses to rates of uptake of pollutants, rather than to express exposure as the product of pollutant concentration and time.

With such large disagreement between experiments in relatively controlled environments (*Table 16.4*), it is hardly surprising that no clear conclusions can be drawn about the influence of SO_2 on plants in the field. One of the complicating factors is that SO_2 may have a dual role, as a plant nutrient as well as a potentially damaging pollutant. It is well known that, in soils where sulphur supplies are limited, atmospheric sulpur plays an important role in plant nutrition (McLaren, 1975) both by leaf absorption of SO_2 and by root absorption of sulphur reaching the ground gaseously or in rainfall. *Figure 16.1* illustrates this point in glasshouse experiments with S23 ryegrass growing in pots of a typical cultivated soil. All pots were fertilized with nitrogen, phosphorus, potassium and magnesium during the experiment, but only half received an additional 100 ppm sulphate. *Figure 16.1* shows two effects. First, plants in pots which had not received additional sulphate in the soil showed progressively reduced yields at successive harvests when growing in low SO_2 concentrations. Second, irrespective of sulphur in the soil, yields of plants growing in 150 ppb SO_2 were reduced compared with those in clean air. In the experimental system described, about 40–80 ppb SO_2 were necessary to alleviate sulphur deficiency. As an extrapolation from this type of experiment it has been suggested that, since many agricultural crops require between 10 and 30 kg S ha^{-1} yr^{-1}, and since roughly this quantity of sulphur reaches plants and soil each year from the atmosphere in many agricultural parts of industrialized countries, SO_2 may be a highly effective supplier of the sulphur that is lacking in modern fertilizers (Prince and Ross, 1972). This oversimplified proposal ignores several facts. For example, much of the annual input of sulphur to soil from the atmosphere is in winter when crops are either growing very slowly or are not planted; Bromfield and

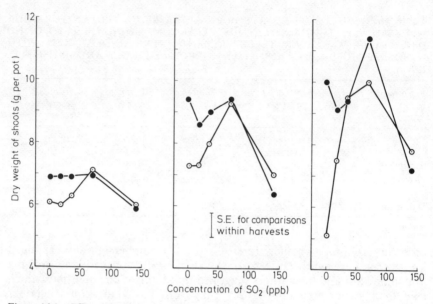

Figure 16.1   *Effects of sulphur dioxide on the yield at three successive harvests of S23 ryegrass plants grown in pots with (●—●) or without (○—○) additions of sulphate to the soil. (After Lockyer, Cowling and Jones, 1976)*

Williams (1974) found no evidence that soils can store substantial quantities of sulphur from the atmosphere for use later by growing plants. Clearly, we need to understand far more about the supply and demand for sulphur in the field and about the capacity of soil for storing and releasing inorganic and organic sulphur before the role of atmospheric sulphur as a crop nutrient can be adequately assessed.

Physiological Responses to Pollutants

If air pollutants influence growth and yield, then clearly they act at the physiological and biochemical level. Consequently, studies of physiological and biochemical responses to pollutants are a powerful tool for identifying mechanisms of action, but of course it is just as difficult to estimate effects on yield of a certain physiological response to a pollutant as it is with any other form of stress! In the space available here I will summarize some recent work at Sutton Bonington and elsewhere on physiological responses to SO_2 as an illustration of some approaches that physiologists adopt to study the stresses of air pollution. Much less will be said of physiological responses to ozone or to NO_x.

SULPHUR DIOXIDE

Much early work which studied physiological responses to SO_2 was at damagingly high concentrations almost entirely irrelevant to present practical

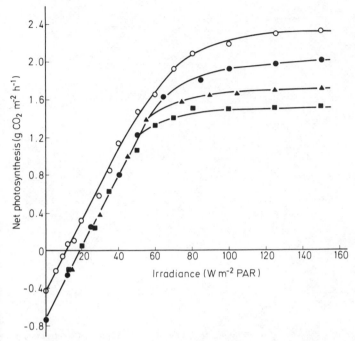

Figure 16.2 *Net photosynthesis versus light response curves for* Vicia faba *plants in clean air* (o—o) *and at three SO₂ concentrations: 35 ppb (●—●), 87.5 ppb (▲—▲) and 175 ppb (■—■). (From Black and Unsworth, 1979a)*

problems. From outstanding work by Thomas and associates (Thomas and Hill, 1937), who used automatic chemical analysers for SO_2 and CO_2, it was concluded that, at SO_2 concentrations above about 400 ppb, photosynthesis was depressed by SO_2 in many species but that, provided no leaf damage occurred, recovery was rapid when the pollutant was removed. Katz (1949) claimed that no inhibition of photosynthesis occurred at SO_2 concentrations below 400 ppb. Modern gas analysis methods allow studies at much lower SO_2 concentrations. *Figure 16.2* shows net photosynthesis versus light response curves for whole plants of *Vicia faba* exposed to a range of SO_2 concentrations. The figure shows that photosynthesis was depressed and dark respiration increased even at SO_2 concentrations as low as 35 ppb. At light saturation, the magnitude of the photosynthetic depression depended on SO_2 concentration, which suggests that SO_2 may compete with CO_2 for binding sites. The large increase in dark respiration was independent of SO_2 concentration and persisted for one photoperiod after the SO_2 was switched off, although net photosynthesis recovered within about one hour. The results suggest that there is more than one mechanism of action for SO_2 at these very low concentrations. A number of possible responses at the cellular and biochemical level has been proposed (e.g. as reviewed by Hallgren, 1978), and it will need cooperation between whole-plant physiologists and biochemists to design experiments to test the range of hypotheses.

Many people have made unwarranted deductions about the presence or absence of stomatal responses to air pollutants in experiments where

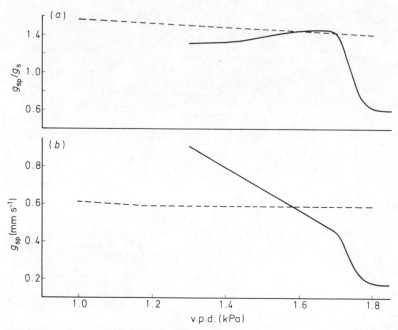

Figure 16.3   *(a) The dependence of the ratio g_{sp}/g_s (conductance in 35 ppb SO_2/conductance in clean air) on vapour pressure deficit (v.p.d.) for* Phaseolus vulgaris *(– – –) and* Vicia faba *(——). (b) The dependence of stomatal conductance (g_{sp}) on vapour pressure deficit for* Phaseolus vulgaris *(– – –) and* Vicia faba *(——) plants exposed to 35 ppb SO_2. (After Black and Unsworth, 1979)*

environmental control was inadequate for physiological studies. Mansfield (1973) reviewed the tangled literature. Majernik and Mansfield (1971) used a viscous flow porometer to study stomatal movements in *V. faba* exposed to SO_2 at concentrations from 250 ppb upwards. They found that at high humidity stomata were opened by SO_2 whereas at low humidity they were closed. Work at Nottingham, initially by Biscoe, Unsworth and Pinckney (1973) and later by Valerie Black, using much lower SO_2 concentrations (from 18 ppb) has demonstrated more completely the influence of SO_2 and vapour pressure deficit (v.p.d.) on stomatal conductances of several species. *Figure 16.3b* shows that conductances of *Phaseolus vulgaris* in 35 ppb SO_2 decreased only slightly with increasing v.p.d. whereas those of *V. faba* decreased linearly at first and then very sharply when v.p.d. was about 1.7 kPa. The ratio (conductance in 35 ppb SO_2/conductance in clean air) is shown in *Figure 16.3a*, which indicates that in *P. vulgaris* conductances were larger in the polluted treatment than in the control at all v.p.d.s, whereas in *V. faba* conductances were larger in the polluted treatment only at v.p.d.s below about 1.7 kPa.

Other species showing v.p.d. sensitivity, e.g. *Helianthus annuus* and *Nicotiniana tobacum*, had similar responses to *V. faba* but the threshold for changeover from opening to closure in response to SO_2 occurred at different v.p.d.s. Over the range of SO_2 concentrations from 18 to 350 ppb, Black found that the magnitude of the change in stomatal conductance was independent of SO_2 concentration. After SO_2 was removed there was no recovery to unpolluted conductance values, and conductances also remained larger than those of controls in the dark.

These and other experiments (Black and Unsworth, 1979b) lead us to the conclusion that stomatal responses to SO_2 arise quite independently of any action of SO_2 on CO_2 exchange. In fact, the common response of stomatal opening is associated with a *reduction* in photosynthesis, contrary to expectation if stomatal conductance were limiting photosynthesis. There is considerable new evidence from studies with light and scanning electron microscopy (Black and Black, 1979a, 1979b) that the mechanism of stomatal response to low concentrations of SO_2 arises from changes in water relationships of the epidermal cells adjacent to stomata. *Figure 16.4* illustrates the difference between the turgid epidermal cells of control plants of *V. faba* in clean air, and the collapsed epidermal cells of plants exposed for two hours to 175 ppb SO_2.

At higher SO_2 concentrations, it is known (Paul and Huynh-Long, 1975; Black and Black, 1979a) that there is damage to chloroplasts in guard cells which leads first to a reversible swelling of the grana thylakoids (Wellburn, Majernik and Wellburn, 1972; Godzik and Sassen, 1974) and, after longer exposures, to plasmolysis. In plants exposed to high concentrations of SO_2, stomatal closure has been reported (Sij and Swanson, 1974; Bonte, de Cormis and Longuet, 1977) consistent with the evidence from microscopy of the loss of turgor in guard cells.

There is still much debate in plant physiology over the mechanism of stomatal responses to v.p.d. (Rawson, Begg and Woodward, 1977). It seems likely that the action of SO_2 on epidermal cells adjacent to stomata leads initially to increased water loss and stomatal opening compared with controls in clean air. In species where cells of the stomatal complex are partially isolated hydraulically from other cells of the epidermis and mesophyll, the rate of water loss at large vapour pressure deficits may reduce guard cell turgor and cause stomatal closure. Damage to epidermal cell walls by SO_2 (Black and Black, 1979a, 1979b) would enhance this response.

The most recent development from our studies of SO_2 responses concerns the measurement of fluxes of SO_2 into plants and simultaneous determination of inhibition of photosynthesis (Black and Unsworth, 1979c). Analysis of fluxes when SO_2 concentrations ranged from 20 to 300 ppb shows that the concentration of gaseous SO_2 in the substomatal cavities of *V. faba* was close to zero. There are two important implications from this. First, plants in our experiments were clearly capable of absorbing SO_2 rapidly even at large fluxes, and this poses interesting new questions concerning the sinks for SO_2 within plants. Second, if it is found generally that SO_2 concentrations within leaves are zero, it is relatively simple to estimate the uptake of SO_2 by leaves from direct measurements of the stomatal conductance for water vapour (Black and Unsworth, 1979c; Unsworth, 1980).

OZONE

Until quite recently it was thought unlikely that high ozone concentrations would be produced photochemically in Europe because there were neither high enough air temperatures nor sufficient sunlight associated with the

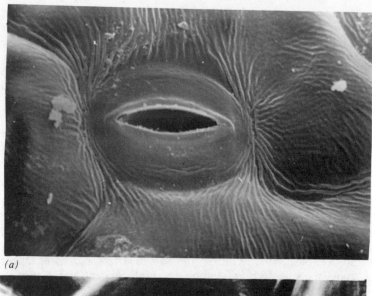

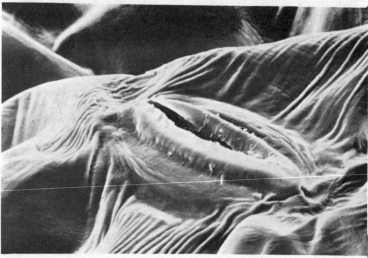

Figure 16.4 *Scanning electron micrographs of samples of leaf tissue from* Vicia faba *plants, prepared by critical point drying and coated with gold. (a) Control plant exposed in clean air showing turgid epidermal and guard cells. (b) Plant exposed for two hours to 175 ppb SO_2, showing collapsed epidermal cells. (From Black and Black, 1979b)*

relevant precursor gases. However, surveys both with gas analysers (Cox *et al.*, 1975) and with biological indicators (Posthumus, 1976; Ashmore, Bell and Reily, 1978) reveal that there is potential for ozone damage to plants over a wide area.

Physiological responses to ozone have been reviewed by Verkroost (1974). In contrast to the usual response to SO_2, ozone generally causes stomatal closure (Hill and Littlefield, 1969; Beckerson and Hofstra, 1979a, 1979b). Several years ago Mansfield (1973) pointed out that there had been

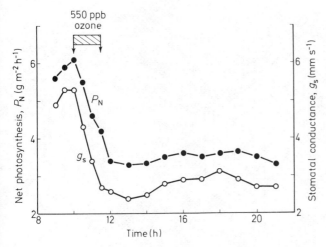

Figure 16.5   *Influence of ozone on net photosynthesis rate, P_N (●—●) and stomatal conductance for water vapour, g_s (○—○) of an attached leaf of Zea mays cv. Coop 3990 exposed for 90 minutes to a mean ozone concentration of 550 ppb. (After K. M. King and P. J. Smith, unpublished data)*

surprisingly few studies by physiologists aimed at revealing the mechanisms of the closure response; the situation still appears unclear. It is known that ozone reduces photosynthesis and increases dark respiration (Hill and Littlefield, 1969). *Figure 16.5* illustrates the responses of net photosynthesis and stomatal conductance in *Zea mays* exposed briefly to a rather high ozone concentration. It is not known whether the stomatal closure observed in this and other work arises because of an increase in internal CO_2 concentration or because of an independent action of ozone on cells of the stomatal complex. Unfortunately, Mansfield's (1973) suggestions for clarifying the matter do not seem to have been taken up.

In the field, pollutants seldom occur singly and so there have been several studies of physiological responses to mixtures of pollutants. Recent work by Beckerson and Hofstra (1979a, 1979b) showed that mixtures of 150 ppb ozone and 150 ppb SO_2 decreased the stomatal conductance on both surfaces of leaves of *P. vulgaris*. Stomata on the upper surface of the leaves responded to the pollutants applied singly by closure with ozone and opening with SO_2. On the lower surface, stomata did not respond to SO_2 or ozone alone, but there was marked closure with the mixture.

OXIDES OF NITROGEN

As a final example of the complexity of responses to gaseous pollutants, we will consider some problems which can arise in glasshouses. An increasingly popular arrangement for providing CO_2 enrichment in glasshouses involves the burning of propane or kerosene and venting the exhaust inside the house. Hand (1973) discussed the air pollutants produced in this process, predominantly unburnt hydrocarbons, SO_2 and oxides of nitrogen. Capron and Mansfield (1975) measured simultaneously concentrations of CO_2, NO_2

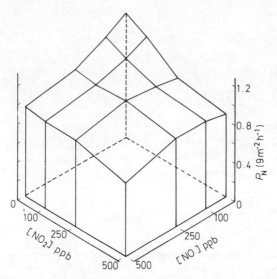

Figure 16.6   *Results of a 4 × 4 factorial experiment to determine the influence of NO and NO$_2$ on rates of net photosynthesis P$_N$ in tomato plants. (After Capron and Mansfield, 1976)*

and NO in a glasshouse which employed a kerosene burner. No SO$_2$ could be detected, presumably, they concluded, because the small quantity generated was very efficiently absorbed by the vegetation and soil. Concentrations of NO$_2$ reached about 100 ppb and of NO about 350 ppb. In the free atmosphere NO is rapidly oxidized to NO$_2$ by ozone and other species, but in glasshouses it appears to be the dominant NO$_x$ species. In subsequent laboratory experiments with tomato plants, Capron and Mansfield (1976) investigated the effects of NO and NO$_2$ on net photosynthesis. *Figure 16.6* summarizes their conclusions that both gases reduced photosynthesis to approximately the same extent, and that the effects were additive rather than synergistic. Wellburn *et al.* (1976) showed that fumigation of tomato seedlings with 400 ppb NO led to a small increase in the activity of nitrite reductase. Capron and Mansfield (1976) speculated that the toxic effects of both NO and NO$_2$ might be related to the inability of the plant to remove nitrite quickly enough by the action of nitrite reductase. Consequently, plant responses to NO$_x$ may be positive or negative. It seems essential to make further studies of the potentially serious problem of NO$_x$ pollution in glasshouses, in particular studying responses to NO$_x$ when CO$_2$ concentrations are high.

Conclusions

In this brief and selective review I have attempted to give an outline of some of the challenging problems for physiologists in understanding plant responses to pollutants. There is increasingly clear evidence that low concentrations of air pollutants influence the growth and yields of plants both positively and negatively. In particular, the old ideas that yields are influenced only by pollution concentrations above some well-defined (and high) threshold, must be re-examined. Experimental design will need to be much more carefully controlled than in the past to avoid interactions between air pollutants and other environmental factors. There is also a need for field experimentation to study the growth and development of plants stressed by air pollutants.

Recent work using techniques well established in other branches of environmental physiology has shown that gas exchange and stomatal responses are influenced by very low concentrations of SO_2 and by other pollutants singly and as mixtures. Further progress in two areas is important. First, responses must be related to pollutant uptake so that the influences of different experimental designs and of pollutant losses to apparatus can be eliminated from analyses. Second, collaboration between biochemists and physiologists is necessary to advance towards an understanding of mechanisms of physiological responses.

Ultimately, physiologists must attack the important practical problem of investigating responses to mixtures of pollutants, but the demands, both intellectual and in terms of facilities required for experimentation, are daunting. The thirty-first Easter School (1980) draws together the wide range of disciplines in air pollution research under the title 'Effects of gaseous air pollution in agriculture and horticulture'. It is hoped that by then more physiologists will have taken up the gauntlet and tackled some of the challenges posed by these complex cases of physiological responses to stress.

Acknowledgements

I am grateful to my colleagues Valerie and Colin Black for many helpful discussions of topics in this paper. Thanks are also due to Professor K. M. King of the University of Guelph for permission to use unpublished results.

References

ASHENDEN, T. W. and MANSFIELD, T. A. (1977). *J. exp. Bot.*, **28**, 729–735

ASHMORE, M. R., BELL, J. N. B. and REILY, C. L. (1978). *Nature*, **276**, 813–815

BECKERSON, D. W. and HOFSTRA, G. (1979a). *Atmos. Env.*, **13**, 533–535

BECKERSON, D. W. and HOFSTRA, G. (1979b). *Atmos. Env.*, **13**, 1263–1268

BELL, J. N. B. (1976). Pollution Paper No. 7, Department of the Environment. HMSO, London

BELL, J. N. B. and CLOUGH, W. S. (1973). *Nature*, **241**, 47–49

BISCOE, P. V., UNSWORTH, M. H. and PINCKNEY, H. R. (1973). *New Phytol.*, **72**, 1299–1306

BLACK, C. R. and BLACK, V. J. (1979a). *J. exp. Bot.*, **30**, 291–298

BLACK, C. R. and BLACK, V. J. (1979b). *Pl., Cell Env.*, **2**, 329–333

BLACK, V. J. and UNSWORTH, M. H. (1979a). *J. exp. Bot.*, **30**, 473–483

BLACK, V. J. and UNSWORTH, M. H. (1979b). *J. exp. Bot.*, **30**, 81–88

BLACK, V. J. and UNSWORTH, M. H. (1979c). *Nature*, **282**, 68–69

BLACK, V. J. and UNSWORTH, M. H. (1980). *J. exp. Bot.*, **31**, 667–677

BLEASDALE, J. K. A. (1973). *Env. Poll.*, **5**, 275–285

BONTE, J., DE CORMIS, L. and LONGUET, P. (1977). *Env. Poll.*, **12**, 125–133

BROMFIELD, A. R. and WILLIAMS, R. J. B. (1974). *Nature*, **252**, 470–471

BROWN, H. T. and ESCOMBE, F. (1902). *Phil. Trans. R. Soc.*, *B*, **195**, 397–413

CAPRON, T. M. and MANSFIELD, T. A. (1975). *J. hortic. Sci.*, **60**, 233–238

CAPRON, T. M. and MANSFIELD, T. A. (1976). *J. exp. Bot.*, **27**, 1181–1186

COWLING, D. W. and LOCKYER, D. R. (1978). *J. exp. Bot.*, **29**, 257–265

COX, R. A., EGGLETON, A. E. J., DERWENT, R. G., LOVELOCK, J. E. and PACK, D. H. (1975). *Nature*, **255**, 118–121

DEPARTMENT OF THE ENVIRONMENT. The Investigation of Air Pollution, Annual Summaries, Warren Spring Laboratory, Stevenage, Herts

DERWENT, R. G. and STEWART, H. N. M. (1973). *Atmos. Env.*, **7**, 385–401

GODZIK, S. and SASSEN, M. M. A. (1974). *Phytopath. Z.*, **79**, 155–159

GUDERIAN, R. (1977). *Air Pollution*. Springer-Verlag, Berlin

HALLGREN, J.-E. (1978). In *Sulphur in the Environment*, Part II: *Ecological Impacts*, pp. 163–209. Ed. by J. O. Nriagu. John Wiley, Chichester

HAND, D. W. (1973). *Sci. Hortic.*, **24**, 142–152

HECK, W. W. and BRANDT, C. S. (1977). In *Air Pollution*, Vol. 2, pp. 157–229. Ed. by Stern, A. C. Academic Press, New York

HILL, A. C. and LITTLEFIELD, N. (1969). *Env. Sci. Tech.*, **3**, 52–56

HORSMAN, D. C., ROBERTS, T. M. and BRADSHAW, A. D. (1979). *J. exp. Bot.*, **30**, 485–493

KATZ, M. (1949). *Ind. Eng. Chem.*, **41**, 2450–2465

LOCKYER, D. R., COWLING, D. W. and JONES, L. H. P. (1976). *J. exp. Bot.*, **27**, 397–409

MAJERNIK, O. and MANSFIELD, T. A. (1971). *Phytopath. Z.*, **71**, 123–128

MANSFIELD, T. A. (1973). *Comment. Pl. Sci.*, No. 2, April 1973

MARTIN, A. and BARBER, F. R. (1973). *Atmos. Env.*, **7**, 17–37

MCLAREN, R. G. (1975). *J. agric. Sci., Cambridge*, **85**, 571–573

ORMROD, D. P. (1978). *Pollution in Horticulture*. Elsevier, Amsterdam

PAUL, R. and HUYNH-LONG, V. (1975). *Parasitica*, **31**, 30–39

POSTHUMUS, A. C. (1976). In *Proceedings of the Kuopio Meeting on Plant Damages caused by Air Pollution*, pp. 115–120. Ed. by Karenlampi, L. Kuopio, Finland

PRINCE, R. and ROSS, F. F. (1972). *Water, Air and Soil Poll.*, **1**, 286–302

RAWSON, H. M., BEGG, J. E. and WOODWARD, R. G. (1977). *Planta*, **134**, 5–10

SIJ, J. W. and SWANSON, C. A. (1974). *J. env. Qual.*, **3**, 103–107

STOKLASA, J. (1923). *Urban und Schwarzenberg*, p. 487. Springer-Verlag, Berlin

THOMAS, M. D. (1951). *Ann. Rev. Pl. Physiol.*, **2**, 293–322

THOMAS, M. D. and HILL, G. R. (1937). *Pl. Physiol., Lanc.*, **12**, 309–383

UNSWORTH, M. H. (1980). *Arch. env. Prot.*, in press

VERKROOST, M. (1974). *Med. Landbouwh. Wageningen*, **74–19**, 1–78

WELLBURN, A. R., CAPRON, T. M., CHAN, H-S. and HORSMAN, D. C. (1976). In *Effects of Air Pollution in Plants*, pp. 105–114. Ed. by T. A. Mansfield. Cambridge University Press, Cambridge

WELLBURN, A. R., MAJERNIK, O. and WELLBURN, F. A. M. (1972). *Env. Poll.*, **3**, 37–49

WISLICENUS, H. (1914). *Sammlung von Abhandlungen über Abgase und Rauchschäden*. Parey, Berlin

17

NUTRIENT INTERCEPTION AND TRANSPORT BY ROOT SYSTEMS

DAVID T. CLARKSON
ARC Letcombe Laboratory, Wantage, UK

Introduction

The publication of the report *Limits to Growth* (Meadows *et al.*, 1972) seems to be a popular landmark in the way in which we have come to think about world resources and the rate at which we use them. The appropriateness of various technologies, not least those employed in agriculture, has come under increasing scrutiny in the last decade; those which are at present economic in crop production in developed countries may become progressively less acceptable because of the large consumption of energy needed to manufacture high-grade fertilizers. It is unlikely that such technology will ever be applied economically in crop production on a significant scale in the developing countries. These bleak assessments raise the questions whether crop species can be better adapted for the efficient exploitation of nutrients in the soil, and what is meant by efficiency. In the context of the present symposium it should be asked whether there are circumstances where the performance of the root system limits plant production overall, and what we can do about it if the answer is affirmative.

Despite popular misuse, efficiency cannot have an absolute measure, only one related to some subjective property of the system considered. A comparison of crop plants with native species reveals that the former are generally distinguished by the vigour of their growth and the high demand they place on soil resources. Many native species exploit the environment conservatively, slowly consuming resources and having slow growth rates. In selecting crop plants we have attempted to maximize *production*, the necessary inputs being freely available, whereas natural selection may maximize the chances of *survival* where the inputs may be restricted. This is, of course, a statement of the obvious, but it is worth making at the outset of this discussion because it is crucial to clarify the objectives in selecting plants which exploit the environment efficiently. A grass or pasture species which thrives in soils with very low available phosphorus may do so by growing so slowly that 10 ha of its production are needed to feed one animal. Such a plant has little to recommend it to the grazier. Although we may learn much of interest from native species, the lesson may not be always one that can be applied fruitfully to food production. What we seek is relatively rapid extraction of nutrients from the soil coupled with a system of rigorous internal economy within the plant itself.

307

The dynamics of nutrient transfer between soil and plant root are more complex than those usually considered by plant physiologists working in the laboratory. The complexity depends on the unstirred nature of the soil solution and the numerous interactions between ions in solution and the solid phases in the soil. The final design of the root system must fit it for the absorption of both relatively immobile nutrients, which must be sought out, and mobile nutrients, which arrive abundantly at the root surface due to water absorption by the plant. In the first instance exploratory root growth with a large surface area is important whereas such properties may be less important for mobile nutrients.

Distribution and Mobility of Nutrients in the Soil

Mineral nutrients in ionic or complexed form move through the soil by convective and diffusive processes. The former, more commonly known as mass flow, may occur by percolation of water through the soil or by water movement to the root surface of transpiring plants. The water absorbed from the soil may bring with it an adequate supply of some nutrients, especially those which are present in the soil solution at a relatively high concentration and which do not interact strongly with exchange sites in the soil. Brewster and Tinker (1970) calculated that mass flow would provide an adequate supply of calcium and magnesium but not potassium to the root surface of plants which grow in a clay loam topsoil. Using much simpler calculations to derive the total delivery of nutrients to plants over an entire growing season, based on the total water transpired and the average soil solution concentration, Barber and Olson (1968) arrived at conclusions similar to those of Brewster and Tinker (1970). *Table 17.1* shows that mass flow could deliver the total nitrogen, sulphur, calcium and magnesium required by robust corn plants, but for phosphate, potassium and some minor nutrients the rate of absorption at the root surface greatly exceeds the rate of supply by mass flow. Thus the concentration in the vicinity of the root will be reduced relative to that in the

Table 17.1   THE SIGNIFICANCE OF MASS-FLOW IN SUPPLYING *ZEA MAYS* (CORN) WITH ITS NUTRIENT REQUIREMENTS FROM A TYPICAL FERTILE SILT LOAM SOIL. (FROM BARBER AND OLSON, 1968)

Nutrient	Amount needed for a yield of 9500 kg ha^{-1}	Approximate amount supplied by mass flow (kg ha^{-1})	Nutrient	Amount needed for a yield of 9500 kg ha^{-1}	Approximate amount supplied by mass flow (kg ha^{-1})
Nitrogen	187	185	Copper*	0.1	0.4
Phosphorus	38	2	Zinc	0.3	0.1
Potassium	192	38	Boron*	0.2	0.7
Calcium*	38	165	Iron	1.9	1.0
Magnesium*	44	110	Manganese*	0.3	0.4
Sulphur	22	21	Molybdenum*	0.01	0.02

*These elements are potentially supplied in excess by mass flow. In some instances, e.g. calcium (Barber and Ozanne, 1970), ions may accumulate at the soil/root interface and diffuse back into the bulk soil.

Table 17.2 TYPICAL VALUES FOR DIFFUSION COEFFICIENTS FOR IONS IN MOIST SOIL

Ion	Diffusion coefficient (cm^2 s^{-1})	Reference
Cl^-	$2-9 \times 10^{-6}$	Rowell, Martin and Nye (1967)
NO_3^-	1×10^{-6}	Nye (1969)
SO_4^{2-}	$1-2 \times 10^{-6}$	Wray (1971)
$H_2PO_4^-$	$0.3-3.3 \times 10^{-9}$	Rowell, Martin and Nye (1967)
Rb^+	$6-16 \times 10^{-8}$	Graham-Bryce (1963)
K^+	$1-28 \times 10^{-8}$	Drew, Nye and Vaidyanathan (1969)

The values given are chosen to illustrate the characteristic range in soils with 20–40 per cent water content; diffusion coefficients vary with soil type, ionic concentration and water status (Nye and Tinker, 1977). Phosphate diffuses much more slowly than the other ions because it makes specific interactions with binding sites on clay minerals.

bulk soil and a diffusion gradient created which will bring additional amounts of nutrient to the root surface. Where the rate of diffusion of a nutrient is slow relative to its rate of absorption by the root, the former process will dominate the supply of the plant. The mobilities of some of the major nutrient ions are listed in *Table 17.2* where it is evident that phosphate, among the major nutrients, has the lowest diffusion coefficient. The phosphate concentration in the soil solution rarely exceeds a mean value of 2 μM, although it may be much greater near a fertilizer granule. The combination of this low concentration with low mobility means that, when a young root enters new regions of the soil, the phosphate concentration at the root surface is rapidly reduced. *Table 17.1* also suggests that plants may depend on diffusion for most of their potassium requirement. The diffusion coefficient for K^+ is, however, 10–100 times greater than that of $H_2PO_4^-$ and there is therefore likely to be a much larger volume of soil around a root from which K^+ can be extracted than around that from which $H_2PO_4^-$ is available. The calculations above are based on fertile soils; with the exception of phosphate and potassium, the nutrition of plants in such a soil is not very different from those grown in a water culture. As fertilizer applications decrease and an approach is made towards natural soils, the concentrations of many nutrients may decrease greatly. Diffusion may then become a much more significant factor in the supply of nitrogen and sulphur. It may also be more important in young plants with a high relative growth rate than in mature ones where uptake of nutrients per unit root length is known to decline (Mengel and Barber, 1974).

Cultivation and fertilizer practices, irrigation and soil profile development can all lead to stratification of nutrients in the soil. This process is again associated with the differential ionic mobility in the soil. There is a strong tendency for nitrate to be leached from the upper horizons by water percolating through the soil profile so that during a growing season the vertical distribution of available nitrogen can change radically. At the other extreme, immobile elements such as phosphate tend to remain close to the point at which they entered the soil, either by fertilizer application or by the decay of plant litter. In an undisturbed soil, either a natural one or one managed by no-tillage practices, the available phosphate becomes concentrated in the upper horizons. When the soil becomes dry, additional problems in phosphate supply might develop. It is clear that stratification of available nutrients and water in the soil can give rise to a certain amount of 'division of labour' within the root system with roots at different depths engaged in transport of different commodities at different rates. This situation cannot be reproduced in solution culture experiments nor in well-mixed soil potting composts and so has been comparatively little studied.

The Plant Root and its Associates

The slow diffusion and low concentration of some nutrients impose restraints on supplies to the plant which can be overcome effectively only by continuous exploration of new soil volume by the root system. This is the sense in which 'interception' is used in this paper, and it can be achieved either by the growth of the root in such a way as to maximize its surface area (Bouldin, 1961) or by developing associations with mycorrhizal fungi, which can extend for considerable distances into the soil and thus tap sources remote from the root surface (Gerdemann, 1968; Mosse, 1973; Sanders, Mosse and Tinker, 1975).

Species which produce roots that are coarse, unbranched or which lack root hairs exploit soil phosphate ineffectively. Baylis (1972) showed that phosphate extraction from phosphate-deficient soil by a variety of species was directly related to the size of their root/soil interface. *Table 17.3* indicates

Table 17.3   THE GROWTH OF FIVE SPECIES WITH CONTRASTING ROOT SURFACE AREAS IN PHOSPHORUS-DEFICIENT STERILIZED SOIL WITH VARIOUS ADDITIONS OF PHOSPHATE. (FROM BAYLIS, 1970, 1972)

Species	Root diameter (mm)	Root hairs (frequency/length)	Mean dry weight (mg) of seedlings Added P*:			
			0	15	45	135
Podocarpus totara	>1.0	None	9	9	11	**29**
Coprosma robusta	0.2–0.3	Few: 0.2 mm	3	3	**5**	**71**
Leptospermum scoparium	0.15–0.2	Moderate: <1 mm	10	16	**38**	61
Solanum nigrum	0.15–0.2	Frequent: 1–2 mm	2	**9**	**60**	**243**
*Lolium perenne***	>0.1	Abundant: 1 mm	4050			4020

Values in bold type differ significantly from those to their immediate left.
*1 mg of phosphate from fertilizer added to pots containing 200 g steam-sterilized soil.
**Results taken from Figure 1 of Baylis (1970). Other results as published in Table 1 of Baylis (1972).

that *Lolium perenne* (ryegrass) which has a finely divided, profusely hairy root system, was able to thrive in soil without phosphate amendment while the coarse hairless root of *Podocarpus totara* required relatively large amounts of phosphate to be added to the soil in order to grow. All the plants used in this work were kept non-mycorrhizal; had they been infected, the outcome of the experiment would have been different (*see* later).

ROOT HAIRS AND THE SOIL/ROOT INTERFACE

It is evident that species differ enormously in the quantity of root hairs which can be produced (Cormack, 1949, 1962), but environmental conditions strongly influence their rate of formation within a given species. There is no evidence that root hairs possess special properties which would make them more effective than other cells of the root in ion absorption. Their advantage over other cells is one of position since they form the effective interface between root and soil. Their presence may increase the surface area of this interface by ten-fold or more (Dittmer, 1937). Roots of *Lolium multiflorum*, grown in a sandy loam, were found to have approximately 68 cm of root hair per cm of root length, the frequency of hairs being approximately 100 mm^{-1} (Drew and Nye, 1969). In *L. perenne* the hairs are longer but at about the same frequency, with 110 cm of hair per cm of root (Reid, personal communication). Calculations based on a diffusion model indicated that potassium uptake by *L. multiflorum* was increased by 70 per cent by the presence of root hairs when plants were grown in a low-potassium-status soil over a four-day period (Drew and Nye, 1969). In the same experiments, it was observed that phosphate uptake was two- to three-fold greater than that predicted from the amount of phosphate which could diffuse to the surface of the root epidermis; uptake could be accounted for only by assuming that the ends of the dense root hairs described a circle which was the effective boundary of the root. Not all experimental findings agree on this matter, however. Bole (1973) found relatively little effect of increasing root hair density on phosphate uptake by wheat genotypes selected for this characteristic. In such conflicting instances the duration of the experiment may be an important consideration; where this is prolonged (four weeks in Bole's, 1973, work) the geometrical benefit of the hair may be lost due to depletion of the soil, competition and to deposition of cuticle over the root hairs (Scott, 1963) which may reduce the effectiveness with which they can absorb ions. In short-term experiments, phosphate extraction rates by non-mycorrhizal roots of *Brassica napus*, with abundant root hairs, was compared with that of *Allium cepa*, with few or none. The former species absorbed 1 µmol phosphate in five days whereas the latter absorbed one fortieth of this amount during the development of a comparable length of root (Brewster, Bhat and Nye, 1976). While some of this difference may be accounted for by relative growth rates of the species, the principal factor was attributed to the root hairs of *B. napus*.

When barley roots with abundant root hairs were compared with those possessing only a few in a nutrient solution, phosphate, potassium and calcium absorption per unit root weight were similar (Crossett and Campbell, 1975); in a well-stirred medium, root hairs cease to have a positional advantage.

MYCORRHIZAS

Both ectotrophic and endotrophic (vesicular-arbuscular, v.a.) mycorrhizas may extend the absorbing surface of the root deep into the surrounding soil. Rhodes and Gerdemann (1975) showed that labelled phosphate could be transported by hyphae of *Glomus faciculatus* associated with roots of *A. cepa* from an injection into soil 8 cm away from the root surface; during the course of the experiment, autoradiographs showed that there was no detectable ^{32}P in the soil at a distance from the injection point greater than 0.75 cm. There is less convincing evidence that v.a. mycorrhizas can also assist in the delivery of sulphur (Rhodes and Gerdemann, 1978) and possibly zinc (Gilmore, 1971; Bowen, Skinner and Bevege, 1974) to the root. The significance of mycorrhizas in mineral nutrition diminishes as the concentration and mobility of a given ion increase.

The great interest in mycorrhizal fungi has developed from the realization that most plants in both natural and agricultural soils are mycorrhizal, and

Figure 17.1   *The growth of* Medicago satira *(lucerne) inoculated with endophytic mycorrhizas in field conditions. The phosphate status of the soil varied in different plots ranging from 14 ppm Olsen-extractable phosphate in the soil supporting the plants in the top line to 8 ppm in that for plants in the bottom line. NI = Not inoculated. IL = Inoculated with a mixed culture of endophytis containing* Glumus mosseae. *IR = Inoculated with a pure culture of* Glomus caledonius. *Note that at all levels of soil phosphate the interaction of the host with* G. caledonius *produced much larger plants. The inoculated endophytes would have had to compete with indigenous fungi in the field soil in order to colonize the roots of the host. (From Owusu-Bennoah and Mosse, 1979)*

that in nutrient-poor soils the incidence of infection is greatest. There are many instances where the formation of an effective association is crucial for the host's survival (Kleinschmidt and Gerdemann, 1972). The extent to which the fungal hyphae are benignly associated with roots can be surprisingly high; values of 80 cm hyphae per cm root length were measured in *A. cepa* by Sanders and Tinker (1973), and they are thus comparable in extent to root hairs (*see* above). There appear to be inverse relationships between the infection of roots by v.a. fungi and phosphate and nitrogen fertilizer application to the soil (Hayman, 1975) and the phosphate-status of the plant. This may diminish the benefits of fertilizers to some extent because efficiency of exploitation of phosphate may go down as the inputs increase. It is known that there are degrees of effectiveness of specific host/fungal strain associations and that it can be maximized by successful manipulation (Bowen, 1977; Mosse, 1977). This suggests one plausible method of improving the efficiency with which crop plants exploit soil resources (*Figure 17.1*).

FREE-LIVING MICROORGANISMS IN THE RHIZOSPHERE

Experiments have shown that hyphae of mycorrhizal fungi draw on the same sources of phosphate in the soil as plant roots (Sanders and Tinker, 1971). There are suggestions, however, that other free-living microorganisms in the rhizosphere can mobilize sources of phosphate normally unavailable to the plant. *Pseudomonas* species and *Aspergillus niger* strains have been isolated from the rhizosphere of coconut plants which release phosphate from insoluble tricalcium phosphate, largely by their production of 2-ketogluconic acid and citric acid (Nair and Subba Rao, 1977). Much speculation and controversy surrounds the role of free-living, nitrogen-fixing organisms in the rhizosphere. A discussion of this is clearly outside the scope of this paper but it has relevance because of attempts which have been made to improve the growth and nutrition of plants by bacterial inoculation of seeds with *Azotobacter* and other organisms. Opinions differ as to whether the somewhat variable benefits of this procedure are due to a contribution to the nitrogen nutrition of the plant or to the production of growth-stimulating substances by the bacteria (Brown, 1974).

EXUDATION/LEAKAGE OF MATERIALS FROM ROOTS

Inorganic and organic substances of many kinds are released from roots into the rhizosphere (Barber and Martin, 1976). Environment and plant species can influence the quantity and composition of this exudate flow (Vancura, 1967; Bowen, 1969). Research on this subject is in its infancy but it is known that production of hydrogen ions and reductants (Brown, 1978) and of acid phosphatase (Clark and Brown, 1974) at the root surface shows marked genotype variation within a given species.

The release of protons from roots tends to acidify the rhizosphere and can bring about marked improvements in the supply of phosphate near the root, particularly in alkaline soils. Extensive work by Brown and colleagues at USDA Beltsville has shown that genotypes which can acidify their rhizosphere soil are efficient at extracting iron from the soil and rarely develop severe chlorosis. Those which are unable to do this, particularly when the plant becomes deficient in iron, are unable to thrive in calcareous soils. This property appears to be an attribute of calcicole species.

The tendency of calcicole or 'phosphate-efficient' genotypes to acidify their rhizosphere is a distinct liability when attempts are made to establish them in acidic soils. Further acidification here may increase the concentration of toxic aluminium (Al^{3+}) in the vicinity of the root (Foy *et al.*, 1967). Some calcifuge plants are able to avoid the toxic effects of Al^{3+} by *raising* the pH of the soil near the root surface, thus causing precipitation and hydrolysis of the Al^{3+} ion; this attribute renders such plants inefficient in acquiring iron and phosphate in neutral and alkaline soils. The evolution of calcicole and calcifuge genotypes can occur rapidly by the selection of individuals from a population which thrive on calcareous or acid soil. This discovery has given much impetus to the work of those with enthusiasm for 'fitting the plant to the soil' by selection (*see* Wright, 1977 for extensive discussion of this matter).

The predominant form of nitrogen in soil, itself pH dependent, can have a marked effect on the release of hydroxyl and hydrogen ions by roots (Miller, 1974). The most striking effects are with NH_4^+ since each ammonium ion assimilated gives rise to one H^+ which must be extruded (Dijkshoorn, 1962). The consequent acidification of the rhizosphere increases phosphate availability and its absorption (Miller, Mamaril and Blair, 1970), and may also increase the uptake of iron, manganese and zinc (Sarkar, 1977). If, however, the pH of the rhizosphere soil is reduced below pH 4.7 the concentration of Al^{3+} will increase rapidly (Magistad, 1925) and may damage the roots.

The release of acid phosphatase from the surface of nutrient-stressed roots seems to be a well-documented phenomenon (Bieleski, 1971) the nutritional significance of which does not seem to have been demonstrated convincingly. It seems probable that organic phosphate close to root surfaces will be degraded by the enzyme (Boero and Thien, 1979). The range over which this process could be expected to operate would be limited, however, by the slow diffusion of a large protein away from the root surface and the high rate of enzyme breakdown to be expected in the rhizosphere. It is easy to envisage this as one of several factors which come into operation in the nutrient-stressed plant so that it may scavenge ions from the depleted environment around the root.

The Root Surface: Distribution of Transporting Capabilities

In any assessment of mineral uptake by root systems a distinction should be made between limitations on the process imposed by the plant and those imposed by the soil. It is important to know how much of the root surface is capable of nutrient absorption and of transferring material to the shoot.

The structure and metabolic activity of cells vary along the length of any root axis or branch and it has been appreciated for many years that these changes may influence ion absorption and long-distance transport. Earlier speculations about these matters (Scott, 1928; Prevot and Steward, 1936) have given way to direct observations using radioactive tracers to measure uptake and translocation from different root zones. Observations from a number of species (*Table 17.4*) reveal that the transport capabilities for various nutrients become segregated along the length of their roots by the anatomical and biochemical changes which accompany ageing.

A hypothesis to explain some of these results depends on root age having a much stronger influence on substances which move by an extracellular, or cell wall, pathway (the *apoplast*) across the cortex, than on those which travel in the cytoplasmic continuum (the *symplast*) which links all living cells. The structure and development of two tissue elements, the endodermal cell walls and the plasmodesmata, are crucial in this hypothesis. It is widely agreed that the free space of the cortex, or the apoplast, has an inner limit marked by the endodermis (Clarkson and Robards, 1975). In younger regions of the root the most important structural feature of this tissue is a band of lignin and suberin (the Casparian band) in the radial walls to which the plasmamembranes of adjacent endodermal cells are firmly attached (Bryant, 1934; Bonnett, 1968; Robards *et al.*, 1973). These properties of the Casparian band introduce a zone of high resistance to flow into the radial walls and ensure that substances in the apoplast have to cross the plasmamembrane of the endodermal protoplast to gain access to the stele and vascular tissue. Since there seems to be frequent confusion on the matter, it must be emphasized that all but a few millimetres of the apical part of the root have the endodermis in this form, including those parts normally regarded as the most permeable to water and solutes. All cells in the endodermis develop a Casparian band at approximately the same time, usually corresponding with the end of the phase of cell elongation (Robards *et al.*, 1973). As time passes, the endodermis changes in such a way that the endodermal protoplast becomes isolated from the apoplast.

At a variable distance from the root tip, frequently but not always corresponding with the emergence of lateral branches, layers of suberin are formed over the entire inner surface of endodermal walls; the layers create a lamellar structure which subsequently becomes buried by further cellulose deposition. The original attachment of the Casparian band and plasmamembrane is disrupted during these developments (Robards *et al.*, 1973). The suberin lamellae thus resemble a damp-proof course in a building, and are interposed between the plasmamembrane of the endodermal cell and the apoplast. This development is accompanied in barley and marrow (Harrison-Murray and Clarkson, 1973; Graham, Clarkson and Sanderson, 1974) by a distinct decrease in water flow and calcium movement across the root (*Figure 17.2*).

With other ions, the changes which occur in the endodermis seem to have little significance (*Table 17.4*); it is believed that these cross the cortex predominantly in the symplast in older parts of the axis. They negotiate the suberized endodermis by entering and leaving it via the abundant plasmodesmata, which connect it to cells in the cortex and pericycle. The plasmodesmata pass through suberin lamellae (Robards *et al.*, 1973) thus maintaining symplasmic continuity across the thickened endodermis. The marked reduction in calcium translocation from such zones (*Figure 17.3*) implies ineffective movement of this ion in the symplast; this is probably due to the very low thermodynamic activity of calcium ions in the cytoplasm. The reduced water uptake in older zones (*Figure 17.3*) may indicate a limited

Table 17.4 A SELECTION OF PUBLISHED ACCOUNTS OF ION UPTAKE, EFFLUX AND TRANSLOCATION ALONG THE LENGTH OF ROOTS OF INTACT PLANTS

Species	Reference	Ions measured	Length of root examined (cm)	Type of root	Conclusions/comments
Barley (Hordeum vulgare)	Wiebe and Kramer (1954)	$H_2PO_4^-$	0–7	Seminal axis	All zones examined translocated, peak value at 3 cm from tip; experiment ran for 6 h
	Clarkson et al. (1968)	$H_2PO_4^-$	1 and >40	Seminal axis	Endodermis suberized in old zone but high rates of uptake/translocation recorded; experiment ran for 24 h
	Clarkson et al. (1978)	$H_2PO_4^-$	>30	Seminal axis	Phosphate stress increased uptake/translocation in zones >15 cm from the tip before 1 and 2 cm zones responded. In phosphate-sufficient plants, peak absorption/translocation 2.5 cm from tip
	Wiebe and Kramer (1954)	K^+	0.7	Seminal axis	Peak translocation around 2 cm from tip
	Russell and Clarkson (1973)	K^+	>30	Seminal and nodal axes	Rate of uptake maximal 2–4 cm from tip; translocation from all zones about 50 per cent of uptake in 24 h
	Clarkson et al. (1975)	NH_4^+	>40	Seminal axis	Uniform rate of uptake along the root; translocation to shoot constant
	Wiebe and Kramer (1954)	SO_4^{2-}	7	Seminal axis	Total absorption variable but amount translocated increased with distance from root tip
	Robards et al. (1973)	Ca^{2+}	>40, 32, 4	Seminal axis, Nodal axis, First order lateral	In each type of axis maximum seen in calcium translocation, probably related to xylem maturation, followed by a very major reduction in translocation as suberized lamellae appeared in the endodermis
	Ferguson and Clarkson (1976b)	Ca^{2+}, Mg^{2+}	30	Seminal axis	Translocation of both ions reduced to trivial amounts where endodermis suberized
	Clarkson and Sanderson (1978)	Fe^{3+}	>30	Seminal axis	Effective translocation found only in the 1–5 cm region of Fe-sufficient and 1–8 cm region of Fe-stressed plants
			6	First order lateral	Only the apical 1.5 cm translocated iron effectively; pattern not altered by Fe stress

Species	Reference	Ion(s)	Root type		Notes
Wheat (*Triticum aestivum*)	Bowen and Rovira (1967)	$H_2PO_4^-$	Seminal axis	7	Described rapid scanning technique after short uptake periods. Compared excised and intact roots
	Rovira and Bowen (1968)	$H_2PO_4^-$, SO_4^{2-}, Cl^-	Seminal axis	>25	Used scanning technique—uptake by laterals combined with main axis absorption. Results stress the significance of lateral roots in absorption/translocation of phosphate
Maize (*Zea mays*)	Yakovlev (1970)	$\left\{ \begin{array}{l} H_2PO_4^- \\ SO_4^{2-}, Cl^- \\ Ca^{2+} \end{array} \right.$	Primary root	16	Measured net absorption by depletion method over 2 d (non-sterile). No measurement of translocation to shoot. Relatively uniform uptake of SO_4^{2-} and Cl^-; $H_2PO_4^-$ uptake large peak in the apical 2 cm; Ca^{2+} uptake declined at >14 cm from tip. Efflux of all ions and of H^+ greatest at root tip
	Marschner and Richter (1973)	K^+	Primary root	18	Relatively uniform rate of uptake but percentage translocated increased with distance from tip
	Richter and Marschner (1973)	K^+	Primary root	18	Rate of efflux very large in apical 3 cm of root, amounting to 40 per cent of ^{42}K received from other zones
	Burley *et al.* (1970)	$H_2PO_4^-$	Primary root	28	Amount translocated relatively constant; increased movement to shoot in mature regions associated with full differentiation of the xylem. (N.B. Endodermis probably suberized in oldest zones tested)
	Ferguson and Clarkson (1975)	$H_2PO_4^-$	Primary root	>30	Suberization of endodermis was without effect on translocation; suberization of hypodermis placed very great restriction on radial movement into cortex/stele
	Marscher and Richter (1973)	Ca^{2+}	Primary root	18	Absorption declined after 3 cm from tip but rate of translocation remained >85 per cent of uptake in 24 h
	Ferguson and Clarkson (1975)	Ca^{2+}	Primary root	>30	Translocation reduced markedly by endodermal suberization. Metabolic dependence of transport declined with age
	Kashirad, Marschner and Richter (1973)	Fe^{3+}	Primary root	18	Small amounts of labelled iron, derived from ionic and chelated sources, reached shoot from all zones examined
Sand sedge (*Carex arenaria*)	Robards, Clarkson and Sanderson (1979)	K^+, $H_2PO_4^-$	Adventitious axis	12	Radial penetration of both ions virtually eliminated beyond 3 cm from tip by development of suberized hypodermis

Table 17.4 (cont.)

Species	Reference	Ions measured	Length of root examined (cm)	Type of root	Conclusions/comments
Marrow (*Cucurbita pepo*)	Harrison-Murray and Clarkson (1973)	K^+, $H_2PO_4^-$, Ca^{2+}	>50	Seminal axis	High rates of translocation of $H_{2.4}PO_4^-$ and K^+ from zones where endodermis suberized and secondary thickening of stele had begun. Ca^{2+} uptake/translocation declines to low level when endodermis suberizes
Pea (*Pisum sativum*)	Grasmanis and Barley (1969)	NH_4^+, NO_3^-	13.5	Primary root	Absorption of both ions maximal in apical 1 cm, minimal in 1–4.5 cm zone but thereafter constant. The only published account of NO_3^- uptake pattern
Pine (*Pinus radiata*)	Bowen (1969) Bowen (1970)	$H_2PO_4^-$ $H_2PO_4^-$	10 10	Primary root Primary root	In phosphate-sufficient plants relatively uniform uptake but marked peak at tip of phosphate-stressed roots. Translocation from all zones
	Bowen (1968)	Cl^-	10	Primary root	Measured loss of $^{36}Cl^-$ from labelled roots using scanning technique. Loss greatest in zone 5–10 cm
Apple and cherry (*Malus* and *Prunus*)	Atkinson and Wilson (1979)	$H_2PO_4^-$	0.5 Not specified	White root Woody root	Uptake and translocation of all three ions occurred in brown, secondarily thickened parts of the root system
	Atkinson and Wilson (1980)	K^+, Ca^{2+}	0.5 Not specified	White root Woody root	Calcium translocated from thickened roots: this contrasts with predictions from work on herbaceous species. Rates of uptake by the two species comparable

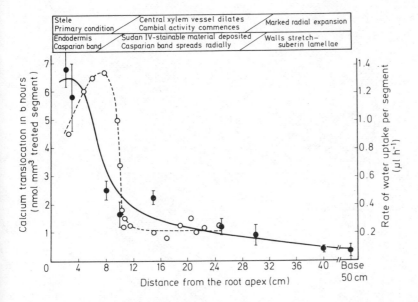

Figure 17.2 *A comparison of the patterns of water absorption and calcium translocation along the length of primary root axes of Cucurbita pepo (marrow) cv. greenbush, in relation to the anatomical development of the stele and the endodermis. The information was obtained from sets of plants growing in slightly varying conditions on different occasions, but both water absorption (– – –) and calcium translocation (——) are greatly reduced in the zone where suberized endodermal cells make their appearance, even though the conductivity of the xylem increases in this zone. (Results taken from Harrison-Murray and Clarkson, 1973; Graham, Clarkson and Sanderson, 1974)*

carrying capacity through the plasmodesmata and a relatively high resistance to water flow in the symplast. On this last point, however, opinion is divided (*see* Newman, 1976 for a contrary view).

The hypothesis above rests heavily on the assumption that suberin lamellae make the endodermal walls relatively impermeable. This assumption is difficult to test directly because the endodermis is surrounded by other tissues and the walls are perforated by numerous plasmodesmata (up to 10^6 mm^{-2} of wall surface). The endodermis can be isolated by enzymatic digestion but unless the plasmodesmatal channels become blocked when the tissue is disrupted, the isolated walls would have sieve-like properties which they probably do not have *in vivo*. This problem applied to a lesser extent in studies of the permeability of isolated, suberized hypodermal walls in *A. cepa* (Clarkson *et al.*, 1978) and *Zea mays* where there appear to be relatively few plasmodesmata. These studies suggest that the presence of suberin lamellae does not necessarily imply low permeability, because suberized hypodermal cells are present in *A. cepa* in a zone characterized by very high rates of water absorption (Rosene, 1941). By contrast, suberin lamellae in the hypodermis of *Z. mays* greatly restrict radial movement of phosphate and water into the root (Ferguson and Clarkson, 1976a and unpublished results). Until the nature of the suberin lamellae in endodermal walls has been characterized some caution is necessary in accepting the above hypothesis.

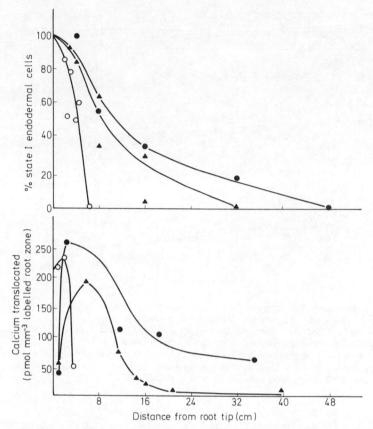

Figure 17.3   *The influence of endodermal suberization on calcium translocation in three types of root member of* Hordeum vulgare *(barley) cv. midas. State I endodermal cells lack suberized lamellae in their walls and possess only a Casparian band. The linear distance over which suberization occurs is related to the elongation rate and meristem size of a given root member. In each type, suberin lamellae appear after approximately the same length of time. Suberization does not occur simultaneously in all of the endodermal cells, those over the phloem being the first and those over the protoxylem poles being the last to develop lamellae in their walls.* ▲ = Seminal axis. ● = Nodal axis. ○ = 1 degree lateral root. *(From Robards* et al., *1973)*

Whether or not the explanation advanced above is correct, it is clear that most of the root surface can absorb the major nutrient ions and transport them to the shoot (*Table 17.4*). The persistence of phosphate transport in old parts of barley root axes (> 0.5 m from the tip) and in secondarily thickened roots of fruit trees (Atkinson and Wilson, 1980) invalidates the idea that xylem loading depends on living xylem cells close to the root tip, the so-called test-tube hypothesis (Hylmö, 1953). A detailed study of long-distance transport in relation to xylem differentiation in barley (Läuchli *et al.*, 1978) indicates that the release of ions into the xylem during vessel differentiation could contribute only a minor fraction of the total xylem flux.

The much more extensive distribution of ion translocation capability than was formerly supposed also disposes of the idea that roots have absorbing zones characterized by the presence of root hairs, unsuberized endodermal cells and high metabolic activity. When roots grow in soil, however, time-dependent factors such as nutrient depletion at the soil/root interface may operate so as to make young parts of roots, which have recently entered a soil zone, more effective absorbers. This type of absorbing zone is determined by properties of the soil, not by those of the root.

Response to Nutrient Stress

ION TRANSPORT

Several responses to stress, such as the release from the root of hydrogen ions, reductants and enzymes, have been mentioned earlier (pages 313–314). They seem to be part of a mechanism to increase the availability of certain nutrients in the soil near the root. A further part of this mechanism concerns the rate of ion transport by the root and the affinity of transport mechanisms for the ions they carry, both of which increase as nutrient deficiency makes itself felt in the plant. In microorganisms there appear to be facultative transport systems of many kinds which are activated by nutrient deprivation; these activities are biochemically distinct from those which commonly operate in 'normal' cells (Ducet, Blasco and Jeanjean, 1977). In higher plants, no comparable systems have been described; this may be because the transport systems for the major nutrients can respond flexibly to a wide range of demand created by growth (Wild and Breeze, this volume). In some species there is an apparently facultative response induced by iron deficiency which may have two components: proton release and enhanced ion transport.

When plants of sunflower were grown in an unbuffered culture solution to which 2 µg Fe^{3+} ml^{-1} had been added either as $FeCl_3$ or as FeEDTA or omitted (−Fe), the pH of the culture solutions altered in a way consistent with a stress-induced hydrogen ion release (Marschner, Kalish and Römheld, 1974). The initial pH of the culture solution, >pH 6, would have precipitated most of the iron added as $FeCl_3$ so that from the outset the plants treated with the −Fe and the $FeCl_3$ solutions would have been equally disadvantaged in comparison with those treated with FeEDTA, which remains in solution at pH 6–7. After 5–6 days, the pH in the −Fe and the $FeCl_3$ solutions began to decrease rapidly to pH 3.9 (*Figure 17.4*). In these acidified cultures any precipitated iron would redissolve and be available for absorption. The consequent correction of Fe-deficiency within the plant seems to 'switch off' the proton release, and the pH of the culture solution increased back to its initial value. A further drop in pH 5–6 days later indicated that this cycle of pH changes can be repeated as long as some iron remains potentially available for absorption. The cycle does not occur when plants receive a continuous supply (FeEDTA) or none (−Fe).

Hydrogen ion release has been accompanied by release of reductants from the roots of some species (Marschner, Kalish and Römheld, 1974), and it is possible that the release of the former promotes the membrane leakage of the latter (Brown, 1978). Since the reduction of Fe^{3+} to Fe^{2+} is a crucial step in

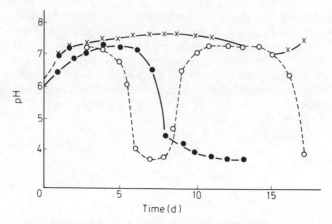

Figure 17.4 *Changes in the pH of an unbuffered culture solution during the development of iron stress in* Helianthus annuus *(sunflower). Plants were supplied with either 2μg Fe^{3+} ml^{-1} as Fe-EDTA (×) or $FeCl_3$ (○), or had no iron supplied (●). The cyclical pH changes in cultures which contained $FeCl_3$ were related to the alternating development and relief of iron stress; the shoots of the plants gave no visual indication of iron deficiency. (From Marschner, Kalish and Römheld, 1974)*

iron transport the increased synthesis of reductant within the cells may be an important part of the facultative response, but most evidence concerns release rather than endogenous levels of reductant.

The second part of the mechanism can be illustrated by some experiments with barley (Clarkson and Sanderson, 1973). The development of iron deficiency is accompanied by a rapid enhancement of translocation of iron to the shoot. This effect was induced before any visual symptoms of iron chlorosis appeared in the plant. During an extended period of iron deprivation (18 days), the tissue iron content of shoots was 78 nmol whereas in Fe-sufficient controls, which were of similar size, the content was 217 nmol. When both sets of plants were placed for 24 hours in a solution containing 5 μM Fe^{3+} (chelated with EDDHA) the greatly increased transport to shoots in Fe-stressed shoots increased the shoot content to 292 nmol, a level slightly greater than the controls, 262 nmol. Thus, in 24 hours, the shortfall in iron supply which had developed over 18 days was made good. Once this had occurred the rate of iron translocation to the shoot declined very rapidly to a rate comparable with the Fe-sufficient control (*Table 17.5*).

It would seem that species can operate either both parts of the above mechanism or the second part only; in barley the development of iron stress does not seem to be correlated with an increased hydrogen ion release by the roots (Marschner, Kalish and Römheld, 1974) but it may be promoted by an increased level of endogenous reductant.

The membrane transport systems for most ions increase in their activity as nutrient deficiency develops in a plant. The maximum velocity of absorption (v_{max}) increases and the affinity of the transport system for the ion it carries may also increase; this is indicated by a decrease in the apparent Michaelis constant, K_m. Studies of nutrient uptake in flowing culture systems have now

Table 17.5 THE INFLUENCE OF IRON STRESS ON THE TRANSLOCATION OF LABELLED IRON TO THE SHOOT OF *HORDEUM VULGARE* (BARLEY) CV. MIDAS. (FROM CLARKSON AND SANDERSON, 1973)

Duration of treatment		*Translocation of labelled iron to shoot during 24 h uptake* $(nmol\ g^{-1}\ root\ fresh\ weight \pm S.E.)$	
Fe stress (d)	*Unlabelled iron* (h)	*Control* (+ *Fe*)	*Iron-stressed*
18	0	31 ± 6	1478 ± 178
18	24	54 ± 9	536 ± 142
18	30	40 ± 7	41 ± 4
18	48	49 ± 4	77 ± 12

The rapid translocation characteristic of iron-stressed plants is quickly lost when the plants are returned to unlabelled 10 µ Fe Eddha before the tracer uptake was measured

made it clear that these two kinetic parameters can vary widely according to the internal demand for nutrients and their supply to the root surface (Wild and Breeze, this volume). The response elicited by nutrient deficiency is, however, a very specific one for the nutrient in short supply. In barley plants, phosphate deficiency greatly increased the rate of phosphate uptake but had no influence on the absorption of Cl^-, NO_3^- or SO_4^{2-} (Lee, 1980).

PATTERN OF NUTRIENT UPTAKE BY THE ROOT SYSTEM

Nutrient deficiency may reinforce an existing pattern of nutrient uptake over the root surface or alter that pattern. The response to iron stress is an example of the former kind (Clarkson and Sanderson, 1978). In barley, iron translocated from roots of iron-sufficient plants came mainly from the zone 1–4 cm from the root tip of main axes and from apical 2 cm of lateral roots. When translocation was greatly enhanced by iron deficiency this pattern was kept and most of the root surface remained ineffective in supplying iron to the shoot. When barley plants became phosphate deficient the most marked early response was a greatly increased absorption and translocation of phosphate from the old and mature zones of root axes (Clarkson *et al.*, 1978). This markedly changed the pattern of translocation which, in phosphate-sufficient plants, showed a maximum value in the zone 2–5 cm from the apex of axes. The youngest zones of the axis showed no uptake response to phosphate deficiency for up to ten days after the older regions had become fully responsive. It has also been shown that the balance of absorption between young and mature root zones can be shifted by growth temperature (Bowen, 1970) and carbohydrate supply.

In the soil where nutrients are not uniformly available it seems likely that a portion of the root system which finds itself in a soil zone where a nutrient is available can increase its rate of absorption from that zone to satisfy the whole plant requirement (Drew and Nye, 1969; Drew and Saker, 1978). As discussed below, this uptake response may be compounded with effects on root proliferation in response to favourable nutrient supply.

ROOT MORPHOLOGY

Variation in the supply of phosphate and nitrogen can have marked effects on the root/shoot weight ratio. According to a model of plant growth proposed by Brouwer and De Wit (1969) a shortage of an essential nutrient such as nitrogen limits shoot growth and utilization of carbohydrate. This results in an increase in the proportion of photosynthate translocated to the root and thus the growth of root relative to shoot increases. This seems a very sensible arrangement, for the larger root system may explore a greater volume of soil and perhaps tap new sources of nutrient. There is a penalty to be paid for this development since the larger root system will consume in its own growth and maintenance a larger proportion of the nutrient resources within the plant. This may, to some extent, offset the advantages.

The fineness or the extensiveness of root branching is also influenced by nutrition, particularly by phosphate supply. In water-cultured barley plants cv. Proctor, Hackett (1968) found a five-fold increase in the number of second-order laterals and a four-fold increase in their volume (weight) when plants were grown in a low phosphate supply as opposed to an adequate one. This response was not found, however, in another cultivar of barley, Maris Badger. In these experiments potassium deficiency had no effects on root morphology of either cultivar. In phosphate-deficient soils, root systems of several species have been described as more finely branched and 'exploratory' (Humphries, 1951; Taylor and Goubran, 1976). Bearing in mind the low mobility of phosphate in soil, this type of response to phosphate stress could be of greater practical effectiveness than alterations in the activity of the phosphate transport system in the root.

Zones of the soil with a favourable nutrient supply are frequently those where roots tend to proliferate (Weaver, 1926). In controlled conditions it has been shown that the first-order laterals, and the second-order laterals borne on them, proliferate greatly in a zone enriched in phosphate, ammonium or nitrate when the remainder of the root system receives a low rate of supply (Drew and Saker, 1975, 1978).

Efficiency of Nutrient Absorption and Utilization

The most effective use of soil nutrient resources by a plant depends on highly efficient interception and transport of nutrients as discussed in this paper, and on their efficient use within the plant. Much less is known about this latter subject where there is no general agreement how this efficiency can be measured. Attempts have been made to compare the efficiency of nutrient utilization of plants on the basis of the amount of growth produced by a given nutrient input (Boken, 1970). This approach is probably not subtle enough for comparisons of the performance of species which differ in growth rate (Blair and Cordero, 1978) and is further complicated by the fact that utilization coefficients may increase with plant age (or size) as in studies of nitrogen, phosphate and potassium nutrition of pea plants by Lastuvka and Minar (1970). Clearly, much more information on intracellular and intraorgan compartmentation and turnover of nutrients is required before we can understand the internal nutrient economy of a plant and make proper comparisons between species and genotypes.

The response to low nutrient availability may represent a balance between the efficiencies of uptake and plant utilization. Different species may accommodate themselves to such conditions by different means. Perennial species usually do so by slow growth and efficient redistribution of nutrients within the plant (Specht and Groves, 1966). In annuals of potential interest in agriculture, White (1972) found that *Phaseolus atropurpureus* exploited phosphate-deficient soil by enlarging its root system relative to the shoot. The disadvantage of this strategy is that a large demand is created in the root system for incoming phosphate, but this was offset by a high utilization coefficient in the shoot tissue. Two other species were also able to thrive in this soil, *Stylosanthes humilis* and *Desmodium intortum*, but in these the ratio of root to shoot was kept at a lower level, thus the absorptive surface was relatively smaller than in *P. atropurpureus*, but this was offset by the very low retention of phosphate in the root system.

Practical Considerations

If we accept that it is a desirable goal to improve crop species so that they can grow in soils less heavily amended with fertilizers than is the case at present, we should consider which courses of action offer the greatest scope for practical manipulation.

ROOT MORPHOLOGY AND BRANCHING

Although emphasis has been given to the importance of root geometry in the interception of nutrients of low soil mobility (pages 310 *et seq.*), the prospects of exploiting varietal differences in root form by plant breeding do not seem very promising. It is certainly known that root characteristics such as relative size, branching pattern (Troughton and Whittington, 1969) and root hair density (Bole, 1973) are heritable but the laborious experimentation necessary to demonstrate these differences in soil-grown plants has doubtless discouraged investigation. A further discouragement is that root form can be so strongly influenced by the chemical and physical condition of the soil (Russell, 1978) that genetic variation can be masked entirely. Extreme differences in form between species such as are considered in *Table 17.3* may be maintained under all circumstances but the more subtle differences between varieties within a species can be easily obscured. Attempts at major modification of the form of root systems within a species have been made using mutagens (Zobel, 1975).

With our present knowledge it is possible to suggest the morphological characteristics we would like a crop species to have, but it is difficult to select for them and probably impossible to ensure their expression in the soil except under ideal conditions.

MYCORRHIZAS

The prospects for successful manipulation of mycorrhizal associations, to allow plants to exploit soils which are of low phosphate status, seem to be excellent now that it has been demonstrated that exotic fungi with valuable attributes can be introduced into soils that contain indigenous mycorrhizal fungi and can compete with them successfully in infecting host plants (Bowen, 1977; Mosse, 1977; Powell, 1977). Practical trials in hill pasture in New Zealand have shown that spores of *Glomus tenuis* can be introduced with seeds of grass and clover and establish mycorrhizas which can exploit crude rock phosphate fertilizer (Powell, 1978). The benefits of mycorrhizas are most evident with phosphate-deficient soils and with species which have a root geometry that is not particularly favourable for exploring the soil. Thus, in experiments where native upland grasses were compared with introduced white clover (*Trifolium repens*), it was only in the latter that the formation of mycorrhizas was positively beneficial in promoting plant growth (Sparling and Tinker, 1978a, 1978b).

NUTRITION AND GENOTYPE INTERACTIONS

The rapid influence of mineral nutrition on heritable characters has been demonstrated by genetic analysis (Durrant, 1962; Durrant and Tyson, 1964), population genetics (Crossley and Bradshaw, 1968) and by the accidental selection of calcicole and calcifuge, salt and heavy metal tolerant genotypes of common crop plants at experimental stations with different soil types in the USA and elsewhere (Wright, 1977). This leads to the uncomfortable suspicion that varieties selected on well-fertilized soils of plant breeding stations may have come to lack the thriftiness necessary for survival in harsher nutritional conditions and may not be of much use for agricultural development in marginal land. A more optimistic view, however, is that this process can be put into reverse. There is much evidence that more purposeful breeding programmes have been initiated in which plants are selected in soils of low fertility or with other problems, for example, aluminium toxicity and salinity.

Selection under specific environmental pressure is perhaps the most promising way of reducing the limitations on crop production sometimes imposed by the soil and its exploitation by plant roots. It seems probable that any practical success in this endeavour will outstrip our scientific understanding of how it has been achieved, and will thus continue to provide a challenge to the physiologists to swap their role as 'followers' in the hunt of plant improvement for that of 'scouts'.

Acknowledgements

I would like to thank Dr Barbara Mosse for providing me with the photograph used for *Figure 17.1* and my colleagues Dr J. V. Lake, Dr M. G. T. Shone and Dr M. C. Drew for reading the manuscript.

References

ATKINSON, D. and WILSON, S. A. (1979). In *The Soil-Root Interface*, pp. 259–271. Ed. by J. L. Harley and R. S. Russell. Academic Press, London

ATKINSON, D. and WILSON, S. A. (1980). In *The Mineral Nutrition of Fruit Trees*. Ed. by D. Atkinson, J. E. Jackson, R. O. Sharples and W. M. Waller. Butterworths, London

BARBER, D. A. and MARTIN, J. K. (1976). *New Phytol.*, **76**, 69–80

BARBER, S. A. and OLSON, R. A. (1968). In *Changing Patterns in Fertilizer Use*, pp. 163–188. Ed. by L. B. Nelson *et al*. Soil Sci. Soc. Am., Madison, Wisconsin

BARBER, S. A. and OZANNE, P. G. (1970). *Soil Sci. Soc. Am. Proc.*, **34**, 635–637

BAYLIS, G. T. S. (1970). *Plant and Soil*, **33**, 713–716

BAYLIS, G. T. S. (1972). *Plant and Soil*, **36**, 233–234

BIELESKI, R. L. (1971). In *Recent Advances in Plant Nutrition*, pp. 143–153. Ed. by R. M. Samish. Gordon and Breach, New York

BLAIR, G. J. and CORDERO, S. (1978). *Plant and Soil*, **50**, 387–398

BOERO, G. and THIEN, S. (1979). In *The Soil-Root Interface*, pp. 231–242. Ed. by J. L. Harley and R. S. Russell. Academic Press, London

BOKEN, E. (1970). *Plant and Soil*, **33**, 645–652

BOLE, J. B. (1973). *Can. J. Soil Sci.*, **53**, 169–175

BONNETT, H. T. JR. (1968). *J. Cell Biol.*, **37**, 109–205

BOULDIN, D. R. (1961). *Soil Sci. Soc. Am. Proc.*, **25**, 476–479

BOWEN, G. D. (1968). *Nature*, **218**, 686–687

BOWEN, G. D. (1969a). *Austr. J. biol. Sci.*, **22**, 1125–1135

BOWEN, G. D. (1969b). *Plant and Soil*, **30**, 139–142

BOWEN, G. D. (1970). *Austr. J. Soil Res.*, **8**, 31–42

BOWEN, G. D. (1977). In *Reviews in Rural Science*, Vol. III: *Prospects for Improving Efficiency of Phosphorus Utilization*, pp. 103–112. Ed. by G. J. Blair. University of New England Press, Armidale, Australia

BOWEN, G. D. and ROVIRA, A. D. (1967). *Austr. J. biol. Sci.*, **20**, 369–378

BOWEN, G. D., SKINNER, M. F. and BEVEGE, D. I. (1974). *Soil Biol. Biochem.*, **6**, 141–144

BREWSTER, J. L., BHAT, K. K. S. and NYE, P. H. (1976). *Plant and Soil*, **44**, 295–328

BREWSTER, J. L. and TINKER, P. B. (1970). *Soil Sci. Soc. Am. Proc.*, **34**, 421–426

BROUWER, R. and DE WIT, C. T. (1969). in *Root Growth*, pp. 224–244. Ed. by W. J. Whittington. Butterworths, London

BROWN, J. C. (1978). *Pl., Cell Env.*, **1**, 249–258

BROWN, M. E. (1974). *Ann. Rev. Phytopath.*, **12**, 181–197

BRYANT, A. E. (1934). *New Phytol.*, **33**, 231

BURLEY, W. J., NWOKE, F. I. O., LEISTER, G. L. and POPHAM, R. A. (1970). *Am. J. Bot.*, **57**, 504–511

CLARK, R. B. and BROWN, J. C. (1974). *Crop Sci.*, **14**, 505–508

CLARKSON, D. T., MERCER, E. R., JOHNSON, M. G. and MATTHAM, D. (1975). *Agric. Res. Council Letcombe Lab. Ann. Report* 1974, pp. 10–13

CLARKSON, D. T. and ROBARDS, A. W. (1975). In *The Development and Function of Roots*, pp. 415–436. Ed. by J. G. Torrey and O. T. Clarkson. Academic Press, London

CLARKSON, D. T., ROBARDS, A. W., SANDERSON, J. and PETERSON, C. A. (1978). *Can. J. Bot.*, **56**, 1526–1532
CLARKSON, D. T. and SANDERSON, J. (1973). *Agric. Res. Council Letcombe Lab. Ann. Report* 1972, pp. 3–5
CLARKSON, D. T. and SANDERSON, J. (1978). *Pl. Physiol.*, **61**, 731–736
CLARKSON, D. T., SANDERSON, J. and RUSSELL, R. S. (1968). *Nature*, **220**, 805–806
CLARKSON, D. T., SANDERSON, J. and SCATTERGOOD, C. B. (1978). *Planta*, **139**, 47–53
CORMACK, R. G. H. (1949). *Bot. Rev.*, **15**, 583–612
CORMACK, R. G. H. (1962). *Bot. Rev.*, **28**, 446–464
CROSSETT, A. N. and CAMPBELL, D. J. (1975). *Plant and Soil*, **42**, 453–464
CROSSLEY, G. and BRADSHAW, A. D. (1968). *Crop Sci.*, **8**, 383–387
DIJKSHOORN, W. (1962). *Nature*, **194**, 165–167
DITTMER, H. J. (1937). *Am. J. Bot.*, **24**, 417–420
DREW, M. C. and NYE, P. H. (1969). *Plant and Soil*, **31**, 407–424
DREW, M. C., NYE, P. H. and VAIDYANATHAN, L. V. (1969). *Plant and Soil*, **30**, 252–270
DREW, M. C. and SAKER, L. R. (1975). *J. exp. Bot.*, **26**, 79–90
DREW, M. C. and SAKER, L. R. (1978). *J. exp. Bot.*, **29**, 435–451
DUCET, G., BLASCO, F. and JEANJEAN, R. (1977). In *Regulation of Cell Membrane Activities in Plants*, pp. 55–62. Ed. by E. Marrè and O. Ciferri. Elsevier/North-Holland, Amsterdam
DURRANT, A. (1962). *Heredity*, **17**, 27–61
DURRANT, A. and TYSON, H. (1964). *Heredity*, **19**, 207–227
FERGUSON, I. B. and CLARKSON, D. T. (1975). *New Phytol.*, **75**, 69–79
FERGUSON, I. B. and CLARKSON, D. T. (1976a). *New Phytol.*, **77**, 11–14
FERGUSON, I. B. and CLARKSON, D. T. (1976b). *Planta*, **128**, 267–269
FOY, C. D., FLEMING, A. L., BURNS, G. R. and ARMINGER, W. H. (1967). *Soil Sci. Soc. Am. Proc.*, **31**, 513–521
GERDEMANN, J. W. (1968). *Ann. Rev. Phytopath.*, **6**, 397–418
GILMORE, A. E. (1971). *J. Am. Soc. hortic. Sci.*, **96**, 35–37
GRAHAM, J., CLARKSON, D. T. and SANDERSON, J. (1974). *Agric. Res. Council Letcombe Lab. Ann, Report* 1973, pp. 9–12
GRAHAM-BRYCE, I. J. (1963). *J. agric. Sci.*, **60**, 239–244
GRASMANIS, V. O. and BARLEY, K. P. (1969). *Austr. J. biol., Sci.*, **22**, 1313–1320
HACKETT, C. (1968). *New Phytol.*, **67**, 287–300
HARRISON-MURRAY, R. S. and CLARKSON, D. T. (1973). *Planta*, **114**, 1–16
HAYMAN, D. S. (1975). In *Endomycorrhizas*, pp. 495–510. Ed. by F. E. Sanders, B. Mosse and P. B. Tinker. Academic Press, London
HUMPHRIES, E. C. (1951). *J. exp. Bot.*, **2**, 344–379
HYLMÖ, (1953). *Physiol. Plant.*, **6**, 333–405
KASHIRAD, A., MARSCHNER, H. and RICHTER, C. H. (1973). *Z. Pfl.-ernähr. Bodenk.*, **134**, 136–147
KLEINSCHMIDT, G. D. and GERDEMANN, J. W. (1972). *Phytopath.*, **62**, 1447–1453
LASTUVKA, Z. and MINAR, J. (1970). *Plant and Soil*, **32**, 189–197
LÄUCHLI, A., PITMAN, M. G., LÜTTGE, U., KRAMER, D. and BALL, E. (1978). *Pl., Cell Env.*, **1**, 217–223
LEE, R. B. (1980). *Agric. Res. Council Letcombe Lab. Ann. Report* 1979, p. 46
MAGISTAD, O. C. (1925). *Soil Sci.*, **20**, 181–226

MARSCHNER, H., KALISCH, A. and RÖMHELD, V. (1974). In *Plant Analysis and Fertilizer Problems*, pp. 273–281. Proc. 7th Int. College, Hanover, September 1974

MARSCHNER, H. and RICHTER C. H. (1973). *Z. Plf.-ernähr. Bodenk.*, **135**, 1–15

MEADOWS, D., *et al.* (1972). *The Limits to Growth*. Earth Island, London

MENGEL, D. B. and BARBER, S. A. (1974). *Agron. J.*, **66**, 399–402

MILLER, M. H. (1974). In *The Plant Root and its Environment*, pp. 643–648. Ed. by E. W. Carson. University of Virginia Press, Charlottesville

MILLER, M. H., MAMARIL, C. P. and BLAIR, G. J. (1970). *Agron. J.*, **62**, 524–527

MOSSE, B. (1973). *Phytopath.*, **11**, 171–196

MOSSE, B. (1977). *New Phytol.*, **78**, 277–288

NAIR, S. K. and SUBBA RAO, N. S. (1977). *J. Plant. Crops*, **5**, 67–70

NEWMAN, E. I. (1976). *Phil. Trans. R. Soc. London, B*, **273**, 463–478

NYE, P. H. (1969). In *Ecological Aspects of Mineral Nutrition*, pp. 105–114. Ed. by I. H. Rorison. Blackwell, Oxford

NYE, P. H. and TINKER, P. B. (1977). *Solute Movement in the Soil-Root System*. Blackwell, Oxford

OWUSU-BENNOAH, E. and MOSSE, B. (1979). *New Phytol.*, **83**, 671–679

POWELL, C. LL. (1977). *N. Z. J. agric. Res.*, **20**, 343–348

POWELL, C. LL. (1978). In *Microbial Ecology*, pp. 310–313. Ed. by M. W. Loutit and J. A. R. Miles. Springer-Verlag, Berlin

PREVOT, P. and STEWARD, F. C. (1936). *Pl. Physiol.*, **11**, 509–534

RHODES, L. H. and GERDEMANN, J. W. (1975). *New Phytol.*, **75**, 555–561

RHODES, L. H. and GERDEMANN, J. W. (1978). *Soil Biol. Biochem.*, **10**, 355–360

RICHTER, C. and MARSHNER, H. (1973). *Z. Pfl.-physiol.*, **70**, 211–221

ROBARDS, A. W., CLARKSON, D. T. and SANDERSON, J. (1979). *Protoplasma*, **101**, 331–347

ROBARDS, A. W., JACKSON, S. M., CLARKSON, D. T. and SANDERSON, J. (1973). *Protoplasma*, **77**, 291–312

ROSENE, H. F. (1941). *Pl. Physiol.*, **16**, 19–38

ROVIRA, A. D. and BOWEN, G. D. (1968). *Trans. 9th Congr. Int. Soil Sci. Soc.*, Vol. II, 209–217

ROWELL, D. L., MARTIN, M. W. and NYE, P. H. (1967). *J. Soil Sci.*, **18**, 204–222

RUSSELL, R. S. (1978). *Plant Root Systems. Their Functions and Interaction with the Soil*. McGraw-Hill, London

RUSSELL, R. S. and CLARKSON, D. T. (1973). In *Potassium in Biochemistry and Physiology*, pp. 79–92. International Potash Institute, Uppsala

SANDERS, F. E., MOSSE, B. and TINKER, P. B. (Eds) (1975). *Endomycorrhizas*. Academic Press, London

SANDERS, F. E. and TINKER, P. B. (1971). *Nature*, **223**, 278–279

SANDERS, F. E. and TINKER, P. B. (1973). *Pest. Sci.*, **4**, 385–395

SARKAR, A. N. (1977). PhD, Thesis. University of Wales, Cardiff

SCOTT, F. M. (1963). *Nature*, **199**, 1009–1010

SCOTT, L. I. (1928). *New Phytol.*, **27**, 141–174

SPARLING, G. P. and TINKER, P. B. (1978a). *J. appl. Ecol.*, **15**, 951–958

SPARLING, G. P. and TINKER, P. B. (1978b). *J. appl. Ecol.*, **15**, 959–964

SPECHT, R. L. and GROVES, R. H. (1966). *Austr. J. Bot.*, **14**, 201–221

TAYLOR, B. K. and GOUBRAN, F. H. (1976). *Plant and Soil*, **44**, 149–162

TROUGHTON, A. and WHITTINGTON, W. J. (1969). In *Root Growth*, pp. 296–313. Ed. by W. J. Whittington. Butterworths, London

VANCURA, V. (1967). *Plant and Soil*, **27**, 319–328
WEAVER, J. E. (1926). *Root Development of Field Crops*. McGraw-Hill, New York
WHITE, R. E. (1972). *Plant and Soil*, **36**, 427–447
WIEBE, H. H. and KRAMER, P. J. (1954). *Pl. Physiol.*, **29**, 342–348
WRAY, F. J. (1971). DPhil. Thesis. Oxford
WRIGHT, M. J. (Ed.) (1976). *Plant Adaptation to Mineral Stress in Problem Soils*. Cornell Univ. Agric. Exp. Stn. Spec. Publ.
YAKOVLEV, A. A. (1970). *Sov. Pl. Physiol.*, **17**, 379–383
ZOBEL, R. W. (1975). In *The Development and Function of Roots*, pp. 261–275. Ed. by J. G. Torrey and D. T. Clarkson. Academic Press, London

18

NUTRIENT UPTAKE IN RELATION TO GROWTH

A. WILD
V. G. BREEZE
Department of Soil Science, University of Reading, UK

Introduction

The two premises on which this paper is based are: (a) that plant growth is the synthesis of development (ontogeny) and dry matter production, and (b) that rate of growth for a particular genotype is determined by environmental conditions. For nutrient uptake, dry matter production is of prime importance; ontogeny is more relevant to nutrient redistribution within the plant, for example during reproductive growth and during the senescence of organs.

The extent to which nutrition is associated with dry matter production may be conveniently described in terms of the carbon balance of the plant. It would be more correct to refer to some of these processes, such as photosynthesis and respiration, by the changes in chemical energy of the components. This is because the carbon balance during photosynthesis does not necessarily indicate the amount of energy absorbed, as some energy-requiring processes such as nitrate reduction may be driven directly by light-generated reducing power. However, the magnitude of the processes which use light-generated energy directly is small in comparison with CO_2 reduction.

An overall scheme of nutrition and growth is given in *Figure 18.1*. The processes illustrated operate with feedback, so that each is dependent on others. For example, low levels of nutrition may lead to a low rate of leaf growth and of synthesis (together with a build-up in soluble carbohydrate content), which lower assimilation rates and might increase the rate of senescence. There may also be a change in the root/shoot ratio. Some of the feedback may operate via the rate of carbon assimilation, and some may be controlled by hormone levels. A more sophisticated form of *Figure 18.1* has been used by Baldwin (1976) to model plant growth and nitrogen uptake.

This scheme of nutrition and growth raises the question of how best to relate the processes quantitatively. There is a better understanding of the growth and carbon balance of certain crops than there is of nutrient uptake. The ideas of synthesis and respiration (de Vries, 1975), in which the growth rate is explained in terms of the conversion of substrate (usually glucose) into plant material of known composition, might be applied to uptake of nutrients by plant roots. Certainly, the two plant processes of carbon assimilation and nutrient uptake need to be collated and both should be treated as rate processes.

331

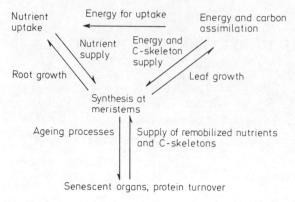

Figure 18.1   *The interdependence of nutrient uptake and growth*

Nutrient Requirement

One way to assess the nutrient requirement of crops is to multiply growth rates by the threshold internal nutrient concentration. The difficulty is that threshold concentrations vary between crops, stage of growth, the part of the plant analysed (Chapman, 1966) and probably growth rates. No single value has general validity, but for nitrogen, phosphorus and potassium the values appear to be about 2, 0.4 and 2 per cent respectively on plant dry matter, at least in young tissues.

Maximum short-term growth rates have been give as 50–54 g m^{-2} d^{-1} for C_4 plants, and 34–39 g m^{-2} d^{-1} for C_3 plants (Monteith, 1978). Short-term growth rates of crops range from 16 to 52 g m^{-2} d^{-1} for reasonably good environmental conditions (Cooper, 1975). Using the values given above for threshold nutrient concentrations, and an average growth rate of 28 g m^{-2} d^{-1} from Cooper's data, the nutrient requirements are 0.56, 0.11, and 0.56 g m^{-2} d^{-1} for nitrogen, phosphorus and potassium, respectively. These requirements agree reasonably well with measurements of nutrient uptake by crops grown under field conditions (Viets, 1965; Bromfield, 1969). Recent measurements with the potato crop have given growth rates and nitrogen uptake rates that reached 30 g m^{-2} d^{-1} and 0.5 g m^{-2} d^{-1} respectively over a five-week period (Asfary, unpublished data). In flowing nutrient solutions, ryegrass takes up about 1 g N m^{-2} d^{-1} during long days with high light intensity (Clement *et al.*, 1978; Clement, Jones and Hopper, 1979).

Results with tobacco have been reported by Raper *et al.* (1977). They grew plants in solutions of three nutrient concentrations and at three temperatures. From four harvests taken between days 7 and 32 after transplanting they calculated the relative growth rate (RGR) and the relative absorption rate (RAR) of each nutrient. Based on whole plant analysis the authors showed that the RAR for phosphate and potassium were equal to the RGR of the plants, that is, the percentage composition was constant, whereas the RAR for nitrogen, calcium and magnesium were lower than the RGR. They considered that a model for the phosphate and potassium requirement of tobacco during this growth period could be developed as a function of plant growth. For nitrogen, the other major nutrient, they observed a relationship between RAR and the RGR of the roots, which they propose to incorporate into a model (Raper *et al.*, 1978).

This method of assessing the nutrient requirements of crops has the advantage that it can be compared directly with the rate of supply of nutrients to the root surface, as has been set out by Nye and Tinker (1977). It has been used by Russell (1978) to compare the required uptake rate (the critical nutrient requirement rate) with the fluxes of nutrient release from, and fixation into, non-labile pools in soils. It is also the basis of a comprehensive model to predict the fertilizer requirements of crops (Barnes, Greenwood and Cleaver, 1976). The limitations are that there is inadequate information on the threshold nutrient concentrations for optimal growth rates and on the extent to which the requirements can be met by redistribution of nutrients within the plants. At present, only broad generalizations can be made about the changes in the required uptake rate throughout ontogeny.

Nutrient Uptake

For high crop yields the soil should maintain nutrient concentrations at the surface of roots which are optimal for plant growth. Here the interests of plant physiologists and soil scientists overlap: both might ask what the concentrations of nutrients should be at the root surface; the physiologist in order to know, for example, what minimum concentrations permit sufficient uptake to achieve potential growth rates, and the soil scientist to ensure that there are sufficient reserves of nutrients in the soil to maintain optimal concentrations. Two recent books, by Nye and Tinker (1977) and by Russell (1977), help to bring together the interests of the two groups of workers and help them to understand each other's problems.

THE ROOT–SOIL SYSTEM

Assuming that at low concentration nutrient uptake into roots varies linearly with solution concentration, for flux across the root surface we can write $F = \alpha C_1$, where $F =$ flux (mol nutrient cm^{-2} s^{-1}), $C_1 =$ concentration of nutrient in solution at the root surface (mol cm^{-3}), and α is described as the root absorption coefficient (Nye and Tinker, 1969). The concentration at the root surface will decrease if the rate of supply is less than the rate of uptake by the root, as usually occurs for nutrients such as phosphate which diffuse slowly through soil. The flux equation then becomes

$$F = \alpha \left(\frac{C_1}{C_{li}} \right) \times C_{li}$$

where C_{li} is the concentration of the bulk solution, and the ratio (C_1/C_{li}) is the fractional reduction of concentration at the root surface. Nye (1966) gave a graphical presentation of values of (C_1/C_{li}) which can be used to calculate the effect of plant properties (root radius and root absorption coefficient), of soil properties (diffusion coefficient and buffer capacity), and time.

Transpiration from leaf surfaces of plants decreases their water potential (Slatyer and Taylor, 1960), which causes mass flow of solution into the roots. The flux equation for mass flow is

$$F = VC_{li}$$

where V is the water flow ($cm^3 cm^{-2} s^{-1}$). Although mass flow and diffusion are not simply additive, an approximate equation for the total flux (Passioura, 1963) is

$$F = (C_{li} - C_l) \frac{bD\gamma}{a} + VC_{li} \tag{18.1}$$

where b = buffer power of the soil for the nutrient considered,
$\quad D$ = diffusion coefficient of nutrient in the soil,
$\quad a$ = root radius,
$\quad \gamma$ is a complex term with a value of about 0.5.

Equation (18.1) shows that flux of nutrient into the root depends on the solution concentration for both mass flow and diffusion.

These concepts have been set out fully by Nye and Tinker (1977), who have also described the less 'ideal' conditions in soil where plants reduce the water content, induce intense bacterial and often fungal activity and secrete bicarbonate ions or protons. Although there are these complicating factors, the relationship between nutrient uptake into roots and the concentration external to the roots remains of central importance. This relationship can best be understood from experiments using nutrient solutions at constant concentrations.

Use of Flowing Nutrient Solutions

Early work with static solutions (Hewitt, 1966) used relatively high nutrient concentrations in order to avoid severe depletion or too frequent replenishment of nutrients. The problem of depletion has been met by using large volumes of well-stirred solutions (Williams, 1961) and by using flowing solutions of constant concentration (Asher, Ozanne and Loneragan, 1965). The latter technique most nearly provides constant nutrient concentrations close to the roots, which can be maintained during growth. Methods have been described for measuring, controlling and recording pH and nitrate concentrations by the use of specific ion electrodes (Clement *et al.*, 1974). The method has been adapted for potassium measured by flame photometry (Woodhouse, Wild and Clement, 1978) and most recently for phosphate using colorimetry (Breeze *et al.*, unpublished data).

Using flowing nutrient solutions Asher and Ozanne (1967) and Wild *et al.* (1974) showed that the external potassium ion concentration required for maximum yield differs between species, and Spear, Asher and Edwards (1978) showed a difference between cultivars (cassava). Differences of threshold nutrient concentration between species have also been reported for phosphate (Asher and Loneragan, 1967), zinc (Carroll and Loneragan, 1968) and calcium (Loneragan, Snowball and Simmons, 1968), and seems likely to apply to all nutrients. A review by Asher (1978) gives the range for phosphate for several crop species as $\leqslant 0.1$ to $\leqslant 12\,\mu M$ (deficient), $\geqslant 0.25$ to $\geqslant 50\,\mu M$ (adequate) and $\geqslant 1$ to $\geqslant 130\,\mu M$ (toxic). These concentrations may be compared with those recommended for static solutions which are about 1 mM.

The flowing nutrient solution technique can be seen to have practical value by four examples:

(1) it is a means of quickly distinguishing cultivars which attain maximum yield at low nutrient concentrations, as was done by Spear, Asher and Edwards (1978) with cassava;
(2) it lends itself well to studies of interactions between ions such as aluminium and phosphate because low concentrations can be used which reduce the likelihood of precipitation (Asher and Edwards, 1978);
(3) it shows unequivocally that plants will attain maximum yields in very dilute solutions, which helps to substantiate the use of concentrations of 0.01–0.4 ppm phosphorus (0.3–13 μM) in soil solutions as optimal for crop growth under field conditions (Juo and Fox, 1978);
(4) it lends itself to investigation of changing nutrient supplies, as will happen under field conditions when fertilizer is applied.

Thus, when the concentration of nitrate in solution is comparatively high, nitrate which is taken up and is surplus to the plant requirements accumulates in the shoots and to a lesser extent in the roots; when the concentration in the nutrient solution decreases towards zero only part of this pool of nitrate appears to be used metabolically (Clement, Jones and Hopper, 1979).

Comparison of results between different groups of workers must be made with caution. Edwards and Asher (1974) recommended a flow rate of about $1\,l\,min^{-1}$ through each pot containing 10 g roots in a 1.5 l solution, in order to limit depletion to 5 per cent from dilute solutions (e.g. $5\,\mu M\,NO_3^-$). Sufficiently high flow rates are therefore needed in order to ensure that the nominal concentration is maintained at the root surface.

The conditions at the root surface are different in some obvious respects for plants grown in flowing nutrient solution and those grown in soil. For solution-grown plants the microbial population in solution is comparatively small (P. J. Harris, private communication); cellular components which are exuded or which leak from the roots are swept away by the flow of liquid, and ions such as calcium and sulphate probably do not accumulate close to the root as they often do in soil (Barber, Walker and Vasey, 1963; Barber and Ozanne, 1970). The value a of flowing nutrient solution is that it isolates the effects of nutrient solution composition and plant parameters from any limitations imposed by the soil. The use of this technique to measure the root absorption coefficient, and to investigate the relationship between solution concentration and plant growth, will be discussed in turn.

Root absorption coefficient

Flowing nutrient solutions are well suited to the measurement of the root absorption coefficient, α, defined as F/C, and which has been discussed by Nye and Tinker (1969, 1977). Using three plant species, Christie and Moorby

336

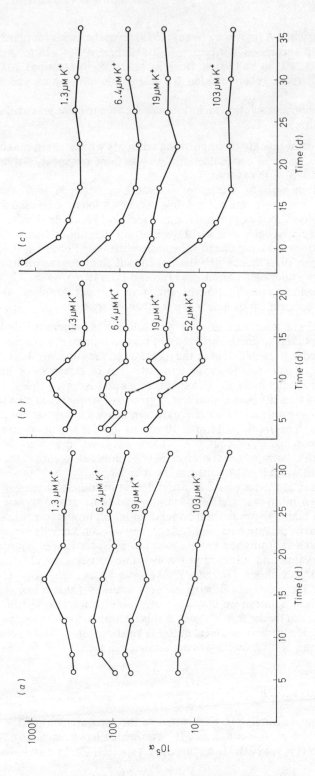

Figure 18.2 *Relationship with plant age of root absorption coefficient, α, for potassium uptake by three species: (a) radish (Raphanus sativus); (b) barley (Hordeum vulgare); (c) ryegrass (Lolium perenne)*

(1975) showed that for phosphate α decreases as plants grow older. *Figure 18.2* shows a similar change for potassium using results of experiments reported by Woodhouse, Wild and Clement (1978). The two sets of results for phosphate and potassium show in general that as plants age, their requirement for nutrient per unit surface area of root becomes less. The relationship in *Figure 18.2* is not a simple one, however, nor should this be expected because α is a resultant from interacting factors determined by plant growth, as discussed below.

Thickness of the unstirred layer

The highest value for α in *Figure 18.2* is 10^{-2} cm s^{-1}, which was the value for ryegrass in the most dilute solution (1.3 μM K$^+$) at eight days. If movement from the bulk solution to the root absorbing sites is by diffusion and this is unaffected by electrical potentials within the root, the thickness of the water film, δ, through which diffusion occurs can be calculated (Nye and Tinker, 1977) from the relationship:

$$F = \frac{D_1(C_1 - C_{int})}{\delta} = \alpha C_1 \qquad (18.2)$$

where C_{int} is the K$^+$ concentration at the interface (absorption sites for K$^+$ within the roots). As $C_{int} \to 0$, $\alpha \to D_1/\delta$. For $\alpha = 10^{-2}$ cm s^{-1} and $D_1 = 10^{-5}$ cm^2 s^{-1}, a value of $\delta = 10$ μm gives zero concentration at the root absorbing sites; that is, for a value for α of 10^{-2} cm s^{-1}, the 'zero sink' within the root is only 10 μm from the outside bulk solution. The average root radius of the ryegrass roots was 100 μm.

The highest values of α for barley and fodder radish were 6×10^{-3} and 7×10^{-3} cm s^{-1} respectively in solutions of 1.3 μM K$^+$, implying values for δ of about 20 μm for these plants. The thickness of the water film in well-stirred solutions has been estimated as 20 μm thick on roots (Levitt, 1957), and between 10 and 500 μm at the interface between membranes and solution (Dainty, 1963). Even without this water film, the sites within the root which absorb K$^+$ and reduce the [K$^+$] almost to zero must be close to the epidermis.

The calculation has not taken into account the mass flow of K$^+$ in the transpiration stream, but its contribution is less than 1 per cent to uptake from solutions of 1 μM K$^+$ and can be neglected. Possibly, in flowing nutrient solutions the water film is thinner than in stirred solutions, although there is no obvious reason why this should be so. The diffusive flux might be increased by the electrical potential of the plasmalemma, but this seems unlikely to affect diffusion over distances of more than a few nanometres.

The calculation has assumed that the flux of K$^+$ into the root is determined by the diffusive flux across the water film (passive movement). If transport within the root is mainly symplasmic (Clarkson, 1974), in the examples given where α has high values, absorption of K$^+$ seems most likely to occur in the outer cells of the cortex. Transport to the xylem through the symplast is likely to be rapid (Robards and Jackson, 1976).

The results imply strong absorption of K$^+$ at the absorbing sites, leaving a solution concentration of K$^+$ close to zero, and certainly well below 1 μM.

It has been noted elsewhere (Wild, Woodhouse and Hopper, 1979) that in terms of Michaelis–Menten kinetics, K_m is below 1 µM for K^+ when plants are grown in solutions which are maintained at constant concentrations.

Nutrient Uptake and Plant Growth

The coefficient, α, being a measure of plant demand, will depend on the growth rate of the crop. Nye and Tinker (1969) wrote:

$$\text{uptake rate per plant} = \frac{\mathrm{d}N}{\mathrm{d}t} = \frac{\mathrm{d}(WK)}{\mathrm{d}t} \tag{18.3}$$

where N = amount of nutrient in plant,
W = plant weight,
K = nutrient concentration in plant,
t = time.

As

$$\frac{\mathrm{d}(WK)}{\mathrm{d}t} = K\frac{\mathrm{d}W}{\mathrm{d}t} + W\frac{\mathrm{d}K}{\mathrm{d}t} \tag{18.4}$$

the mean rate of uptake per unit root surface area is given by

$$\text{flux} = F = \left(K\frac{\mathrm{d}W}{\mathrm{d}t} + W\frac{\mathrm{d}K}{\mathrm{d}t}\right) \times \frac{1}{2\pi\bar{r}L} \tag{18.5}$$

where $\bar{r}$ = mean root radius and L = total root length.

To maintain the potential growth rates the critical concentration of potassium within the plant requires a K^+ flux of $\mathrm{d}N/\mathrm{d}W > K_{\mathrm{crit}}$. Since $\mathrm{d}N/\mathrm{d}W = \mathrm{d}N/\mathrm{d}t \times \mathrm{d}t/\mathrm{d}W$, and from equations (18.3), (18.4) and (18.5) $\mathrm{d}N/\mathrm{d}t = F \cdot 2\pi\bar{r}L$, the condition $\mathrm{d}N/\mathrm{d}W > K_{\mathrm{crit}}$ becomes

$$F \times 2\pi\bar{r}L\,\frac{\mathrm{d}t}{\mathrm{d}W} > K_{\mathrm{crit}}, \quad \text{or} \quad F > K_{\mathrm{crit}} \times \frac{\mathrm{d}W}{\mathrm{d}t} \times \frac{1}{2\pi\bar{r}L}$$

The minimum flux, F_m, to maintain potential growth rates is thus

$$F_m = K_{\mathrm{crit}}\,\frac{\mathrm{d}W}{\mathrm{d}t} \times \frac{1}{W} \times \frac{W}{2\pi\bar{r}L} \tag{18.6}$$

where $(\mathrm{d}W/\mathrm{d}t) \times (1/W)$ is the potential relative growth rate, and $W/2\pi\bar{r}L$ is the ratio of plant weight to root surface area.

Using flowing nutrient solutions, Woodhouse, Wild and Clement (1978) found the same RGR in solutions of the two highest K^+ concentrations used, and substituted the mean value in equation (18.6). As they found no effect of K^+ concentration on the ratio $W/2\pi\bar{r}L$, the mean value obtained at all K^+ concentrations was used in equation (18.6). The value of K_{crit} was calculated from a fitted regression of the concentration of potassium in the plant against per cent RGR (using the mean of the two highest values as 100 per cent RGR), and is the concentration at 95 per cent potential RGR. The value used for all species was 0.08 mol K^+ g^{-1} fresh weight, which corresponds to approximately 2 per cent potassium in the dry matter. A fresh weight basis

Table 18.1   PARAMETERS (*see* EQUATION 18.6) USED TO
CALCULATE THE MINIMUM FLUX (F_m) REQUIRED TO
ACHIEVE OPTIMUM GROWTH RATES

Age (days)	RGR (day^{-1})	$\dfrac{W/2\pi\bar{r}L}{(\text{g cm}^{-2})}$	F_m (mol K^+ cm^{-2} s^{-1} × 10^{-12})	Required [K^+] (μM)
Radish (Raphanus sativus, L., cv. Slobolt)				
4	0.32	0.130	38.7	>100?
6	0.30	0.041	11.3	19
8	0.29	0.033	8.9	6
12	0.28	0.026	6.8	2
17	0.27	0.019	4.8	<1
21	0.26	0.020	4.8	<1
25	0.22	0.020	4.1	<1
29	0.16	0.024	3.6	<1
35	0.05	0.020	0.9	<1
Barley (*Hordeum vulgare*, L.)				
3	0.41	0.059	22.5	19
5	0.29	0.026	6.9	6
7	0.24	0.022	4.8	3
9	0.21	0.017	3.3	<1
11	0.19	0.015	2.7	<1
13	0.18	0.013	2.2	<1
15	0.18	0.012	2.0	<1
17	0.17	0.011	1.8	<1
21	0.17	0.011	1.7	<1
25	0.16	0.010	1.7	<1
Ryegrass (*Lolium perenne*, cv. S23)				
8	0.19	0.040	7.2	<1
11	0.19	0.018	3.2	<1
13	0.19	0.014	2.5	<1
17	0.19	0.012	2.2	<1
22	0.19	0.010	1.8	<1
26	0.19	0.011	1.9	<1
30	0.18	0.012	2.0	<1
34	0.16	0.014	2.1	<1
39	0.13	0.016	1.9	<1

was preferred because it accounted for a greater proportion of the variance, probably because early growth rates depend on the surface area of turgid leaves, that is, they depend on leaf fresh weight rather than on dry weight.

The values used to calculate F_m for three species are given in *Table 18.1*. To achieve potential growth rates, a high flux is required by the three species during early growth and F_m becomes less in older plants. In the early period, RGR is high, especially for barley and radish, but as the plants grow older the most marked change is the increase of root surface area to plant weight. Comparison between species shows that in the early period, when F_m is highest for radish and least for ryegrass, the ratio of root surface area to plant weight is generally lowest for radish and highest for ryegrass, but differences in RGR also exist. As the plants grow older, F_m remains higher for radish than for the other two species, but for all three the value is in the narrow range of 2 to 4×10^{-12} mol cm^{-2} s^{-1}. (A value of 2×10^{-12} mol K cm^{-2} s^{-1} corresponds to 0.9 g K m^{-2} soil surface per day for a root radius of 0.02 cm and a root length of 100 cm per cm^2 of soil surface.)

The measured flux is related to external concentration in *Figure 18.3*. For the appropriate plant age, the K$^+$ concentration that is required to give F_m can be obtained by interpolation from these, and other, plots. Only during the early period (12 days for radish, seven days for barley) was a concentration above 1 μM K$^+$ needed to achieve potential growth rates (*Table 18.1*). With ryegrass a concentration of 1 μM K$^+$ was sufficient throughout growth. Dry matter yields from these experiments have been given by Woodhouse, Wild and Clement (1978).

These experimental results show that there is no single value for the external concentration of K$^+$ close to the root which can be regarded as generally critical for plant growth. The value depends on three plant parameters: the critical internal concentration of K$^+$, the RGR, and the ratio of root surface area to plant weight. The values for the two latter parameters vary between species and change with plant age. The value of critical concentration of K$^+$ within the plant probably also varies with plant age but the experiments were with comparatively young plants and yielded no evidence on this point. This emphasis on the close relationship between K$^+$ uptake and growth supports the evidence of Pitman (1972), who postulated that a feedback mechanism between roots and shoots regulated K$^+$ uptake.

Discussion of *Table 18.1* has been in relation to the K$^+$ flux required to achieve the potential RGR. The actual K$^+$ flux for each species, measured from K$^+$ uptake between successive harvests, has been given by Woodhouse, Wild and Clement (1978). The values were shown to increase with nutrient concentration during the early period of growth, but after about 20–30 days the flux became independent of K$^+$ concentration (within the range 1–100 μM K$^+$), and became about the same for all three species.

We can ask whether the measured K$^+$ flux was determined solely by the three plant parameters incorporated in equation (18.1), or whether, irrespective of these parameters, the effectiveness of the uptake mechanism differed between species. As F_m is the required flux, F/F_m is a measure of the effectiveness of the plant in meeting its requirements for K$^+$. In *Figure 18.4* F/F_m is plotted against F_m for all values of (F/F_m) less than 1, that is, where K$^+$ uptake was less than that required. Only for radish and barley, and for these species only during early stages of growth, were there values of F/F_m less than 1. The data are therefore few, but indicate no difference between the two species in their effectiveness of uptake. Because values of (F/F_m) less than 1 occurred only during early growth (up to 12 days with radish in the most dilute K$^+$ solution), the results shed no light on whether the effectiveness of uptake changes with time.

Two points of clarification should be noted. First, the conlusion that there is no difference in effectiveness of K$^+$ uptake relates to plant populations and not to individual plants which, even in an apparently uniform batch, may differ markedly in susceptibility to potassium deficiency (Asher and Ozanne, 1977). Second, there is strong evidence that the efficiency ratio (mg dry weight per mg element in plant) does differ between crop species and between

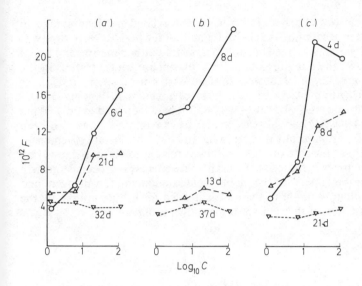

Figure 18.3 *Relationship between K^+ flux (mol cm^{-2} s^{-1}) and external K^+ concentration (μM) for three species: (a) radish* (Raphanus sativus); *(b) ryegrass* (Lolium perenne); *(c) barley* (Hordeum vulgare)

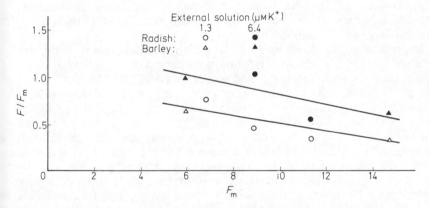

Figure 18.4 *Effectiveness of uptake mechanisms of K^+ (as F/F_m) for two plant species*

strains (Gerloff, 1976); that is, the value of the threshold nutrient concentration in plants (K_{crit} in equation 18.6) differs between species and strains. Thus crops selected for growth in nutrient-deficient soils should have a low threshold concentration; root distribution in relation to nutrient supply in soil is also important (Clarkson, this volume).

Values of (F/F_m) greater than 1 occurred more commonly in the experiments. Thus, F became temporarily higher after periods of insufficient uptake, as shown in Figure 3 of Woodhouse, Wild and Clement (1978). Apart from the early period at low concentrations, the plants always took up more K^+ than they needed by the definition of F_m. These higher ratios are not related to requirements for growth and will not be discussed.

Other workers (e.g. Barber, 1979) have also related nutrient uptake to growth. Brewster, Bhat and Nye (1976) found that the high RGR of rape was compensated by a high surface area of root, so that although rape and onion differed in RGR, their requirements for a phosphate concentration in the nutrient solution of about 10 μM were similar. With nitrate, Clement, Hopper and Jones (1978) found a maximum yield of ryegrass in flowing culture solutions at 1400 μM NO$_3^-$, but at 14 μM NO$_3^-$ it had decreased to only 90 per cent and at 1.4 μM NO$_3^-$ to 70 per cent of the maximum. It has also been shown that defoliation results in a rapid fall in NO$_3^-$ uptake rate (Clement *et al.*, 1978), and that the rate of NO$_3^-$ uptake is related to diurnal, day-to-day and seasonal changes in radiation. Overall, these effects arise because growth of the photosynthetic parts of the plants creates a demand for nitrogen, and supply of the photosynthate to the root is needed for nitrate uptake by the root. The integration of these processes is shown in *Figure 18.1*.

Conclusions

The conclusion that nutrient uptake is determined by plant growth is not new (Williams, 1946). It is also widely recognized that nutrient uptake by plant roots is regulated by the activity of the photosynthetic parts of plants (Hatrick and Bowling, 1973). Less is known about the regulating mechanisms themselves, or the sites of active uptake in the root, but progress is being made (Lüttge and Pitman, 1976).

In gross terms, the requirement of crop plants for nutrients from the soil solution for the achievement of potential growth rates is now becoming clear, as has been shown in this paper and by Barber (1979). Research of the last few years has also developed a better understanding of root development (Russell, 1977). Transport of nutrients through the soil, and the processes whereby the solution is replenished in soil, have been set out by Nye and Tinker (1977). In the soil, some major problems remain, especially in unravelling the complex interactions close to the root/soil interface and in describing the effects of water depletion on nutrient supply. Further, the application of principles to particular problems has so far been on a very limited scale. Nevertheless, the possibility of predicting nutrient requirements now exists. When this is realized there will be the opportunity to improve our understanding of the concept of 'available' nutrients and to improve the efficiency of the use of fertilizers.

References

ASHER, C. J. (1978). In *Nutrition and Food*, Sect. G: *Diets, Culture Media, Food Supplements*, Vol. III. Ed. by R. Rechcigl, Jr. CRC Press, Cleveland
ASHER, C. J. and EDWARDS, D. G. (1978). In *Mineral Nutrition of Legumes in Tropical and Subtropical Soils*, pp. 131–152. Ed. by C. S. Andrew and E. J. Kamprath. CSIRO, Melbourne
ASHER, C. J. and LONERAGAN, J. F. (1967). *Soil Sci.*, **103**, 225–233
ASHER, C. J. and OZANNE, P. G. (1967). *Soil Sci.*, **103**, 155–161

ASHER, C. J. and OZANNE, P. G. (1977). *Austr. J. Pl. Physiol.*, **4**, 499–503
ASHER, C. J., OZANNE, P. G. and LONERAGAN, J. F. (1965). *Soil Sci.*, **100**, 149–156
BALDWIN, J. P. (1976). *J. agric. Sci., Cambridge*, **87**, 341–356
BARBER, S. A. (1979). In *The Soil-Root Interface*, pp. 5–20. Ed. by J. L. Harley and R. Scott Russell. Academic Press, London
BARBER, S. A. and OZANNE, P. G. (1970). *Soil Sci. Soc. Am. Proc.*, **34**, 635–637
BARBER, S. A., WALKER, J. M. and VASEY, E. H. (1963). *J. agric. Food Chem.*, **11**, 204–207
BARNES, A., GREENWOOD, D. J. and CLEAVER, T. J. (1976). *J. agric. Sci., Cambridge*, **86**, 225–244
BREWSTER, J. L., BHAT, K. K. S. and NYE, P. H. (1976). *Plant and Soil*, **44**, 279–293
BROMFIELD, A. R. (1969). *Exp. Agric.*, **5**, 91–100
CARROLL, M. D. and LONERAGAN, J. F. (1968). *Austr. J. agric. Res.*, **19**, 859–868
CHAPMAN, H. D. (Ed.) (1966). *Diagnostic Criteria for Plants and Soils.* University of California Press
CHRISTIE, E. K. and MOORBY, J. (1975). *Austr. J. agric. Res.*, **26**, 423–436
CLARKSON, D. T. (1974). *Ion Transport and Cell Structure in Plants.* McGraw-Hill, London
CLEMENT, C. R., HOPPER, M. J., CANAWAY, R. J. and JONES, L. H. P. (1974). *J. exp. Bot.*, **25**, 81–99
CLEMENT, C. R., HOPPER, M. J. and JONES, L. H. P. (1978). *J. exp. Bot.*, **29**, 453–464
CLEMENT, C. R., HOPPER, M. J., JONES, L. H. P. and LEAFE, E. L. (1978). *J. exp. Bot.*, **29**, 1173–1183
CLEMENT, C. R., JONES, L. H. P. and HOPPER, M. J. (1979). In *Nitrogen Assimilation of Plants*, pp. 123–133. Ed. by E. J. Hewitt and C. V. Cutting. Academic Press, London
COOPER, J. P. (1975). In *Photosynthesis and Productivity in Different Environments*, pp. 593–621. Ed. by J. P. Cooper. Cambridge University Press, Cambridge
DAINTY, J. (1963). *Adv. bot. Res.*, **1**, 279–326
DE VRIES, F. W. T. P. (1975). In *Photosynthesis and Productivity in Different Environments*, pp. 459–480. Ed. by J. P. Cooper. Cambridge University Press, Cambridge
EDWARDS, D. G. and ASHER, C. J. (1974). *Plant and Soil*, **41**, 161–175
GERLOFF, G. C. (1976). In *Plant Adaptation to Mineral Stress in Problem Soils*, pp. 161–173. Ed. by M. J. Wright. Cornell University, Ithaca
HATRICK, A. A. and BOWLING, D. J. F. (1973). *J. exp. Bot.*, **24**, 607–613
HEWITT, E. J. (1966). Commonw. Bur. Hort. Plantn Crops Tech. Commun., No. 22. Commonw. Agric. Bur., Farnham Royal
JUO, A. S. R. and FOX, R. L. (1978). *Soil Science*, **124**, 370–376
LEVITT, J. (1957). *Physiol. Plant.*, **10**, 882–888
LONERAGAN, J. F., SNOWBALL, K. and SIMMONS, W. J. (1968). *Austr. J. agric. Res.*, **19**, 845–857
LÜTTGE, U. and PITMAN, M. G. (Eds) (1976). *Transport in Plants*, Vol. II, Part A: *Cells*; Part B: *Tissues and Organs*. Encycl. Pl. Physiol. N.S. Vol. 2. Springer-Verlag, Berlin
MONTEITH, J. L. (1978). *Exp. Agric.*, **14**, 1–5

NYE, P. H. (1966). *Plant and Soil*, **25**, 81–105

NYE, P. H. and TINKER, P. B. (1969). *J. appl. ecol.*, **6**, 293–300

NYE, P. H. and TINKER, P. B. (1977). *Solute Movement in the Soil-Root System.* Blackwell, Oxford

PASSIOURA, J. B. (1963). *Plant and Soil*, **18**, 225–238

PITMAN, M. G. (1972). *Austr. J. biol. Sci.*, **25**, 905–919

RAPER, C. D., OSMOND, D. L., WANN, M. and WEEKS, W. W. (1978). *Bot. Gaz.*, **139**, 289–294

RAPER, C. D., PATTERSON, D. T., PARSONS, L. R. and KRAMER, P. J. (1977). *Plant and Soil*, **46**, 473–486

ROBARDS, A. W. and JACKSON, S. M. (1976). In *Perspectives in Experimental Biology*, Vol. 2: *Botany*, pp. 413–422. Ed. by N. Sunderland. Pergamon Press, Oxford

RUSSELL, J. S. (1978). In *Mineral Nutrition of Legumes in Tropical and Subtropical Soils*, pp. 75–92. Ed. by C. S. Andrew and E. J. Kamprath. CSIRO, Melbourne

RUSSELL, R. S. (1977). *Plant Root Systems: Their Function and Interaction with the Soil.* McGraw-Hill, London

SLATYER, R. O. and TAYLOR, S. A. (1960). *Nature*, **187**, 922–924

SPEAR, S. N., ASHER, C. J. and EDWARDS, D. G. (1978). *Field Crops Res.*, **1**, 347–361

VIETS, F. G. (1965). In *Soil Nitrogen*, pp. 503–549. Ed. by W. V. Bartholomew and F. E. Clark. Am. Soc. Agron., Madison

WILD, A., SKARLOU, V., CLEMENT, C. R. and SNAYDON, R. W. (1974). *J. appl. Ecol.*, **11**, 801–812

WILD, A., WOODHOUSE, P. J. and HOPPER, M. J. (1979). *J. exp. Bot.*, **30**, 697–704

WILLIAMS, D. E. (1961). *Plant and Soil*, **15**, 387–399

WILLIAMS, R. F. (1946). *Ann. Bot., New Ser.*, **10**, 41–72

WOODHOUSE, P. J., WILD, A. and CLEMENT, C. R. (1978). *J. exp. Bot.*, **111**, 885–894

19

SYMBIOTIC NITROGEN FIXATION

ANTHONY HAYSTEAD
Hill Farming Research Organization, Midlothian, UK

JANET I. SPRENT
Department of Biological Sciences, University of Dundee, UK

Introduction

Symbiotic associations which fix atmospheric nitrogen occur between a wide variety of organisms. This paper will be restricted to a discussion of symbiotic associations between vascular plants and microorganisms because they are generally of greater importance in agricultural systems. Four types of symbiotic association occur:

(1) those involving cyanobacteria (blue-green algae) as microsymbionts, which are found in cycads and in the angiosperm genus *Gunnera*. In some cases nodules are formed at the site of infection and in all cases the endosymbiont, although potentially photosynthetic, restricts its activity to nitrogen fixation;

(2) those involving actinomycetes (filamentous bacteria). Associations of this type are called non-legume nodulated plants or actinorrhizal associations. The nodules formed on the roots of infected plants have a central stele and a cortex which contains the endophyte. The actinomycete takes on a characteristic morphology inside the cortical cells to form vesicles which may be the site of nitrogen fixation (van Straten, Akkermans and Roelofsen, 1977). After many years of effort, the endosymbiont in non-legume root nodules has at least been obtained in pure culture (Callaham, Tredici and Torrey, 1978);

(3) those involving species of the bacterial genus *Rhizobium* as microsymbiont. Rhizobia associate with legumes and with one non-legume genus, *Parasponia* (Ulmaceae). Apart from this exception, legume nodules have peripheral vascular tissue and a central area of infected cells. Inside the nodule, nitrogen-fixing rhizobia take on characteristic shapes and are known as bacteroids;

(4) associative symbioses occur between the roots of angiosperms, notably graminaceous plants, and soil bacteria. These looser associations have had considerable publicity and will be considered at the end of the chapter.

Clearly, the field of symbiotic nitrogen fixation is extremely large and it is necessary to restrict a discussion of this length to particular aspects of current research. We have chosen to concentrate on rhizobium-legume associations since they have been most intensively studied and are of considerable importance in a wide variety of agricultural systems. A comprehensive review of the biology of nitrogen fixation has been published by Sprent (1979) and detailed references to the current literature on aspects not dealt with in this chapter can be found there.

The interactions between rhizobia and their legume hosts are many and complex. Nitrogen fixation may be limited by the absence of suitable rhizobia, by suboptimal conditions for infection and nodule development, and by physiological factors that affect both symbionts. We shall consider principally the fully developed nodule and examine limitations to the rate at which nitrogen is fixed.

Legume-Rhizobium Associations

LIMITATIONS IN THE RHIZOBIUM

The limitations relate mainly to the efficiency or otherwise of the actual reduction of nitrogen to ammonia. The enzyme complex 'nitrogenase' has two parts which join and separate at each catalytic cycle (Hageman and Burris, 1978). The first part, nitrogenase reductase, is an Fe protein which accepts electrons from a low redox potential reductant (about -430 mV) such as ferredoxin, which enables it to combine with Mg-ATP. Simultaneously the second part, nitrogenase, an Fe-Mo protein, combines with the dinitrogen molecule. Dinitrogen reduction occurs when the two parts join to form the active enzyme complex and electrons flow singly from nitrogenase reductase to nitrogenase with the concomitant hydrolysis of 2ATP. The complete reduction of dinitrogen to ammonia is thought to take place in three two-electron steps. Thus we have a minimum of 12ATP being hydrolysed for each $2NH_3$ formed: the system is generally thought not to be 100 per cent efficient, and 15ATP per $2NH_3$ are thought to be more realistic.

Associated with the reduction of nitrogen is the production of hydrogen gas, by reduction of protons (H^+). Opinions differ as to whether or not proton reduction is an inevitable part of nitrogenase action, but it is agreed that it occurs after the hydrolysis of ATP and hence is a waste of energy as far as fixation of nitrogen is concerned. Thus hydrogen evolution can be considered as a factor which limits potential nitrogen fixation. However, in addition, legume root nodules can take up hydrogen: the properties of the hydrogenase, which is located in the nitrogen-fixing bacteroids, have been studied by Dixon (1972). Various functions have been suggested for this uptake hydrogenase. One is that it is oxidized to produce ATP, which offsets some of that wasted by hydrogen production. Recent work has shown that ATP from hydrogen oxidation can support nitrogenase activity in soya bean bacteroids (Emerich *et al.*, 1979). The concomitant utilization of oxygen may help to protect the oxygen-sensitive nitrogenase in isolated bacteroids, but whether this role is necessary *in vivo* remains to be determined. Net hydrogen evolution is inversely related to uptake hydrogenase activity. Schubert and Evans (1977) first suggested that hydrogen evolution is a measure of the inefficiency of nitrogen fixation and proposed the term 'relative efficiency' (RE), which is defined as the proportion of the total energy available to the nitrogenase complex which is actually used for nitrogen reduction. Experimentally this is calculated from the following expression:

$$1 - \frac{\text{hydrogen evolved in air}}{\text{hydrogen evolved in N}_2\text{-free air}}$$

The arguments are as follows: in nitrogen-free air, all electrons are available for proton reduction and thus the hydrogen evolved under these conditions is a measure of total electron flux. On the other hand, hydrogen evolved in air represents that part of the total electron flux which is wasted by not being used for nitrogen reduction. A relative efficiency of 1 is achieved when all the hydrogen produced is reused, so that none is evolved. Values of less than 1 indicate inefficiency in this sense. Schubert and Evans (1977) found a range of relative efficiencies which varied with species. Many of the non-legumes are very efficient. Lower efficiency in some of the highly selected legumes may reflect an unwitting selection for less efficient systems, since cultivars are usually selected on their performance on nitrogen-fertilized soils.

Over the last two years, studies on hydrogen evolution by legume nodules have been carried out on various species, at different stages of the growth cycle and under different environmental conditions, all of which affect the extent to which hydrogen is evolved. The important question from an agricultural point of view, however, is does hydrogen evolution affect plant productivity? Schubert, Jennings and Evans (1978) have examined effects on dry matter production and nitrogen content of soya beans (*Glycine max*) and cowpea (*Vigna unguiculata*). Symbiotic systems which recycled hydrogen were compared with those that did not. Data for soya bean cv. Anoka are reproduced in *Figure 19.1*. Plants inoculated with the *Rhizobium* strain USDA 110 had virtually no hydrogen evolution, produced more dry matter, and fixed more nitrogen *in toto* and per unit of dry matter than those nodulated with strain USDA 31, which had a relative efficiency of 0.66–0.74. The results look very convincing, but those for cowpea were much less so and as Schubert, Jennings and Evans (1978) point out, the strains of *Rhizobium* used may have varied in characteristics other than uptake hydrogenase activity. However, mutants of *Rhizobium japonicum*, apparently isogenic apart from the presence or absence of uptake hydrogenase, have now been tested and plants inoculated with the hydrogenase-producing strain contained significantly more nitrogen than those inoculated with the strain which did not synthesize an uptake hydrogenase (Albrecht *et al.*, 1979). It thus seems desirable to use rhizobia which possess an uptake hydrogenase, as far as possible, for legume inoculation.

Another limitation in *Rhizobium* is that nitrogen fixation is reduced in the presence of combined nitrogen in excess of about 3 mM. The two main theories to account for nitrate inhibition are: (a) that photosynthate is diverted to growing root tips where nitrate reduction is occurring, and (b) that nitrate is reduced in nodules, so producing levels of nitrite which inhibit nitrogen fixation. Further, since nitrate reductase has a molybdenum moiety similar to that of nitrogenase, it has been argued that there may be competition for molybdenum or the molybdenum complex between the two enzymes. On the other hand, it has recently been suggested that nitrate

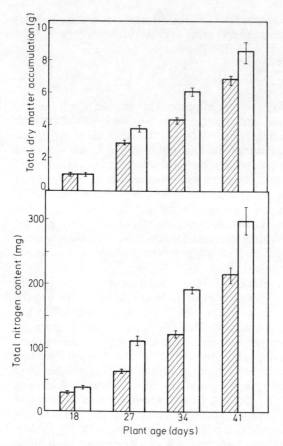

Figure 19.1   *Total dry matter and total nitrogen content of soya bean grown with a strain of Rhizobium which does not evolve hydrogen (open histogram) or with a hydrogen-evolving strain (hatched histogram). S.E. of means indicated by vertical bars. (From Schubert, Jennings and Evans, 1978)*

reduction in nodules is an essential part of the overall nitrogen metabolism in soya beans (Randall, Russell and Johnson, 1978).

Ammonia, supplied as fertilizer or from root nitrate reduction, may inhibit nitrogenase synthesis. Normally it does not accumulate in sufficient quantity in bacteroids, because it is passed to the host for assimilation. Rhizobia apparently do not fix nitrogen in the free-living state, although they have all the necessary genetic material (at least the slow growing types) but they can be made to do so under strictly controlled laboratory conditions. It has been suggested that the reason they do not fix nitrogen in nature is that they are unable to assimilate ammonia while in the nitrogen-fixing state. Effectively, there may be a simultaneous derepression of the nitrogen fixing (*nif*) genes and a repression of the genes for ammonia assimilation (Shanmugam *et al.*, 1978). One of the controlling agents appears to be glutamate, provided by host cells to the bacteroids. Thus there is normally a strict division of labour between the host cells and bacteroids, with the former assimilating the nitrogen reduced by the latter. Mutants of the free-living nitrogen-fixing bacteria *Azotobacter vinelandii* (Bishop *et al.*, 1977) and *Klebsiella pneumoniae* (Anderson, Shanmugam and Valentine, 1977) have been isolated which will fix more nitrogen than they can assimilate, and hence they excrete large quantities into the growing medium. In these mutants the normal repression of nitrogenase synthesis by ammonia must be overcome. If similar mutants of *Rhizobium* were available it would be possible for nodulated plants to assimilate combined nitrogen at the same time as nitrogen fixation was occurring. This would be useful in those crops or under those conditions where nitrogen fixation on its own may not be sufficient; for example, in *Phaseolus vulgaris* in Columbia where the life cycle is too short for some cultivars to fix sufficient nitrogen for their needs (Graham and Halliday, 1977).

LIMITATIONS DUE TO THE LEGUME HOST

Production and distribution of photosynthate

The growth of a legume, like that of any other green plant, is limited by genetically and environmentally determined factors irrespective of its source of nitrogen. Factors which restrict the production of photosynthetically active tissue and its presentation to the light will in turn depress symbiotic development and activity. If we consider a symbiosis which has had access to the special nutrient requirements necessary for nodule growth and functional development, for example molybdenum and cobalt (Evans and Russell, 1971), the rate at which nitrogen is fixed by the nodules is dependent on the host plant's capacity to: (a) supply photosynthate for reductant and ATP production, (b) rapidly remove NH_4^+ from the site of fixation and transport it throughout the host plant, and (c) provide the bacteroids with the nutrients they require, particularly a source of combined nitrogen probably in the form of glutamate (Shanmugam *et al.*, 1978). The plant's capacity to supply photosynthate is dependent on the relative rates of CO_2 fixation and respiratory loss in the shoot. Factors which increase respiration, for example higher temperatures, or which increase CO_2 fixation, for example high light intensity and CO_2 concentration, will affect net photosynthesis and ultimately increase or decrease the rate at which nitrogen is fixed in the nodules. *Figure 19.2* shows the diurnal cycle of nitrogen-fixing activity of white clover grown under two different light/temperature regimes. The data show that plants which grow at high light and a relatively low temperature fix nitrogen at a higher rate and show a minimal diurnal variation. At a lower light intensity and higher temperature, fixation is reduced and is subject to pronounced diurnal variation.

The difference between these two groups of plants probably lies in the size of the carbohydrate reserve pool under the two growth regimes. High light and a relatively low temperature increase net photosynthesis and permit the building up of carbohydrate reserves. In clover, such reserves are located principally in the stolon and can be semiquantitatively measured by counting and measuring starch grains in stolon sections (Haystead and Marriott,

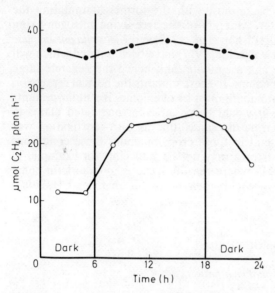

Figure 19.2   *Diurnal rhythms of acetylene reduction in white clover (cv. S184 and Rhizobium strain HFRO-4) grown in a 12 hour photoperiod of 160 Wm² (400–700 nm) at 16°C, with a night temperature of 12°C (●) and the same association grown in a 12 hour photoperiod of 80 Wm² (400–700 nm) at 20°C with a night temperature of 16°C (○)*

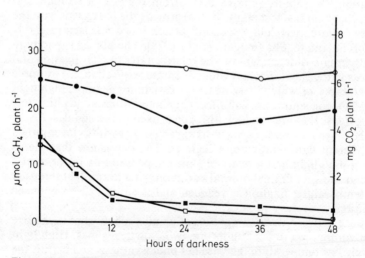

Figure 19.3   *Effect of prolonged darkness on acetylene reduction (○, □) and root respiration (●, ■) of white clover (cv. S184 and Rhizobium HFRO-4) grown in a 12 hour photoperiod of 160 Wm² (400–700 nm) at 16°C and a night temperature of 12°C (○, ●) and plants grown at 80 Wm² at 20°C with a night temperature of 16°C (□, ■)*

unpublished data). *Figure 19.3* shows that when plants from each group are placed in the dark, those with high carbohydrate reserves can continue to fix nitrogen at an undiminished rate for 48 hours while those with low reserves lose most of their nitrogen-fixing activity within 24 hours of darkening. It is an indication of the priority of the nitrogen-fixing tissue in a nodulated root system that, although nitrogen fixing continues in plants with high carbohydrate reserves, root respiration, and by inference, other energy-utilizing root processes slow down more quickly. In some legumes, however, nitrogen fixation and root respiration clearly do compete for reserve carbohydrate. Peas, for example, grow faster and fix more nitrogen in a fluctuating day/night temperature regime in which the night temperature is quite low (Roponen, 1970; Minchin and Pate, 1974), probably because if the temperature falls during the night, respiratory competition for substrates is reduced and nitrogen fixation is higher in the dark.

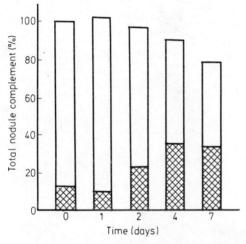

Figure 19.4   *Effect of moving white clover plants (cv. S100 and Rhizobium strain ESA-P3) from high light (120 Wm², 400–700 nm) to low light (40 Wm², 400–700 nm), both under a 12 hour photoperiod at a constant 15°C, on relative proportions of active (red □) and senescing (green ▓ pigment-containing) nodules. The total nodule complement is 100 per cent at the pretreatment harvest*

A legume that grows under conditions in which the supply of photosynthate to the nodules is adequate will support nitrogen-fixing tissue sufficient to utilize the available substrate and in the short term, as we have already shown, a reduction in carbohydrate supply will reduce fixation only until the supply has been restored. A more drastic or prolonged reduction in photosynthate production, as would occur if the shoot were partially defoliated or shaded, has the effect of reducing the amount of active nodule tissue present.

Figure 19.4 shows the change in the relative proportion of active leghaemoglobin-containing nodules and green non-nitrogen fixing nodules in white clover moved from near saturating light to a subsaturating light intensity. The plant, in effect, reduces its nodule demand to match the current rate of supply and so, in turn, reduces the amount of fixed nitrogen produced to balance a lower rate of growth.

The responses to changes in photosynthate supply described so far have been measured on plant material growing vegetatively under constant environmental conditions in a growth cabinet. In an agricultural context, however, if we are to determine whether the useful production of a legume is limited by its capacity to fix nitrogen symbiotically, account must be taken of the seasonal pattern of growth and utilization characteristic of the particular crop we are considering. To illustrate how these factors affect the productivity of nodulated legumes two contrasting examples will be considered: an annual grain legume, the garden pea, and a pasture legume,

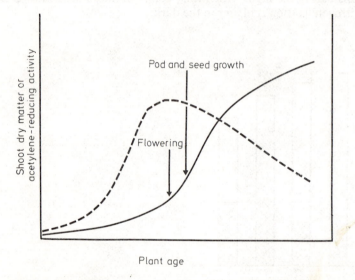

Figure 19.5   *Total shoot dry matter production (———) and acetylene-reducing activity (– – –) of peas grown in the field*

white clover. In the former, a precisely synchronized annual sequence of events occurs in the shoot and the nodulated root, which culminates in the production of reproductive structures and ultimately production and filling of the pod. By contrast, in white clover it is vegetative production of shoot material which is required in pasture systems and to some extent flowering is undesirable. *Figure 19.5* shows the pattern of growth and nitrogen fixation characteristic of the garden pea. Perhaps the most important points to note are that in peas nitrogen fixation rises to a maximum before flowering and continues during pod fill, albeit at a decreasing rate (La Rue and Kurz, 1973).

Inoculating peas with rhizobia selected for high effectivity should demonstrate whether the growth of plants nodulated by a soil population is limited by symbiotic incompetence. In general, trials in the UK have shown little or no response to inoculation, possibly because in the UK peas are usually harvested immature. In countries where peas are harvested fully mature the effects of impaired symbiotic activity are more likely to be apparent and in such circumstances yield increases have resulted from inoculation with highly effective rhizobia.

White clover is a perennial pasture legume sown for the high nutritional value of its herbage to grazing animals and because of the contribution it makes to the nitrogen economy of the associated grasses in the sward. Under grazing, growth is essentially vegetative so seasonal variations in symbiotic nitrogen fixation tend to be determined more by environmental factors than by changes in sink activity concomitant with flowering and seed production. *Figure 19.6* shows seasonal changes in nitrogen-fixing activity measured by acetylene reduction in two contrasting environments. In spring the onset of nitrogen-fixing activity is strongly correlated with soil temperature. Swards

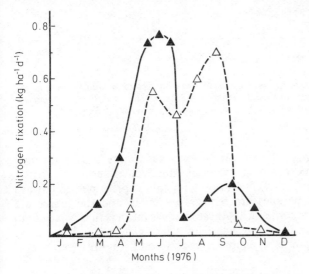

Figure 19.6   *Seasonal pattern of nitrogen fixation (acetylene reduction) of a continuously grazed ryegrass-white clover sward in two environments: peat (△) and brown earth (▲). (Data from Newbould and Haystead, 1978)*

supplied with combined nitrogen at this time of the year produce more herbage dry matter than those dependent entirely on fixed nitrogen (Haystead, unpublished data). An unavoidable consequence of such a strategy, however, is poor nodulation of stolon roots produced in spring and a lower nitrogen-fixing activity later in the year.

The midseason decrease in activity which occurs in the drier of the two environments is probably due to water stress. Sprent (1976) has shown symbiotic activity to be particularly sensitive to water stress. Recent studies (Haystead and Marriott, 1981) have shown that root and nodule senescence occur principally during late summer and at this time clover nodules are extremely sensitive to grazing defoliation. During spring growth, as *Figure 19.7* shows, the grazed plant is relatively insensitive to grazing defoliation, much less so for example than erect plants which are growing in an ungrazed environment. The difference between continuously grazed and infrequently cut clover lies in the morphological changes which occur in response to

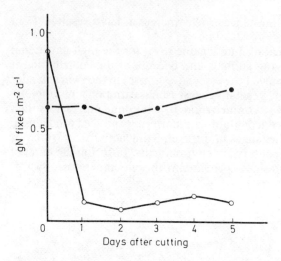

Figure 19.7   *Effect of cutting to 1.5 cm above soil level in early summer on nitrogen-fixing activity of white clover-ryegrass swards subjected to continuous grazing (●) or cut four times in the previous year (○). Grazing defoliation of the grazed swards was stopped two weeks before the cut*

grazing defoliation (King, 1963). Grazed clover has a more branched stolon system and so has more stolon material (and consequently a greater carbohydrate reserve capacity) and also more stolon apices per unit sward area. In addition, the petioles are reduced in length and the leaves are smaller. In their simplest terms, what these changes mean is that when cut to the same height a grazed clover sward loses less material and has more reserves and growing points to enable it to carry on fixing nitrogen and to regenerate rapidly.

Utilization of photosynthate

The efficiency with which carbohydrate supplied to the nodule is utilized in nitrogen fixation affects the rate at which nitrogen is fixed by the symbiosis. As previously discussed, an important factor which regulates efficiency is the extent to which energy lost to the system in hydrogen gas evolution is recycled by an uptake hydrogenase located in the microsymbiont. Other factors such as temperature (Minchin and Pate, 1973) affect the efficiency of carbohydrate utilization and it is not clear whether these factors are determined by the host, the microsymbiont or are a function of the interaction of the two.

The efficiency of photosynthate utilization is most conveniently expressed in terms of the amount of carbon respired by the nodules or the nodulated root per unit of nitrogen fixed (g C g N^{-1}). In practice, it is relatively difficult to make concurrent measurements of nitrogen fixation and the proportion of CO_2 output by the nodulated root, which is directly associated with the provision of energy for nitrogen fixation. A number of experimenters have made comparative measurements of root respiration in nodulated plants growing exclusively on N_2 and plants assimilating NO_3^- in order to determine

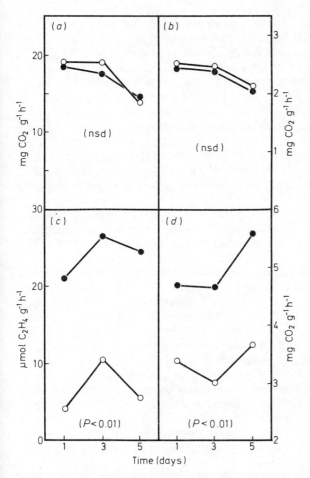

Figure 19.8 *Net photosynthesis (a), canopy dark respiration (b), acetylene reduction (c), and nodulated root respiration (d) of white clover (cv. S184 and Rhizobium strain HFRO-4) grown in a 12 hour photoperiod (80 Wm², 400–700 nm) at a temperature of 15°C supplied with 150 ppm nitrogen as NH_4NO_3 (o) or growing exclusively on N_2 (•). nsd = no significant difference. (Data from Haystead, King and Lamb, 1979)*

the relative efficiency of the two processes. In cases where there is little or no CO_2 loss from the root system associated with NO_3^- reduction (i.e. when NO_3^--reductase activity is located in the shoot), a higher rate of respiration in the nitrogen-fixing nodulated root can be taken as an estimate of the overall energy cost of the processes of nitrogen fixation, product assimilation and nodule maintenance. In many legumes, however, particularly when growing on low levels of combined nitrogen, root-located NO_3^--reductase activity is considerable and such a straightforward interpretation of the experimental data is not possible.

Figure 19.8 shows the results of an experiment conducted on single nodulated plants of white clover grown exclusively on N_2 or supplied with

150 ppm NO_3^-. Under these conditions, over 90 per cent of the NO_3^--reductase activity is located in the shoots of plants assimilating nitrate while N_2-grown plants show no detectable NO_3^--reductase activity. An estimate of the efficiency with which N_2 is fixed is obtained from these data by comparing the rate of acetylene reduction exhibited by the two sets of plants with the measured difference in root CO_2 output. Using an experimentally determined ratio of 3 mol acetylene reduced per mol of N_2 fixed, an efficiency ratio of 4.1 g of carbon respired per g of nitrogen can be calculated. A ratio as low as this implies a high level of efficiency in the white clover-rhizobium association (S184/HFRO-4) used in our experiment. This finding is consistent with our observations that S184 white clover plants nodulated exclusively with HFRO strain 4 do not evolve hydrogen during vegetative growth.

Previous reports of the respiratory cost of symbiotic nitrogen fixation have varied considerably depending on the species under investigation and on the state of development of the host plant. In general, however, during the period of most vigorous nitrogen fixation, values between 4.5 and 9 g C g N^{-1} have been obtained for the nodulated root system of a variety of legume species. Such estimates are the sum of the respiratory costs of growth and maintenance of the nodule and of the subtending root. Concurrent measurements of CO_2 efflux and nitrogen-fixing activity in detached nodules have proved rather unreliable, the range of published figures varying between 1.0 and 8.0 g C g N^{-1}. Ryle, Powell and Gordon (1979), however, have measured nodulated root respiration in cowpea before and after excising nodules and have calculated a value of 3–4 g C g N^{-1} for nodules alone. This figure agrees quite well with our own data for white clover. The theoretical minimum requirement for the nitrogenase reaction is about 2 g C g N^{-1} (Bulen and Le Compte, 1966; Hardy and Havelka, 1975) so it appears that about half of nodule respiration is associated directly with the nitrogenase reaction and half with product assimilation and nodule growth and maintenance.

Haystead *et al.* (1979) have shown that in the presence of acetylene, nodulated root respiration is depressed by 20 per cent during vegetative growth, probably because there is no assimilatory energy requirement when nitrogenase is producing ethylene. If this is the case then it implies that slightly more than 25 per cent of nodule CO_2 efflux can be associated with the assimilation and export of NH_3, the product of the nitrogenase reaction.

The nitrate-assimilating plants in *Figure 19.8* grew at least as well as did the plants growing on atmospheric nitrogen and possessed an active nitrate reductase in the shoots which, as shown in *Table 19.1*, is light dependent. Nitrate assimilation in the light could in theory proceed either at the expense of photochemically produced reductant or, alternatively, energy could be supplied by the oxidation of photosynthate within the leaf. In the latter case CO_2 efflux in the light would be expected to be greater in nitrate-assimilating leaves. Further, unless CO_2 fixation was increased as a result of the greater sink strength of nitrate-reducing leaves (Alderfer and Eagles, 1976), net photosynthesis would be reduced. *Figure 19.8* shows that net photosynthesis in plants which assimilate nitrate (and some nitrogen) is not significantly different from that in plants which grow exclusively on nitrogen. *Table 19.2* shows that in the light, $^{14}CO_2$ efflux from detached leaflets from the two

Table 19.1 ¹⁵NO₃⁻ ASSIMILATION IN 1.5 HOURS OF LIGHT OR DARKNESS BY LEAFLETS OF WHITE CLOVER GROWN ON N_2 AND ON $N_2 + NH_4NO_3$

	Nitrogen source during plant growth	
	$N_2 + NH_4NO_3$ NO_3–N reduction (g)	N_2 NO_3–N reduction (g)
Light*	6.8×10^{-8}	N.D.
Dark	N.D.	N.D.

The values shown are the mean of five separate determinations, each on 25 leaflets.
N.D. = Not detectable. *80 Wm², 400–700 nm at 15°C.

groups of plants is not higher in those that assimilate nitrate. Details of the experimental procedures used to obtain these data are presented in Haystead *et al.* (1979, 1980).

On the basis of these data it would appear that under the growth conditions of this experiment at least, the photoproduction of reductant and ATP within the leaf is sufficient to meet the requirements of the simultaneous reduction of CO_2 and NO_3^-.

Table 19.3 shows the daily carbon budget of single nodulated white clover plants growing exclusively on dinitrogen or supplied with nitrate. The greater CO_2 output from the nodulated roots of plants which fix all their nitrogen

Table 19.2 RATIO OF ¹⁴CO_2 OUTPUT IN LIGHT AND DARK IN LEAFLETS OF WHITE CLOVER GROWN ON N_2 AND $N_2 + NH_4NO_3$

Nitrogen added to nutrient medium*	Nitrogen source during growth	
	$N_2 \pm S.E.$**	$N_2 + NH_4NO_3 \pm S.E.$
0	3.07 ± 0.10	2.38 ± 0.10
150 ppm N as NH_4NO_3	3.08 ± 0.11	2.45 ± 0.15

*Quarter strength Dart and Pate's solution (Dart and Pate, 1959) pH 6.5.
**Each datum is the mean of determinations made on ten sets of leaflets for two successive light/dark cycles, i.e. 20 ratio determinations.

Table 19.3 PHOTOSYNTHESIS, DARK SHOOT RESPIRATION, ROOT RESPIRATION AND GROWTH OF WHITE CLOVER PLANTS GROWING ON N_2 AND ON $N_2 + NH_4NO_3$*

	$N_2 \pm S.E.$	$N_2 + NH_4NO_3 \pm S.E.$
Net canopy photosynthesis (mg CO_2 per plant d⁻¹)	639 ± 36	788 ± 37
Dark canopy respiration (mg CO_2 per plant 12 h dark)	88 ± 9	106 ± 8
Root respiration (mg CO_2 per plant d⁻¹)	154 ± 10	141 ± 12
Growth rate calculated from CO_2 exchange (mg C per plant d⁻¹)	108	148
Calculated increase in total plant** (mg N d⁻¹)	9.0	11.7

*150 ppm N as NH_4NO_3 supplied at a rate of 30 ml d⁻¹.
**Calculated assuming 40 per cent carbon in the DM, using measured nitrogen contents of 3.35 and 3.15 per cent for N_2 and $N_2 + NH_4NO_3$ plants, respectively.

results in about 10 per cent less of the net carbon fixed by the shoots being available for plant growth. If, as the results of *Tables 19.1* and *19.2* indicate, there is no comparable respiratory carbon loss associated with nitrate reduction in the shoots then, compared with plants growing on nitrate, a limitation to the growth of nitrogen-fixing plants derives from the biochemical process in which energy is supplied to the microsymbiont.

Transport of fixed nitrogen

In all species examined so far, ammonia assimilation takes place in the host cell cytosol. For example, Boland, Fordyce and Greenwood (1978) examined 12 species from eight genera of the Papilionoideae and found that although the absolute enzyme contents varied, all species apparently used the glutamine synthetase-glutamate synthase system for assimilating ammonia

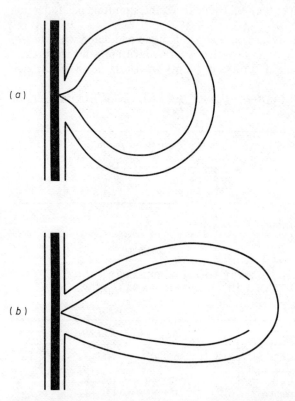

Figure 19.9   *The two basic types of vascular system found in legume nodules. (a) Determinate nodules as found in Phaseolus and Glycine. The vascular system is a closed (in reality also branched) loop of that of the subtending root. Water flow is largely confined to the low resistance pathway of the xylem. (b) Indeterminate nodules as found in Vicia and Trifolium. Water could enter and leave the nodules by different routes and/or by different methods (root pressure and transpirational pull). Water may also enter via the apical meristem. Whichever system obtains, the resistance to water flow is likely to be greater than in the determinate type of nodule*

in the cytosol. Subsequent to this stage there are variations. In some legumes, amides (glutamine and asparagine) are the principal products and are known to be exported from the nodules near the limits of their solubility (Pate, Gunning and Briarty, 1969). Scott, Farnden and Robertson (1976) have described the synthesis of asparagine in the cytosol of lupin nodules. The other major substances exported from nodules are the ureides (allantoin and allantoic acid), which are less soluble than amides (Sprent, 1980). Ureides are synthesized *de novo* from purines in the host cytosol (Herridge *et al.*, 1978; Triplett *et al.*, 1980). In species examined so far, ureides are restricted to the tropical/subtropical genus Phaseoleae which includes important grain (cowpea, soya bean, snap bean) and forage legumes (Macroptilium) (one report of ureides in Pisum (Pate, 1973) is an important exception). The same genera have nodules of determinate growth whose vascular systems fuse at the apical end when meristematic growth ceases. This may present a lower resistance path to water flow than the vascular systems of indeterminate nodules (all of which appear to be amide exporters) which remain open at the growing ends (*Figure 19.9*). At the higher temperatures of tropical/subtropical areas and with a low resistance to water flow, members of the Phaseoleae may be in a position to make best use of carbon-conserving compounds such as allantoin ($C:N = 1:1$), even though it requires 2.6 times as much water to export a similar amount of nitrogen as in asparagine. On the other hand, more temperate types such as the Vicieae and Trifolieae, which also include important grain (peas, broad beans, lentils) and forage (clovers) legumes, may have to forego the $C:N$ advantage in favour of better solubility because of lower temperatures and a more resistant water flow pathway. Once the export products have left the nodule and joined the main stream of xylem flow into the shoot, the concentration of nitrogenous compounds falls (Pate, Gunning and Briarty, 1969) and limitations due to solubility do not arise.

LIMITATIONS DUE TO THE INTERACTION BETWEEN RHIZOBIUM AND LEGUME HOST

Nutman, in 1946, first demonstrated that within a single legume species the symbiotic response of individual genotypes varies with a single strain of Rhizobium and that relative differences in symbiotic performance observed between host genotypes were also bacterial strain-specific. While some progress has recently been made in describing the fundamental processes involved in specific host-bacterium recognition (Brougton, 1978), there are many interactive aspects of symbiotic performance which are poorly understood. In field beans, El-Sherbeeny, Mytton and Lawes (1977) have shown that over 75 per cent of the variation in a range of host genotype/rhizobial strain combinations were due to the interaction between the two sources of variability. Masterson and Sherwood (1974) have shown, however, that while clover appears to select those strains which form highly effective associations from a mixed population of rhizobia (*see also* Robinson, 1969; Jones and Russell, 1972), and Mytton (1975) has demonstrated that in general more productive symbioses are found when white clover genotypes

Table 19.4 PRODUCTION OF WHITE CLOVER PLANTS IN POTS INOCULATED WITH RHIZOBIAL STRAINS, HFRO-15, HFRO-6 (EFFECTIVE) AND ESA-11 (POORLY EFFECTIVE) SEPARATELY AND IN PAIRS (1:1 MIXTURE). VALUES IN g DRY MATTER PER POT

Strain	HFRO-15	HFRO-6	ESA-11
HFRO-15	4.75	3.82	1.27
HFRO-6	—	4.82	0.93
ESA-11	—	—	1.39

LSD 5 per cent = 0.47.

are inoculated with strains of rhizobia taken from their own nodules. Conversely, as *Table 19.4* shows, virulence or competitive ability is not always combined with effectivity, some rhizobial strains being highly competitive and essentially ineffective.

At the extremes of the effectivity spectrum it is frequently possible to ascribe the genetic basis of ineffectivity to either the host or, more frequently, to the microsymbiont. Totally ineffective and highly effective strains of rhizobia frequently display these characteristics over a very wide range of host genotypes. The interactive aspects of symbiotic compatibility are more marked in strains of intermediate effectivity. Processes which limit symbiotic performance, apparently as a result of incompatibility of the host-rhizobium interface, remain a mystery. Two main possibilities exist to explain different degrees of effectivity in symbiotic associations: (a) less effective associations may be less efficient in energy terms, or (b) they may have a lower maximum nitrogen-fixing potential.

The large differences in nitrogen-fixing potential which exist between species and which enable some species, cowpea for example, to grow almost as rapidly on fixed nitrogen as they do when supplied with combined nitrogen while others invariably grow better when supplied with combined nitrogen, do not appear to be due to differences in the efficiency with which nitrogen is fixed (Ryle, Powell and Gordon, 1979).

Associative Symbiosis

The bacteria involved in associative symbiosis are capable of fixing nitrogen in the free-living state. Free-living bacteria which are not photosynthetic (or chemolithotrophic) need a source of respirable carbon from the soil and this they must obtain in open competition with other soil bacteria. Since they use a significant proportion of their energy to fix nitrogen, they are not in a good competitive position unless soil nitrogen is very low and carbon comparatively high. These conditions obtain rather rarely and thus nitrogen-fixing organisms in the rhizosphere do not normally contribute significant amounts of nitrogen to the general ecosystem. However, some appear to be firmly attached to, and even inside, root tissue and it is these which may be important. Most associations studied have involved grasses, but this does not mean that dicotyledonous plants cannot form effective associations. Rather, it reflects the dream of many agricultural scientists to couple the high dry matter yield of grasses with the nitrogen-fixing ability of the legumes.

The range of bacteria associated with roots is large and includes both aerobic and anaerobic species (Balandreau *et al.*, 1978). In many reports, nitrogenase (acetylene-reducing) activity *per se* is measured and the bacteria concerned are not fully identified. Most of the specific types studied in detail have been aerobes, in particular species of *Azotobacter* and *Azospirillum*.

Azotobacter paspali associates closely and specifically with the roots of the cultivar batatais of the tropical Bahia-grass, *Paspalum notatum*, where it may show considerable nitrogenase activity (Dobereiner, 1970). Specificity is a very desirable property because in this way the prokaryotic member of an association can avoid its competitors. How the specificity is achieved is not known and there is little information as to the extent of fixation in the field. *A. paspali* also produces growth-regulating substances which stimulate *P. notatum* growth, often in the absence of detectable nitrogenase activity (Brown, 1976).

Spirillum lipoferum has been found to associate with many grasses and other flowering plants, especially in the tropics (Dobereiner, 1978). This organism has now been reclassified as *Azospirillum* with two species, *lipoferum* and *brasiliense* (Tarrand, Krieg and Dobereiner, 1978). These species differ in various ways such as nutritional requirements and plants with which they associate, temperate species in the case of *A. lipoferum* and tropical in the case of *A. brasiliense*. Many of the latter are C_4 plants and Dobereiner has suggested that their high photosynthetic rates, coupled with efficient water use, makes them ideal for associative symbiosis. Thus, it is claimed, these associations are likely to be of most significance in the tropics. There is abundant evidence that associations with *Azospirillum* can fix nitrogen and studies with ^{15}N have shown that the fixed nitrogen can be passed on to the host plant (de Polli *et al.*, 1977). The extent of fixation varies with host genotype, which indicates that there is room for genetic improvement (von Bulow and Dobereiner, 1975).

Evidence is slowly forthcoming as to the exact location of associative bacteria in the roots. For example, Patriquin (1978) showed tetrazolium-reducing bacteria within acetylene-reducing roots of *Spartina alterniflora*. This root environment is presumed to provide the necessary microaerophilic conditions for nitrogen fixation. Even the aerobic nitrogen-fixing bacteria (including *A. paspali* in association symbiosis; Dobereiner, Day and Dart, 1972) fix more nitrogen at oxygen concentrations below atmospheric. In the case of *Azospirillum*, maximum growth on N_2 as sole nitrogen source was found to occur between 0.5 and 0.7 kPa (Burris, 1978). The generation time was over 5.5 hours compared with one hour for the same strain grown aerobically in ammonia as nitrogen source. Further, the bacteria grown on N_2 produced poly-beta-hydroxybutyrate (PHB) in quantities of up to 25 per cent of total dry weight, whereas those grown on ammonia had less than 1 per cent of PHB. Detailed studies with *Azotobacter* have shown that PHB accumulates under conditions where oxygen limits ATP synthesis (Jackson and Dawes, 1976). If the same is true of *Azospirillum*, growth under nitrogen-fixing conditions is ATP- rather than carbon-limited, because of the low partial pressure of oxygen (Po_2) required. This rather obviates the supposed advantage of C_4 plants in being able to spare more photosynthate to support fixation.

Some strains of *Azospirillum* are deficient in nitrate reductase. These can use nitrate as a terminal electron acceptor, in the absence of oxygen, to support nitrogenase activity (Scott and Scott, 1978). Presumably they would only do this as a last resort.

Very careful studies of the environmental factors that affect associative symbiosis have been carried out by French workers in Africa and France. Balandreau *et al.* (1975, 1978) have recently reviewed this work. The data are based on *in situ* measurements of acetylene reduction and are thus not subject to the criticism levelled at many other reports that plants are removed and incubated for about 24 hours before assays are made, which gives time for the rhizosphere populations to change appreciably. All the major environmental factors, temperature, soil and atmospheric water deficit and photosynthetic potential (leaf dry weight), as well as soil and nutritional factors, were found to affect nitrogenase activity. The sensitivity to water, found in maize (C_4) and rice (C_3) is another argument against C_4 plants being in a particularly favourable position to harbour associative nitrogen-fixing bacteria.

Balandreau *et al.* (1978) point out that, because of the sensitivity to environmental factors, field inoculation trials are not likely to show stimulatory effects unless care is taken to optimize growing conditions. This may be the reason why inoculation generally has not led to marked improvements in fixation (e.g. Burris, 1978). Another reason is specificity. Now that *Azospirillum* is known to have two species and many strains, it is important to establish that a particular strain is effective with the desired host before inoculation can be fairly tested. One interesting possibility is that antibiotic resistance may be involved in specificity. Dobereiner and Baldani (1979) found a 1000-fold increase in streptomycin resistance in populations of *Azospirillum lipoferum* isolated from maize roots compared to those isolated from soil. There are also problems of competition between added and native bacteria, both N_2-fixing and non-fixing types. It is perhaps unfortunate that many people have jumped on to the associative nitrogen-fixing bandwagon without appreciating all the problems. It may not be the answer to the world's protein shortage as implied by the media a year or two ago (e.g. Anon. 1975), but it could still be of considerable significance in poor soils, for example in South America and Africa. Experiments are now under way in Brazil and elsewhere to compare legume and associative symbiosis, and a more realistic assessment of the latter's potential may be expected within the next few years.

References

ALBRECHT, S. L., MAIER, R. J., HANUS, F. J., RUSSELL, S. A., EMERICH, D. W. and EVANS, H. J. (1979). *Science*, **203**, 1255–1257
ALDERFER, R. G. and EAGLES, C. F. (1976). *Bot. Gaz.*, **137**, 351–355
ANDERSON, K., SHANMUGAM, K. T. and VALENTINE, R. C. (1977). In *Genetic Engineering for Nitrogen Fixation*, pp. 95–110. Ed. by A. Hollaender, *et al.* Plenum Press, New York

ANON. (1975). The nitrogen solution. *Sunday Times*, Sept. 21

BALANDREAU, J. (1975). *Rev. Ecol. Biol. Sol.*, **12**, 273–290

BALANDREAU, J., DUCERF, P., HAMAD-FARES, I., WEINHARD, P., RINAUDO, G., MILLIER, C. and DOMMERGUES, Y. (1978). In *Limitations and Potentials for Biological Nitrogen Fixation in the Tropics*, pp. 275–302. Ed. by J. Dobereiner, R. H. Burris and A. Hollaender. Plenum Press, New York

BISHOP, P. E., GORDAN, J. K., SHAH, V. K. and BRILL, W. J. (1977). In *Genetic Engineering for Nitrogen Fixation*, pp. 67–80. Ed. by A. Hollaender, *et al.* Plenum Press, New York

BOLAND, M. J., FORDYCE, A. M. and GREENWOOD, R. M. (1978). *Austr. J. Pl. Physiol.*, **5**, 553–559

BROUGHTON, W. J. (1978). *J. appl. Bact.*, **45**, 165–194

BROWN, M. E. (1976). *J. appl. Bact.*, **40**, 341–348

BULEN, W. A. and LE COMPTE, J. R. (1966). *Proc. NAS*, **56**, 979–986

BURRIS, R. H. (1978). In *Developments in Industrial Microbiology*, Vol. 19, pp. 1–15. Ed. by L. A. Underkofler. Soc. Ind. Microbiol., Arlington

CALLAHAM, D., TREDICI, P. D. and TORREY, J. G. (1978). *Science, N.Y.*, **199**, 899–902

DART, P. J. and PATE, J. S. (1959). *Austr. J. biol. Sci.*, **12**, 427–444

DE POLLI, H., MATSUI, E., DOBEREINER, J. and SALATI, E. (1977). *Soil Biol. Biochem.*, **9**, 119–123

DIXON, R. O. D. (1972). *Arch. Mikrobiol.*, **85**, 193–201

DOBEREINER, J. (1970). *Zentrabl. Bak. Parasit.*, **124**, 224–230

DOBEREINER, J. (1978). *Ecol. Bull. (Stockholm)*, **26**, 343–352

DOBEREINER, J. and BALDANI, V. L. D. (1979). *Can. J. Microbiol.*, **25**, 1264–1269

DOBEREINER, J., DAY, J. M. and DART, P. J. (1972). *J. gen. Microbiol.*, **71**, 103–116

EL-SHERBEENY, M. H., MYTTON, L. R. and LAWES, D. A. (1977). *Euphytica*, **26**, 149–156

EMERICH, D. W., RUIZ-ARGUESO, T., CHING, T. M. and EVANS, H. J. (1979). *J. Bact.*, **137**, 153–160

EVANS, H. J. and RUSSELL, S. A. (1971). In *The Chemistry and Biochemistry of Nitrogen Fixation.*, pp. 191–244. Ed. by J. R. Postgate. Plenum Press, London

GRAHAM, P. H. and HALLIDAY, J. (1977). In *Exploiting the Legume Rhizobium Symbiosis in Tropical Agriculture*, pp. 313–334. Ed. by J. M. Vincent, A. S. Whitney and J. Bose. Agric. Univ. Hawaii, Pub. No. 145

HAGEMAN, R. V. and BURRIS, R. H. (1978). *Proc. Natl Acad. Sci., USA*, **75**, 2699–2702

HARDY, R. W. F. and HAVELKA, V. D. (1975). *Science*, **188**, 633–643

HAYSTEAD, A., KING, J. and LAMB, W. I. C. (1979). *Grass Forage Sci.*, **34**, 125–130

HAYSTEAD, A., KING, J., LAMB, W. I. C. and MARRIOTT, C. (1980). *Grass Forage Sci.*, **35**, 123–128

HAYSTEAD, A. and MARRIOTT, C. (1981). In prep.

HERRIDGE, D. F., ATKINS, C. A., PATE, J. S. and RAINBIRD, R. M. (1978). *Pl. Physiol.*, **62**, 495–498

JACKSON, F. A. and DAWES, E. A. (1976). *J. gen. Microbiol.*, **97**, 303–312

JONES, D. G. and RUSSELL, P. E. (1972). *Soil Biol. Biochem.*, **4**, 277–282

KING, J. (1963). *Pl. Soil*, **18**, 221–224

LA RUE, T. A. G. and KURZ, W. G. W. (1973). *Can. J. Microbiol.*, **19**, 304–305
MASTERSON, C. L. and SHERWOOD, M. T. (1974). *Irish J. agric. Res.*, **13**, 91–99
MINCHIN, F. R. and PATE, J. S. (1973). *J. exp. Bot.*, **24**, 259–271
MINCHIN, F. R. and PATE, J. S. (1974). *J. exp. Bot.*, **25**, 295–308
MYTTON, L. R. (1975). *Ann. appl. Biol.*, **80**, 103–107
NEWBOULD, P. and HAYSTEAD, A. (1978). Hill Farming Research Organization Report No. 7, pp. 49–68
NUTMAN, P. S. (1946). *Nature*, **157**, 463
PATE, J. S. (1973). *Soil Biol. Biochem.*, **5**, 109–119
PATE, J. S., GUNNING, B. E. S. and BRIARTY, L. G. (1969). *Planta*, **85**, 11–34
PATRIQUIN, D. G. (1978). *Ecol. Bull. (Stockholm)*, **26**, 20–27
RANDALL, D. D., RUSSELL, W. J. and JOHNSON, D. R. (1978). *Physiol. Plant.*, **44**, 325–328
ROBINSON, A. C. (1969). *Austr. J. agric. Res.*, **20**, 1053–1060
ROPONEN, I. E. (1970). *Physiol. Plant.*, **23**, 457–460
RYLE, G. J. A., POWELL, C. E. and GORDON, A. J. (1979). *J. exp. Bot.*, **30**, 135–144
SCHUBERT, K. R. and EVANS, H. J. (1977). In *Recent Developments in Nitrogen Fixation*, pp. 469–485. Ed. by W. Newton, J. R. Postgate and C. Rodriguez-Barrueco. Academic Press, London
SCHUBERT, K. R., JENNINGS, N. T. and EVANS, H. J. (1978). *Pl. Physiol.*, **61**, 398–401
SCOTT, D. B., FARNDEN, K. J. F. and ROBERTSON, J. G. (1976). *Nature*, **263**, 703–705
SCOTT, D. B. and SCOTT, C. A. (1978). In *Limitations and Potentials for Biological Nitrogen Fixation in the Tropics*, pp. 350–357. Ed. by J. Dobereiner, R. H. Burris and A. Hollaender. Plenum Press, New York
SHANMUGAM, K. T., O'GARA, F., ANDERSON, K., MORANDI, C. and VALENTINE, R. C. (1978). In *Nitrogen in the Environment*, Vol. 2, pp. 393–416. Ed. by D. R. Nielsen and J. G. MacDonald. Academic Press, London
SPRENT, J. I. (1976). In *Water Deficits and Plant Growth*, Vol. IV, pp. 291–315. Ed. by T. T. Koslowski. Academic Press, New York
SPRENT, J. I. (1979). *The Biology of Nitrogen-fixing Organisms.* McGraw-Hill, London
SPRENT, J. I. (1980). *Pl., Cell Env.*, **3**, 35–43
TARRAND, J. J., KRIEG, N. R. and DOBEREINER, J. (1978). *J. Microbiol.*, **24**, 967–980
TRIPLETT, E. W., BLEVINS, D. G. and RANDALL, D. D. (1980). *Pl. Physiol.*, **65**, 1203–1206
VAN STRATEN, J., AKKERMANS, A. D. L. and ROELOFSEN, W. (1977). *Nature*, **266**, 257–258
VON BULOW, J. F. W. and DOBEREINER, J. (1975). *Proc. Natl Acad. Sci.*, **72**, 2389–2393

20

TOWARDS THE ABOLITION OF LIMITING FACTORS

W. J. WHITTINGTON
University of Nottingham School of Agriculture, Sutton Bonington, UK

Introduction

Objectives for the agricultural scientist in abolishing limiting factors can be easily stated, namely, to identify those cultural methods that give the greatest economic return and to advise on their use. Modification to current practice can come through the use of improved varieties, the greater control of pests, diseases and weeds, the invention of new machinery and new techniques, such as direct drilling, and finally through the use of growth regulators, fertilizers and irrigation. An experimental result which illustrates the effects of some choices available to agriculturalists is that of Brain (1978). In an experiment with swedes he studied the effects of four sowing dates on two varieties which did or did not receive the fungicide Persulon. Expressing the yield of swedes in relation to a measure of the photosynthetic area, it was evident that the highest yields came from early sowing of either a mildew-resistant variety Ruta Øtofte or a susceptible variety Ne Plus Ultra, sprayed with the fungicide to control powdery mildew (*Erysiphe cruciferarum*) which otherwise results in leaf loss and reduced yield. Lower yields were obtained where the crop was planted too late or mildew was not controlled on the susceptible variety (*Figure 20.1*). The results show the benefits obtained by the choice of the best environment, the better genotype or through eliminating a genotype-environment interaction.

It is of course obvious that responses by crops to variation in individual environmental components may be a poor guide to the optimum environments when many factors vary together. The environmental factors inevitably interact, as for example where increased nitrogen applications in wheat may lower yields by increasing the prevalence of mildew.

In formal terms, Engledow and Wadham (1923) advised physiologists and breeders that 'Theoretically the procedure should be to find out the plant characters which control yield per acre and by a synthetic series of hybridizations to accumulate into one plant-form the optimum combination of yield-controlling factors. The optimum combination would doubtless be different for different localities and seasons, a fact which causes attention to revert to the adaptation problem already discussed.' There has been little difficulty in recognizing factors of importance and Vavilov (1951) listed some 40 characters in cereals. Donald (1968) believed that attention had been particularly devoted to defect elimination and that it was necessary to work towards an improved ideotype by incorporating appropriate characteristics.

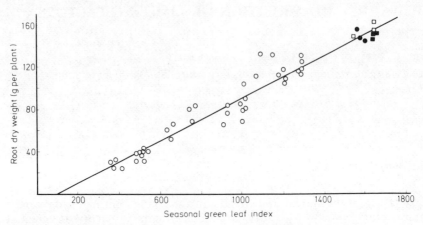

Figure 20.1   *Mean dry matter yield (g) per plant for each of three replicate plots of the varieties Ne Plus Ultra and Ruta Øtofte sown at four dates and treated or not treated with fungicide, in relation to total seasonal green leaf index over nine successive measurements,*

$$green\ leaf\ index = \frac{100 - \% \ mildew\ infection}{10} \times crop\ canopy\ size\ (scale\ 0\text{--}10)$$

Earliest sowings of Ne Plus Ultra receiving fungicide (□) and Ruta Øtofte with (■) and without (●) fungicide are indicated. All later sown plots for either variety or fungicide treatments are undifferentiated (○)

Davies (1977), however, pointed out that much depends upon one's definition of 'defects' and that any improvement may be regarded as coming from the elimination of a defect. Whatever one's views about the semantics of the problem, the requisite action is clearly the identification of factors important to yield among current varieties or primitives or among the individuals of segregating populations. There is, of course, increasing interest in exciting new breeding techniques such as protoplast fusion and genetic engineering, which will widen the genetic base for selection and allow the incorporation of specific genes into species at present without them. Thus the opportunity is available to many types of scientist to benefit agriculture, through the reduction in the number or the severity of limiting factors. It would therefore be possible to consider in this paper all aspects of the improvements of plants, improvement of the environment in which they are grown and the economic considerations which lead us either to fail to produce that which we might, or to waste or destroy deliberately a proportion of that which we do produce. In fact, the paper will consider some aspects of the elimination of limiting factors through a study of physiological genetics and plant breeding.

The ideal situation is that in which the variety used is never exceeded in performance by another and its performance is not altered by changing environments. The level of any environmental input can, however, be easily lowered to that at which yield is reduced and frequently varieties do not respond equally to the changing environments, i.e. genotype-environment interaction occurs. This effect was that which Engledow and Wadham (1923) considered under the heading of 'adaptations'. Thus we are more likely to have positive and possibly differential responses to improvement in the environment. It may then become a matter of importance to the grower to choose his variety to suit the environment or to modify the latter to suit the variety. Investigations into the extent of genotype-environment interaction and its analyses have attracted much interest (Yates and Cochran, 1938; Finlay and Wilkinson, 1963; Perkins and Jinks, 1968), but interaction was considered much earlier by R. A. Fisher as he introduced variance analysis into agricultural research. Fisher and Mackenzie (1923) published results of fertilizer trials on potatoes, which they believed showed that all varieties reacted similarly to the changing environments and thus allowed wider generalizations to be made on fertilizer usage. However, genotype-environment interaction effects are common and their analysis is often instructive. The early history of the Plant Breeding Institute (Bell, 1976) was much concerned with the introduction of two wheat varieties, Little Joss and Yeoman. They were widely used but reanalysis of early results from the National Institute of Agricultural Botany trials, using the Finlay and Wilkinson (1963) method of analysing the regression of individual varietal means against the overall mean at each trial centre, shows that Little Joss, on the basis of total yield, should have been replaced by Iron III (Panser) on the most fertile sites (*Figure 20.2*). The trials were, of course, done at a time when the amount of fertilizer used was lower than today for economic reasons and because the crop would have lodged if excessive nutrient had been applied.

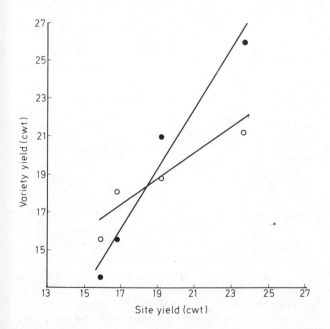

Figure 20.2   *Regression of individual yields for the varieties Little Joss* (○) *and Iron III* (●) *at each of four sites against site mean yield over all varieties under trial. (Data from Parker, 1927)*

One might believe that breeders would now be producing varieties less affected by environments, i.e. varieties with high yields and great stability, but the comparison of the relatively recent Maris Huntsman with the earlier variety Cappelle-Desprez shows that response to better environments is as evident as before (*Figure 20.3*). While it should be noted that in simple analyses such as these the results are entirely related to the varieties and environment chosen, it may yet be that we have done little to reduce

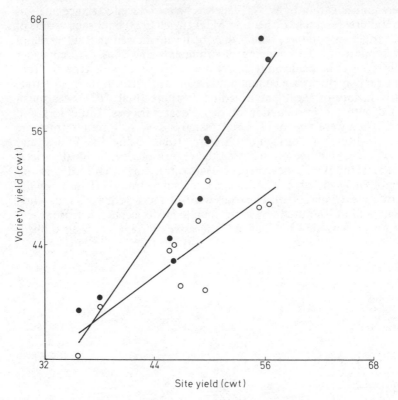

Figure 20.3   *Analysis as in* Figure 20.2 *for the varieties Cappelle-Desprez* (o) *and Maris Huntsman* (•). *(Data from Fiddian, 1970)*

responsiveness of varieties over environments even if we have come to recognize those environments capable of giving higher yields. Indeed, if one chooses to include in an analysis results from environments which give little or no yield, the highest yielding varieties must be the most responsive. It may of course be argued that with increasing concern about energy sources, natural resources and pollution, agriculturalists should be more concerned to achieve maximum production by biological rather than chemical means. Although the range of environments achieved under the latter system might be less than under the former, there is no doubt that genotype-environment interaction would still be of importance.

Selection for Characters and Processes

Although one may make broad distinctions between styles of physiology, as for example between growth analysis or biochemistry, all branches of the subject may contribute to our understanding of why plants vary in their productivity. The analytical techniques used in plant growth analysis have often been and indeed remain simple (Watson, 1968) but the studies become more detailed at the next level, which one can illustrate with reference to the study of variation in the rates of, for example, photosynthesis and respiration. The techniques now begin to involve more complex apparatus, although the time and labour spent in analysis may be little greater. Finally, at a third level are detailed studies of the biochemistry of growth, with even greater sophistication of equipment. Thus, there are three levels of analysis which may be employed to indicate through which processes different environments are having their effects and which show where old varieties might be at fault and new ones introduced by breeding. The time required for each analytical procedure is important because it will determine whether the technique might be used by breeders and used either on potential parents or segregating populations.

If one is to assume that improvements can be introduced by breeding, then genetic variation must be shown to exist for the characters in question. Furthermore, the recognition that a character could be of importance in determining yield implies that a correlation exists between variation in the character and the harvested product. On the first point there is abundant evidence, from the valuable review by Wallace, Ozbun and Munger (1972), that genetic variation exists for all characters that physiologists have been able to measure.

For the purposes of further consideration it will be helpful to examine some evidence which respresents the various levels outlined previously. It should first perhaps be noted that the character of 'yield' may be regarded as an entity despite the fact that it may be the product of several components, e.g. grain weight, spikelets per ear, grains per spikelet, etc.

As a simple character the roots of sugar beet and swedes are harvested and it is apparent that the physiological deficiency in swedes is clearly that in late autumn it is unable, as in the sugar beet, to continue accumulating root dry weight (*Figure 20.4*). This is perhaps partially due to the loss of leaf area from mildew infection followed by the regrowth of new leaves, but it is probably also related to an inherently low dry matter content when compared to the sugar beet.

In terms of yield components, Bingham (1969) concluded that in breeding for higher grain yield in cereals efforts should be concentrated on increasing the sink capacity of the grain and the photosynthetic capacity of the crop after anthesis. There has indeed been much emphasis on the photosynthetic organs, particularly those above the flag leaf node, since results suggested that lower organs made little contribution to the final weight of grain (Thorne, 1966; Biscoe et al., 1975) except under extreme conditions (Gallagher, Biscoe and Scott, 1975; Bidinger, Musgrave and Fischer, 1977). Simpson (1968) obtained significant positive correlations between grain yield and components above the flag leaf node, and flag leaf size was found to be correlated with grain yield (Monyo and Whittington, 1973; Chowdhry, Saleem and Alam,

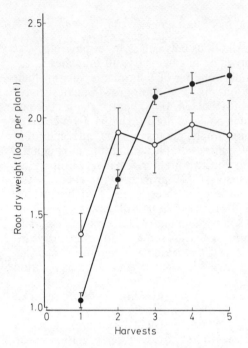

Figure 20.4   *Dry weight yields (*log g*) per plant for sugar beet (*●*) and swede (*○*) over successive harvests. (From Denton, 1974)*

1976; Sharma *et al.*, 1977). The genetics of flag leaf area has been investigated in wheat and found to show overdominance (Hsu anu Walton, 1970; Monyo and Whittington, 1973) while flag leaf area showed almost full dominance in rice (Ganashan and Whittington, 1975).

Although much attention has been centred on the flag leaf area and ear structure in supplying photosynthate to the developing grain, work by Williams and Hayes (1977) has suggested that events before anthesis are of crucial importance. They found that complete leaf excision 15 days before ear emergence affected grain size and number, but defoliation at ear emergence had little effect on ear weight in the two- and six-row Zephyr and Clermont barley varieties. Further, they found (Williams and Hayes, 1979) that dry weight of the third leaf beneath the ear (F^{-2} leaf) was positively correlated with grain number and grain yield per main culm and ear dry weight, and could thus be a useful guide in selection. The importance of the awn as a photosynthetic organ has been recognized in barley for many years. Schmid (1898) emphasized its importance with respect to yield. Zoebl and Mikosch (1892) found that awned ears took up four or five times as much water as ears from which awns were removed. Perlitus (1903) demonstrated that the length of the awns and the duration of the vegetative period were inversely proportional to each other. Vervelde (1953) found that the contribution of awns to yield was 10 per cent in spring barley and Qualset, Schaller and Williams (1965) showed that with four isogenic lines of barley with varying lengths of awns, the yield of awnless types and types with the very shortest awns were lower than those with longer awns. They also found that kernel weight was linearly correlated to awn length and this was regarded as an indication of a pleiotropic effect of genes for awn length. Patterson *et al.* (1962) had previously shown an advantage of awned over awnless types for kernel weight.

Some unpublished results (Humphries, 1973) have shown that in an F_1 diallel analysis (Mather and Jinks, 1971) the inheritance of awn length was largely additive. An examination of 79 two-row varieties from the world collection showed no correlation between awn length and flag leaf area, while a significant correlation existed among 160 six-row types. Mean two- and six-row awn lengths were 11.91 ± 0.33 and 11.37 ± 0.20 cm, respectively, but these values were not significantly different from each other. Flag leaf size was, however, significantly greater in the six-row than in the two-row types (15.08 ± 0.41 and 9.76 ± 0.47 cm^2). No correlation was found between grain weight and mean awn length of the subtending awn of 23 commercial and 11 land race varieties. In terms of number of stomata per unit area, significant variation was found between four varieties in the number of stomata per awn, in the distribution of these stomata and the genotype-position interaction. Over the varieties studied, no correlation was found between flag leaf size, awn length and stomatal densities. In a study of the capacity of awns to fix $^{14}CO_2$ a slight difference was found between the awns of Proctor and Midas, the former variety fixed more on an awn, awn area or chlorophyll content basis, but slightly less on a unit awn length basis. Since barley varieties all possess awns, there is doubtful advantage despite clear genetic differences in considering their attributes in a selection procedure.

On the other hand, Donald (1968) has suggested the incorporation of the awned character into wheat and thus comparison between the extremes of awned and awnless wheats is of great interest. The presence of awns, however, was found to be only beneficial under stress conditions (Atkins and Norris, 1955; Evans *et al.*, 1972; Olugbemi, Bingham and Austin, 1976).

Another easily recognized character which might be altered or modified in the search for a crop ideotype is leaf arrangement. The importance of leaf distribution and angle was reviewed by Wallace, Ozbun and Munger (1972) with particular reference to cereal crops. In grasses, also, the extent of light penetration through the canopy is affected by leaf arrangement (Cooper, Rhodes and Sheehy, 1971). Prostrate leaf arrangements were found to be beneficial in the early stages of growth and under frequent cutting regimes, while erect leaves were preferable for less frequent cutting (Rhodes, 1973).

In the pea crop, Davies (1977) has reviewed the current interest in 'restructuring' the plant to meet the requirements for mechanical harvesting. Essentially these are high yield, uniformity in the extent of ripeness and a high harvest index. Excessive foliage, particularly from an indeterminate habit, is most undesirable and the genetic control for alternative leaflet characteristics is simple. Thus three recessive alleles are at issue: tl, which converts the terminal tendrils of the leaf to leaflets; af, which converts the leaflets to tendrils; and st, which reduces the large leafy stipules to vestigial structures. Two particular types have been considered, namely the semileafless (af af) or the leafless forms (af af, st st). In the former the stipules represent the leaf lamina while in the latter the photosynthetic organs are the tendrils. The advantages of the model are reduced lodging, more open foliage and reduced susceptibility to disease. Although Gritton (1972), using single plants supported on wires, found the mutant types to be lower in yield than conventional types, Snoad and Gent (1974) reported results in which semileafless and leafless types performed as well as or better than leafy types. It is still too early to assess the success of the model since care has to be taken to compare varying plant populations and to use reasonably carefully selected and improved genotypes of the mutant types.

A most thorough analysis of physiological and morphological characters has recently been carried out by Kolawole (1979), who studied 20 characters in 20 barley genotypes but found correlation of yield only with plant weight, fertile tiller number, straw weight and grain to straw ratio. One suspects that one of the problems with this or earlier correlation studies (Lupton, Ali and Subramaniam, 1967) is that, at least in cereals, the yield estimate is made on relatively small numbers of plants or plants grown with wider than normal spacings so that yield often proves to be related only to numbers of tillers, grain numbers or size.

As an example of a physiological character which requires more complex methods of analysis one may take that of dark respiration rate (Jackson and Volk, 1970; Zelitch, 1971; Ludwig, 1972). Although an essential process to growth, it nevertheless results in a loss of carbon assimilates during the night period and one might imagine that overactivity might be disadvantageous. Dark respiration may be measured relatively easily by manometric methods (Zelitch, 1971) or with an infra-red gas analyser.

Genetic variation in dark respiration rates was reported by Izhar and Wallace (1967) and others (Wallace, Ozbun and Munger, 1972) but authors disagree on the relationship between dark respiration rate and net carbon exchange. Wilson (1974) found that dark respiration rates were negatively correlated with plant growth. He also found (Wilson, 1975) that genetic variation in dark respiration rate was additive and heritability estimates were high. Kolawole (1979) noted that the majority of the improved cultivars which he studied had low dark respiration rates and high yields. Further, a genotype with slow dark respiration rate had a higher dry weight, net assimilation rate and relative growth rate than the fast respiring genotype. However, in a genetic analysis the narrow sense heritability on the basis of F_3–F_2 correlations was 29 per cent (broad sense) but zero on a narrow sense basis. In certain of his experiments dark respiration rate did not appear to be correlated with some characters but in others it was negatively correlated with leaf strength, leaf length and width, and positively related to stomatal pore lengths.

Finally, as an example of physiological analysis of enzyme activity, one might consider the enzyme nitrate reductase (NR) in relation to genetic variation and yield. It has on several occasions been suggested that nitrate reductase activity might be used as a selection criterion in plant breeding (Hageman, Long and Dudley, 1967; Warner *et al.*, 1969; Croy and Hageman, 1970). The attraction in studying NR is that the reduction of nitrate has been regarded as rate limiting in the utilization of nitrogen (Beevers and Hageman, 1969) and positive correlations have been found between activity (NRA) of the enzyme and the rate of nitrogen assimilation in wheat (Croy, 1967; Eilrich, 1968; Brunetti and Hageman, 1976). NRA has also been positively correlated with grain yield (Croy, 1967; Eilrich, 1968), grain protein yield

(Croy, 1967; Edwards, 1973; Eilrich and Hageman, 1973), grain protein content (Eilrich, 1968; Dalling, Halloran and Wilson, 1975) and flour quality (Hernandez, 1972). Certainly NRA varies within species (Zieserl and Hageman, 1962) and its level within maize is highly inherited (Warner *et al.*, 1969). Activity of the enzyme can be relatively easily assessed by *in vivo* and *in vitro* techniques but it is usual to find that estimated values are higher from the latter studies (Eck, Wilson and Martinez, 1975). Inevitably there is some controversy about the merits of the two methods and their relationship between estimated and real values in the plants. It is certainly true that variation in the conditions in which the plants are raised and techniques of estimation produce markedly different results.

For the purpose of concentrating results upon one physiological process but without implying that NR activity is the key to a better future, results from recent work on grasses and wheat will be used as illustration of the diversity available to breeders. Osborne (1979), using an *in vivo* technique (Stewart, Lee and Orebamjo, 1973), studied variation in the Agrostis genus. Osborne's results (*Figure 20.5*) showed clearly the variation in NRA between species and also in the responses to nitrogen. *A. tenuis* had higher values than *A. stolonifera*, but *A. stolonifera* had, in this experiment, very similar and elsewhere superior yields and lower levels of NR than *A. tenuis*. There thus appears to be some evidence of a negative relationship between NRA and

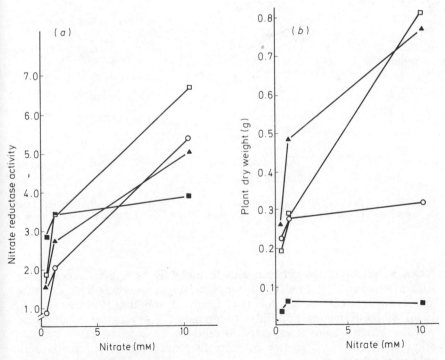

Figure 20.5   (a) Nitrate reductase activity ($\mu M\ NO_2\ h^{-1}\ g\ fwt^{-1}$) for plants of A. canina (○), A. setacea (■), A. stolonifera (▲) and A. tenuis (□) grown at 0.3, 1.0 and 10 mM nitrate. (b) Variation in dry weight for the plants of (a)

vegetative yield which is certainly unexpected and at variance with results of Bowerman and Goodman (1971). Their positive correlation led them to conclude that 'it means that selection for increased enzyme activity does not exclude simultaneous selection for increased yield'. In Osborne's results it was, however, interesting to note that NR values were lower for Italian and perennial ryegrass than for *A. tenuis*, while dry weights were greater. Over different species it appears that NR activity and vegetative productivity are not necessarily closely related.

Osborne also found considerable variation between eight *A. tenuis* plants taken from sand dunes and eight *A. stolonifera* plants from a nearby farm. The individuals and the populations were separated at 1.0 mM nitrogen both with regard to their dry weights and to their NRA values (*Figure 20.6*).

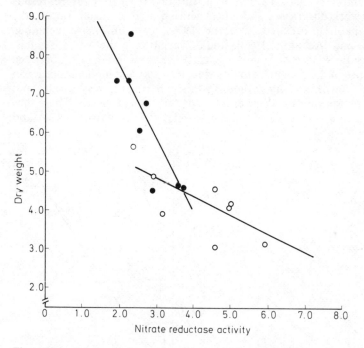

Figure 20.6 *Relationship between mean dry weight (g × 10⁻¹) of eight isolates of* A. tenuis (o) *and* A. stolonifera (•) *and nitrate reductase activity (μM NO₂ h⁻¹ g fwt⁻¹) when grown at 1.0 mM nitrate*

Negative relationships between growth and NR have been reported by Austin, Rossi and Blackwell (1978) for plant weight using Aegilops, Triticum and Triticale seedlings and for plant height in sorghum (Vaishnav *et al.*, 1978). Sinha, Rajagopal and Balasubramanian (1974) also showed that wild wheats, which would be expected to produce the smallest plant weights, often had a two-fold greater capacity to reduce nitrate compared with cultivated wheat.

Genetic control of NR in maize and wheat has been found to be relatively simple to demonstrate (Zieserl and Hageman, 1962; Croy, 1967; Eilrich,

1968; Warner, 1968). Edwards (1973), Deckard, Williams and Hammond (1975), Warner and Konzak (1975) and Sherrard *et al.* (1976), using aneuploid and substitution lines in wheat, have further shown that certain chromosomes or chromosome arms have effects on NRA. Using ditelosomic and other stocks supplied by Dr C. N. Law of the Cambridge Plant Breeding Institute, Jones has carried these investigations further using both *in vitro* and *in vivo* techniques.

It was apparent that the loss of several chromosome arms resulted in an enhanced nitrate reductase activity but the number of significant chromosomal effects were less noticeable using the *in vitro* technique (*Table 20.1*).

Table 20.1 NITRATE REDUCTASE ACTIVITY ($nM\,h^{-1}\,mg^{-1}$ fwt) FOR A GENOME DITELOSOMIC WHEAT STOCKS ASSAYED BY *IN VIVO* AND *IN VITRO* TECHNIQUES (JONES, 1979) COMPARED WITH EUPLOID (CONTROL)

Stock	In vivo	Control (%)	In vitro	Control (%)
DT-1A^L	1.46	74.1	4.70	92.5
DT-1A^S	3.76	191.0*	4.82	93.7
DT-2A^S	3.45	175.1*	5.46	103.4
DT-3A$^\alpha$	2.33	118.3	5.30	99.1
DT-3A^B	3.53	179.2*	5.67	107.4
DT-4A$^\alpha$	2.92	148.2*	6.25	114.3*
DT-5A^L	2.34	118.8	5.62	105.2
DT-6A$^\alpha$	2.15	109.1	5.72	106.3
DT-7A^L	4.69	238.1*	6.30	120.5*
DT-7A^S	4.55	231.0*	4.83	93.4

* Significant increase ($P<0.001$).

Thus it is possible to obtain enhanced enzyme activity, although not necessarily enhanced yields, through the loss of genetic material as well as attempting this by incorporation of foreign genetic material.

In these results (Jones, 1979) it was evident that chromosome 7B plays an important and complex role in the regulation of nitrate reductase. It was one of the chromosomes which exhibit a 'twin arm effect', in that the removal of either arm resulted in an approximately similar increase in NRA, and for the earlier parts of the growing season he could recognize the greater NR activities of the 7B^L + 7B^S lines. Both stocks exhibited a bimodal pattern of seasonal fluctuation in total leaf NRA, but differing rates of development produced a situation in which the peaks and troughs fell at different times. All stocks, however, exhibited the second NRA maximum at the fifth harvest.

When seedling NRA values were related to activities integrated over part of or the entire season, it was evident that whereas little relationship was found with NRA over the reproductive stage, significant positive correlations were obtained between the first harvest NRA and the activities over the vegetative stage and over the season as a whole. Those stocks which, as seedlings, exhibited NRA values greater than that of the euploid maintained the difference over the vegetative stage. Little interstock variation in NRA occurred over the reproductive stage. These results in general agree with those of Croy and Hageman (1970), who reported that significant wheat genotype differences in NRA only occurred up to 32 days after sowing. Thus as Croy (1967), Eilrich (1968) and Edwards (1973) have suggested, it is the seedling tissue that should be assayed if NRA is to be used as a selection criterion.

Problems in Selection

It is obvious from published results that one can find many morphophysiological or biochemical characters to measure and that there may sometimes be significant correlations with yield. Further work may give heritability estimates, such as those quoted, allowing the prediction of response to selection or demonstrating that the character actually does respond to selection with a resulting increase in yield. The most rapid response will occur where environmental, dominance and gene interaction effects are minimal and additive effects are maximal. In a paper for an interdisciplinary meeting such as this, it might not be considered incorrect to present the selection problem in a simpler manner than in texts on population genetics, to emphasize those genetic situations which affect the likely rate of advance. In making such models it is necessary to stipulate the characteristics of the basic population and it is easiest to assume a Hardy-Weinberg population at equilibrium with equal allelic frequencies. An F_2 generation represents such a population.

Let it be assumed that a gene A controls the level of expression of an enzyme which determines yield together with two other genes B and C, which control other physiological characters that are not being observed. In the resultant population the yield characters for each individual genotype may be plotted against the level of enzyme activity determined by the A gene. Essentially, the segregation ratios for two genes are being plotted against the segregation ratio for the other gene. The numbers in *Figures 20.7* and *20.8* represent the frequency of each genotype. Many different models are then possible depending on the values accorded to the effects of the genes and the extent of dominance or gene interaction. In the simplest example one may assume that the three genes all have equal effects on yield such that AA (etc.) = 2, Aa (etc.) = 1, and aa (etc.) = 0 units of yield. If the effects of all three genes are additive then the simplest of all situations for the breeder occurs (*Figure 20.7a*). Here, the high enzyme segregants do contain the highest yielder but this plant would not be recognized as such without consideration also of the yield figures. If the entire high enzyme group were to be selected then the population mean would have advanced, but in the following generation further selection on the basis of high enzyme activity would be ineffective since the selected population would be AA homozygotes. Indeed, one of the problems in the use of specific characters as guides to selection is that once selected, no further variation may remain. The reduction of genetic variation might be described as 'genetic erosion' or better, in relation to the criterion itself, as 'character assassination'.

If A were dominant but B and C still additive then the enzyme classes would be widely separated and selection for enzyme activity would still raise the overall mean but the aa segregants which occur in the following generation would revert to lower yield values (*Figure 20.7b*) so that, as usual

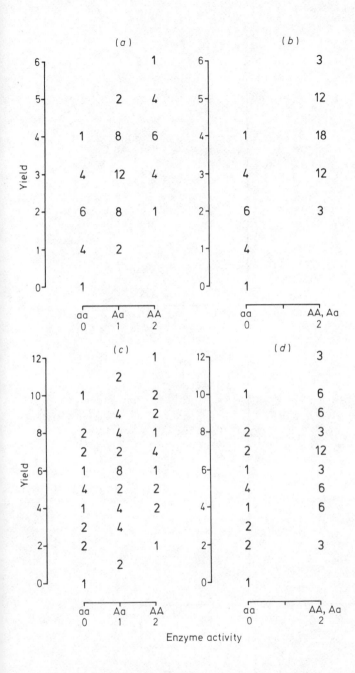

Figure 20.7 *(a) The relationship between the expression of a gene (A, a) controlling enzyme activity and numbers of individuals falling within a series of yield classes determined by three genes (A, a; B, b; C, c) acting additively and equally (further details in text). (b) As in (a) except that the enzyme-controlling gene shows dominance. (c) As in (a), with all genes acting additively but with increasing effect on yield (C > B > A). (d) As for (c) but A shows dominance*

378

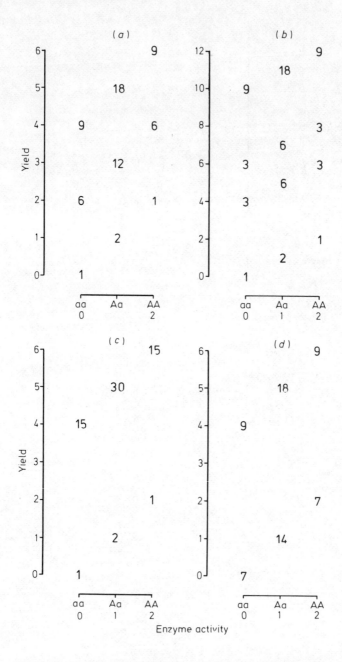

Figure 20.8 *(a) As in Figure 20.7 but A acts additively and B and C have equal effects but show dominance. (b) A acts additively but B and C have unequal effects and show dominance. (c) A acts additively but B and C show duplicate gene interaction. (d) A acts additively but B and C show complementary gene interaction*

with dominance, advance is slowed. Even the highest yielders would segregate if these were selected, for while they would contain the AABBCC homozygote they would also contain two AaBBCC individuals. The two models emphasize the guidance for breeders which are more normally inferred from high or low narrow sense additive heritability estimates obtained by formal genetic analysis and they emphasize the slower advances to selection when dominance is present.

Obviously the success of recognizing either separate yield classes or separate enzyme classes depends on the size of the differences caused by allelic substitution, the effects of the environment (including also the vagaries of the assay systems). It would be possible to make the process of selection appear easier or more difficult according to the assumption made. It should, however, be noted that the models presented here do not include any disturbing environmental effects; all the variation is genetic and that is an ideal but unreal situation.

Other assumptions make the problems for selection even more difficult, for example, where additivity for A exists (AA = 2, Aa = 1, aa = 0) but B (bb = 0, Bb = 2, BB = 4) and C (0, 3, 6) show additive but unequal effects (*Figure 20.7c*). More genetic classes and thus greater yield variation is shown by the high enzyme group because B and C have different but also greater effects than A. Again, dominance for A but not for B and C (*Figure 20.7d*) would reduce the rate of advance for higher yields and the necessity for considering in detail the other two characters and overall yield is obvious. Familiar F_2 ratios occur where A is additive and B and C show dominance but have equal (*Figure 20.8a*) or unequal effects (*Figure 20.8b*). In these examples there is an increase in the number of top yielders but many will segregate in further generations. Further, there are many relatively high yielders which have only intermediate enzyme levels.

To these examples may be added duplicate gene interaction, where A is additive and the presence of a single B or C allele is enough to give full yield expression (*Figure 20.8c*). There would be 15 top yielders but many are heterozygotes and a high proportion of the enzyme heterozygotes are of almost equal yield. Where A acts additively and both B and C are required as single dominants for full expression, segregation will yield a 9:7 ratio (complementary dominance) among the top enzyme group (*Figure 20.8d*). Triplicate gene interaction would of course result in a situation in which nearly all plants (63:1) are high yielders.

In relation to physiological characters we can pursue three courses, namely to select on the basis of the final product (yield) alone, on the basis of the measurement of physiological characters among parents or segregating populations or to combine these approaches.

Selection for single characters, including yield, may be easy if variation in the character can be readily assessed. It is often difficult to do this, however, unless the crop is grown or handled in the usual commercial way, particularly with regard to spacing. Thus cereal breeders have brought forward small plot trials to the earliest possible moment.

The problems in using physiological criteria in selection are:

(1) the number and accuracy of the measurements that can be made. It may not be technically possible to analyse large numbers of segregating

individuals and it should be remembered that some cereal programmes may include 500 000 to 1 million F_2 plants from individual crosses. Inevitably it has been considered more valuable to use physiological criteria to discriminate between potential parents or late generation families, using earlier generations for the elimination of the most obviously defective segregants;

(2) the possibility of a rapid decline in genetic variation as the character is fixed in the population. Some morphological characters can be rapidly fixed, e.g. leafless peas, the presence of awns in wheat, but thereafter further genetic control of these characters can only be by further selection for possibly polygenically controlled quantitative variation and the rate of advance in yield must be lower since the main influence of e.g. awns will be through their presence, not by their size;

(3) the possibility that positive responses made in one character are offset by negative responses in another. Selection for leaf size, for example in ryegrass, was found to be negatively related to the rate of leaf appearance (Edwards, 1970).

Inevitably all three approaches are in use and will continue to be used for they are not mutually exclusive and there is currently great interest in specific problems concerned with screening techniques (Association of Applied Biologists Meeting, Edinburgh, 1979).

In relation to the growth of cereals it appears that high yielding varieties achieve this yield by different combinations of characters. There might, however, be value in reversing current trends towards early maturing varieties which select for short plants with large individual ears. Past selection has also tended to redistribute dry matter from shoot tissue to the grain without increasing the total dry matter produced per unit area. There is now a need to select for an enhanced biomass which will require further studies to increase knowledge of the photosynthetic and respiration processes. The advances shown to be possible in the past encourage the belief that achievements will be no less spectacular in the future.

References

ATKINS, I. M. and NORRIS, M. J. (1955). *Agron. J.*, **47**, 218–220

AUSTIN, R. B., ROSSI, L. and BLACKWELL, R. D. (1978). *Ann. Bot.*, **42**, 429–438

BEEVERS, L. and HAGEMAN, R. H. (1969). *Ann. Rev. Pl. Physiol.*, **20**, 495–522

BELL, G. D. H. (1976). The Massey-Ferguson Papers, No. 9. Massey-Ferguson (UK) Ltd

BIDINGER, F., MUSGRAVE, R. B. and FISCHER, F. A. (1977). *Nature*, **270**, 431–433

BINGHAM, J. (1969). *Agric. Prog.*, **44**, 30–42

BISCOE, P. V., GALLAGHER, J. N., LITTLETON, E. J., MONTEITH, J. L. and SCOTT, R. K. (1975). *J. appl. Ecol.*, **12**, 295–318

BOWERMAN, A. and GOODMAN, P. J. (1971). *Ann. Bot.*, **35**, 353–366

BRAIN, P. W. (1978). PhD thesis. University of Nottingham

BRUNETTI, N. and HAGEMAN, R. H. (1976). *Pl. Physiol.*, **58**, 583–587

CHOWDHRY, A. R., SALEEM, M. and ALAM, K. (1976). *Exp. Agric.*, **12**, 411–415

COOPER, J. P., RHODES, I. and SHEEHY, J. E. (1971). *Ann. Rep. Welsh Plant Breeding Station (1970)*, pp. 57–69

CROY, L. I. (1967). PhD thesis. University of Illinois, Urbana

CROY, L. I. and HAGEMAN, R. H. (1970). *Crop Sci.*, **10**, 280–285

DALLING, M. J., HALLORAN, G. M. and WILSON, J. H. (1975). *Austr. J. agric. Res.*, **26**, 1–10

DAVIES, D. R. (1977). In *Applied Biology*, Vol. 2, pp. 87–121. Ed. by T. H. Coaker. Academic Press, London

DECKARD, E. L., WILLIAMS, N. D. and HAMMOND, J. J. (1975). *Agron. Abstr.*, p. 106

DENTON, O. A. (1974). PhD thesis. University of Nottingham

DONALD, C. M. (1968). *Euphytica*, **17**, 385–403

ECK, H. V., WILSON, G. C. and MARTINEZ, T. (1975). *Crop Sci.*, **15**, 557–561

EDWARDS, I. B. (1973). PhD thesis. University of Illinois

EDWARDS, K. J. R. (1970). *Genet. Res.*, **16**, 17–28

EILRICH, G. L. (1968). PhD thesis. University of Illinois

EILRICH, G. L. and HAGEMAN, R. H. (1973). *Crop Sci.*, **13**, 59–66

ENGLEDOW, F. L. and WADHAM, S. M. (1923). *J. agric. Sci., Cambridge*, **13**, 390–439

EVANS, L. T., BINGHAM, J., JACKSON, P. and SUTHERLAND, J. (1972). *Ann. appl. Biol.*, **70**, 67–76

FIDDIAN, W. E. H. (1970). *J. Nat. Inst. agric. Bot.*, **12**, 165–194

FINLAY, K. W. and WILKINSON, G. N. (1963). *Austr. J. agric. Res.*, **14**, 742–754

FISHER, R. A. and MACKENZIE, W. A. (1923). *J. agric. Sci., Cambridge*, **13**, 311–320

GALLAGHER, J. N., BISCOE, P. V. and SCOTT, R. K. (1975). *J. appl. Ecol.*, **12**, 319–336

GANASHAN, P. and WHITTINGTON, W. J. (1975). *Euphytica*, **24**, 775–784

GRITTON, E. T. (1972). *Pea Newslett.*, **4**, 11–12

HAGEMAN, R. H., LONG, E. R. and DUDLEY, J. W. (1967). *Adv. Agron.*, **19**, 45–86

HERNANDEZ, H. H. (1972). PhD thesis. North Dakota State University

HSU, P. and WALTON, P. D. (1970). *Euphytica*, **19**, 54–60

HUMPHRIES, I. (1973). BSc Dissertation. University of Nottingham

IZHAR, S. and WALLACE, D. H. (1967). *Crop. Sci.*, **7**, 457–460

JACKSON, W. A. and VOLK, R. J. (1970). *Ann. Rev. Pl. Physiol.*, **21**, 385–432

JONES, P. W. (1979). PhD thesis. University of Nottingham

KOLAWOLE, K. B. (1979). PhD thesis. University of Wales

LUDWIG, L. J. (1972). In *Crop Processes in Controlled Environment*, p. 391. Ed. by A. R. Rees. Academic Press, London

LUPTON, F. G. H., ALI, M. A. M. and SUBRAMANIAM, S. (1967). *J. agric. Sci., Cambridge*, **69**, 111–123

MATHER, K. and JINKS, J. L. (1971). *Biometrical Genetics*, 2nd edn. Chapman and Hall, London

MONYO, J. H. and WHITTINGTON, W. J. (1973). *Euphytica*, **22**, 600–606

OLUGBEMI, L. B., BINGHAM, J. and AUSTIN, R. B. (1976). *Annl. appl. Biol.*, **84**, 231–240

OSBORNE, B. A. (1979). PhD thesis. University of Nottingham

PARKER, W. H. (1927). *J. Nat. Inst. agric. Bot.*, **1**, (6), 6–7

PATTERSON, F. L., COMPTON, L. E., CALDWELL, R. M. and SCHAFER, J. F. (1962). *Crop Sci.*, **2**, 199–200

PERKINS, J. M. and JINKS, J. L. (1968). *Heredity*, **23**, 525–535

PERLITUS, L. (1903). Inaugural Dissertation. University of Breslau

QUALSET, C. O., SCHALLER, C. W. and WILLIAMS, J. C. (1965). *Crop Sci.*, **12**, 531–535

RHODES, I. (1973). *Herb. Abstr.*, **43**, 129–133

SCHMID, B. (1898). *Bot. Zentralbl.*, **76**, 1

SHARMA, R. K., DASHORA, S. L., TIKKA, S. B. S. and MATHUR, J. R. (1977). *Z. Pfl.-zücht.*, **79**, 315–319

SHERRARD, J. H., GREEN, D. L., SWINDEN, L. B. and DALLING, M. J. (1976). *Biochem. Genet.*, **14**, 905–912

SIMPSON, G. M. (1968). *Can. J. Pl. Sci.*, **48**, 253–260

SINHA, S. K., RAJAGOPAL, V. and BALASUBRAMANIAN, V. (1974). *Curr. Sci., India*, **43**, 617–618

SNOAD, B. and GENT, G. P. (1974). Rep. John Innes Inst., pp. 22–23

STEWART, G. R., LEE, J. A. and OREBAMJO, T. O. (1973). *New Phytol.*, **72**, 539–546

THORNE, G. N. (1966). In *The Growth of Cereals and Grasses*, pp. 88–105. Ed. by F. L. Milthorpe and J. D. Ivins. Butterworth, London

VAISHNAV, P. P., BHATT, K., SINGH, Y. D. and CHINOY, J. J. (1978). *Austr. J. Pl. Physiol.*, **5**, 39–43

VAVILOV, N. I. (1951). *Chron. Bot.*, **13**

VERVELDE, G. J. (1953). *Neth. J. agric. Sci.*, **1**, 2–9

WALLACE, D. H., OZBUN, J. L. and MUNGER, H. M. (1972). *Adv. Agron.*, **24**, 97–146

WARNER, R. L. (1968). PhD thesis. University of Illinois

WARNER, R. L., HAGEMAN, R. H., DUDLEY, J. W. and LAMBERT, R. J. (1969). *Proc. Natl. Acad. Sci., USA*, **62**, 785–792

WARNER, R. L. and KONZAK, C. F. (1975). *Wheat Newslett.*, **21**, 160–161

WATSON, D. J. (1968). *Ann. appl. Biol.*, **62**, 1–9

WILLIAMS, R. H. and HAYES, J. D. (1977). *Cer. Res. Commun.*, **5**, 113–118

WILLIAMS, R. H. and HAYES, J. D. (1979). *Ann. appl. Biol.*, **91**, 391–395

WILSON, D. (1974). *Ann. Rep. Welsh Plant Breeding Station (1974)*, p. 83

WILSON, D. (1975). *Ann. appl. Biol.*, **80**, 323–338

YATES, F. and COCHRAN, W. G. (1938). *J. agric. Sci., Cambridge*, **28**, 556–580

ZELITCH, I. (1971). *Photosynthesis, Photorespiration and Plant Productivity*, p. 347. Academic Press, New York

ZIESERL, J. F. and HAGEMAN, R. H. (1962). *Crop Sci.*, **2**, 512–515

ZOEBL, A. and MIKOSCH, C. (1892). *Bot. Zentralbl.*, **54**, 240

LIST OF PARTICIPANTS

Alagarswamy, Dr G.	International Crops Research Institute for the Semi-Arid Tropics, 1–11–256 Begumpet, Hyderabad 500016, India
Alcock, Dr M. B.	Department of Agriculture, University College of North Wales, Bangor, Gwynedd, UK
Alden, Dr T.	Department of Forest Genetics and Plant Physiology, Swedish University of Agricultural Sciences, S-90183 Umea, Sweden
Alderson, Dr P. G.	University of Nottingham, Department of Agriculture and Horticulture, Sutton Bonington, Loughborough LE12 5RD, UK
Al-Kummer, M.	University of Nottingham, Department of Physiology and Environmental Studies, Sutton Bonington, Loughborough LE12 5RD, UK
Atherton, Dr J. G.	University of Nottingham, Department of Agriculture and Horticulture, Sutton Bonington, Loughborough LE12 5RD, UK
Avery, Dr D. J.	East Malling Research Station, East Malling, Maidstone, Kent ME19 6BJ, UK
Bayliss, Dr M.	Imperial Chemical Industries Limited, Corporate Laboratory, P.O. Box 11, The Heath, Runcorn, Cheshire, UK
Bengtson, Dr C.	Swedish Water and Air Pollution Research Institute, P.O. Box 5207, S-402 24 Gothenburg, Sweden
Bink, Dr J. P. M.	Department of Tropical Crop Science, Agricultural University, 6700 AH Wageningen, The Netherlands
Black, Dr C.	University of Nottingham, Department of Physiology and Environmental Studies, Sutton Bonington, Loughborough LE12 5RD, UK
Black, Dr V.	University of Nottingham, Department of Physiology and Environmental Studies, Sutton Bonington, Loughborough LE12 5RD, UK
Blackwood, Dr G. C.	University of Nottingham, Department of Applied Biochemistry and Nutrition, Sutton Bonington, Loughborough LE12 5RD, UK
Bonhomme, R.	INRA Bioclimatologie, Estrees-Mons 80200, Peronne, France

Bridges, Dr I.	Imperial Chemical Industries Limited, Corporate Laboratory, P.O. Box 11, The Heath, Runcorn, Cheshire, UK
Burrage, Dr S. W.	Wye College, Wye, Ashford, Kent TN25 5AH, UK
Burrows, Dr F. J.	School of Biological Sciences, Macquarie University, North Ryde, NSW 2113, Australia
Ceulemans, R. J. M.	Department of Biology, University of Antwerp, Universiteitsplein 1, B-2610, Wilrijk, Belgium
Charles, Dr S. A.	Department of Plant Sciences, University of London, King's College, 68 Half Moon Lane, London SE24, UK
Christ, Dr R. A.	Ciba Geigy Limited, Agrochemicals Division, CH-4002 Basel, Switzerland
Clarkson, Dr D. T.	ARC Letcombe Laboratory, Wantage, Oxfordshire, OX12 9JT, UK
Cockshull, Dr K. E.	Glasshouse Crops Research Institute, Worthing Road, Rustington, Littlehampton, West Sussex, UK
Combe, Mrs L.	Station De Bioclimatologic, CNRA–INRA, 7800 Versailles, France
Cormack, Miss S.	University of Nottingham, Department of Applied Biochemistry and Nutrition, Sutton Bonington, Loughborough LE12 5RD, UK
Cox, Dr B. J.	Imperial Chemical Industries Limited, Corporate Laboratory, P.O. Box 11, The Heath, Runcorn, Cheshire, UK
Crawford, Dr D. V.	University of Nottingham, Department of Physiology and Environmental Studies, Sutton Bonington, Loughborough LE12 5RD, UK
Day, Dr W.	Physics Department, Rothamsted Experimental Station, Harpenden, Hertfordshire, UK
Dekhuijzen, Dr H.	Centre Agrobiological Research, P.O. Box 14, 6700AA Wageningen, The Netherlands
Eagles, Dr C. F.	Welsh Plant Breeding Station, Plas Gogerddan, Nr. Aberystwyth, Dyfed, UK
Edwards, Dr M. M.	University of Nottingham, Department of Physiology and Environmental Studies, Sutton Bonington, Loughborough LE12 5RD, UK
Evans, M.	Jealott's Hill Research Station, Bracknell, Berkshire, UK
Fenton, Dr R.	Shell Biosciences Laboratory, Sittingbourne Research Centre, Sittingbourne, Kent, UK
Fernandez, J.	Junta De Energia Nuclear, Apartado 3055, Madrid, Spain
Fitter, Dr A.	Department of Biology, University of York, York YD1 5D3, UK
Folkes, Prof. B. F.	School of Biological Sciences, University of East Anglia, Norwich NR4 7TJ, UK

Fyson, A.	Department of Biological Sciences, Dundee University, Dundee, Scotland
Garrod, Dr J. F.	The Boots Company Limited, Research Department, Lenton Research Station, Lenton House, Nottingham, UK
Glass, Dr A. D.	Department of Botany, University of British Columbia, Vancouver, BC, Canada V6T 1W5
Gosse, G.	Station De Bioclimatologic, INRA–CNRA, 78000 Versailles, France
Gregory, Dr P. J.	University of Nottingham, Department of Physiology and Environmental Studies, ODM Unit, Sutton Bonington, Loughborough LE12 5RD, UK
Grierson, Dr D.	University of Nottingham, Department of Physiology and Environmental Studies, Sutton Bonington, Loughborough LE12 5RD, UK
Hardingham, N. R.	Monsanto Technical Center, Parc Scientifique, Rue Laid Burniat, B-1348 Louvain La Neuve, Belgium
Harper, Dr F.	The Edinburgh School of Agriculture, West Mains Road, Edinburgh, Scotland
Harvey, Dr B. M. R.	Department of Agricultural Botany, The Queen's University of Belfast, Newforge Lane, Belfast, N. Ireland
Hay, Dr R. K. M.	Department of Environmental Sciences, University of Lancaster, Bailrigg, Lancaster, UK
Hayes, Dr P.	Department of Agricultural Botany, Queen's University of Belfast, Newforge Lane, Belfast, N. Ireland
Haystead, Dr A.	Hill Farming Research Organisation, Bush Estate, Penicuik, Midlothian, Scotland
Hebblethwaite, Dr P. D.	University of Nottingham, Department of Agriculture and Horticulture, Sutton Bonington, Loughborough LE12 5RD, UK
Hipkin, Dr C. R.	Department of Botany and Microbiology, University College of Swansea, Singleton Park, Swansea, UK
Hucklesby, Dr D. P.	Department of Agriculture, Long Ashton Research Station, Long Ashton, Bristol B518 9AF, UK
Hughes, B. C.	Department of Science, Bristol Polytechnic, Coldharbour Lane, Frenchay, Bristol, UK
Hughes, J. E.	University of Nottingham, Department of Physiology and Environmental Studies, Sutton Bonington, Loughborough LE12 5RD, UK
Hunter, N. J.	University of Nottingham, Department of Physiology and Environmental Studies, Sutton Bonington, Loughborough LE12 5RD, UK

Ivins, Prof. J. D.	University of Nottingham, Department of Agriculture and Horticulture, Sutton Bonington, Loughborough LE12 5RD, UK
James, D. M.	Department of Agriculture, Long Ashton Research Station, Long Ashton, Bristol B518 9AF, UK
Jarvis, Prof. P. G.	Department of Forestry and Natural Resources, University of Edinburgh, Mayfield Road, Edinburgh, Scotland
Johnson, Dr C. B.	Department of Botany, University of Reading, Reading, Berkshire, UK
Joseph, Mrs M. E.	University of Nottingham, Department of Physiology and Environmental Studies, Sutton Bonington, Loughborough LE12 5RD, UK
Khurana, S. C.	University of Nottingham, Department of Agriculture and Horticulture, Sutton Bonington, Loughborough LE12 5RD, UK
King, Dr R. W.	Division of Plant Industry, CSIRO, Canberra, Australia
Koch, Dr D. W.	Plant Science Department, University of New Hampshire, Durham, NH 03824, USA
Lambers, H.	Department of Plant Physiology, University of Groningen, P.O. Box 14, 9750 AA Haren, The Netherlands
Landsberg, Dr J. J.	Long Ashton Research Station, Long Ashton, Bristol B518 9AF, UK
Larcher, Dr W.	Institut für Botanik der Universität Innsbruck, ABT Resistenzokologie, Sternwartestrasse 15, A-6020 Innsbruck, Austria
Leach, Dr J. E.	Physics Department, Rothamstead Experimental Station, Harpenden, Hertfordshire, UK
Leafe, Dr E. L.	The Grassland Research Institute, Hurley, Maidenhead, Berkshire, UK
Lefroy, R. D. B.	Department of Biology, University of York, Heslington, York, UK
Lemeur, Dr R. J. P.	Laboratory of Plant Ecology, University of Ghent, Coupure Links 533, B-9000 Ghent, Belgium
Leverenz, Dr J. W.	Department of Forestry and Natural Resources, University of Edinburgh, Mayfield Road, Edinburgh, Scotland
Linder, Dr S.	Department of Forest Ecophysiology, Swedish University of Agricultural Sciences, Fack, S-750 07 Uppsala, Sweden
Louwerse, W.	Centre for Agrobiological Research, CABO, Bornsesteeg 65, Wageningen, The Netherlands
Ludwig, Dr L. J.	Glasshouse Crops Research Institute, Worthing Road, Rustington, Littlehampton, West Sussex, UK
Lyne, Dr R. L.	Shell Biosciences Laboratory, Sittingbourne Research Centre, Sittingbourne, Kent, UK

Mangianti, Dr C.	Ufficio Centrali Di Ecologia Agraria, Via Del Caravita 7/A, 00186 Roma, Italy
Mansfield, Prof. T. A.	Department of Biological Sciences, University of Lancaster, Bailrigg, Lancaster, UK
Marshall, Dr B.	University of Nottingham, Department of Physiology and Environmental Studies, ODM Unit, Sutton Bonington, Loughborough LE12 5RD, UK
Marshall, Dr C.	School of Plant Biology, University College of North Wales, Bangor, Gwynedd, UK
McLaren, Dr. J. S.	University of Nottingham, Department of Agriculture and Horticulture, Sutton Bonington, Loughborough LE12 5RD, UK
McNair, Dr D. J.	Department of Biological Sciences, Portsmouth Polytechnic, King Henry 1st Street, Portsmouth, UK
Monteith, Prof. J. L.	University of Nottingham, School of Agriculture, Sutton Bonington, Loughborough LE12 5RD, UK
Naylor, Dr R. E. L.	Botany Division, School of Agriculture, 581 King Street, Aberdeen AB9 14D, Scotland
Norton, Dr G.	University of Nottingham, Department of Applied Biochemistry and Nutrition, Sutton Bonington, Loughborough LE12 5RD, UK
Ong, Dr C.	University of Nottingham, Department of Physiology and Environmental Studies, ODM Unit, Sutton Bonington, Loughborough LE12 5RD, UK
Oquist, Prof. G.	Department of Plant Physiology, University of Umea, S-90187 Umea, Sweden
Palmer, Dr J. W.	East Malling Research Station, East Malling, Maidstone, Kent ME19 6BJ, UK
Parkinson, Dr K. J.	Physics Department, Rothamsted Experimental Station, Harpenden, Hertfordshire, UK
Parsons, A. J.	The Grassland Research Institute, Hurley, Maidenhead, Berkshire, UK
Paul, Miss E. M.	Botany Department, University of Leicester, Adrian Building, Leicester, UK
Peacock, Dr J. M.	The Grassland Research Institute, Hurley, Maidenhead, Berkshire, UK
Pearson, Mrs C.	University of Nottingham, Department of Physiology and Environmental Studies, Sutton Bonington, Loughborough LE12 5RD, UK
Pechan, P. M.	Department of Applied Biology, University of Cambridge, Cambridge, UK
Penny, Dr M. G.	Department of Forestry and Natural Resources, University of Edinburgh, Mayfield Road, Edinburgh, Scotland
Pitman, Prof. M. G.	School of Biological Sciences, University of Sydney, Sydney, NSW 2006, Australia

Raschke, Prof. K.	Michigan State University, MSU-DOE Plant Research Laboratory, East Lansing, Michigan 48824, USA
Raven, Dr J. A.	Department of Biological Sciences, University of Dundee, Dundee DD1 4HN, Scotland
Rees, Dr T. A. V.	Department of Botany and Microbiology, University College of Swansea, Singleton Park, Swansea, UK
Reilly, Dr M. L.	Department of Agricultural Biology, University College, Dublin, Ireland
Robertson, N.	University of Nottingham, Department of Physiology and Environmental Studies, Sutton Bonington, Loughborough LE12 5RD, UK
Robson, Dr M. J.	The Grassland Research Institute, Hurley, Maidenhead, Berkshire, UK
Rode, J. C.	Station De Bioclimatologic, INRA–CNRA, 78000 Versailles, France
Roussopoulos, D.	University of Nottingham, Department of Physiology and Environmental Studies, Sutton Bonington, Loughborough LE12 5RD, UK
Rowland, A. O.	Welsh Plant Breeding Station, Plas Gogerddan, Nr Aberystwyth, Dyfed, UK
Russell, Dr G.	Edinburgh School of Agriculture, West Mains Road, Edinburgh, Scotland
Ryle, Dr G. J. A.	The Grassland Research Institute, Hurley, Maidenhead, Berkshire, UK
Schulze, Prof. E. D.	Lehrstuhl für Pflanzenokologie der Universität Bayreuth, Am Birkengut, D-8580 Bayreuth, West Germany
Sestak, Prof. Z.	Institute of Experimental Botany, Czechoslovak Academy of Sciences, Flemingovo n.2., 160 00 Praha 6, Czechoslovakia
Siva Kumar, Dr M. V. K.	International Crops Research Institute for the Semi-Arid Tropics, 1-11-256 Begumpet, Hyderabad 500016, India
Smith, Prof. H.	Department of Botany, University of Leicester, Leicester, UK
Spiertz, Dr H. J.	Centre for Agrobiological Research, P.O. Box 14, Wageningen, The Netherlands
Squire, Dr G.	University of Nottingham, Department of Physiology and Environmental Studies, ODM Unit, Sutton Bonington, Loughborough LE12 5RD, UK
Stanhill, Dr G.	Institute of Soils and Water, Agricultural Research Organisation, The Volcani Centre, Bet-Dagan, Israel
Strophair, Dr B. A.	Department of Biological Sciences, Portsmouth Polytechnic, King Henry 1st Street, Portsmouth, UK

Thornley, Dr J. H. M.	Glasshouse Crops Research Institute, Worthing Road, Rustington, Littlehampton, West Sussex, UK
Troeng, E.	Swedish Coniferous Forest Project, Jadraas 4485, S-816 00 Ockelbo, Sweden
Unsworth, Dr M. H.	University of Nottingham, School of Agriculture, Sutton Bonington, Loughborough LE12 5RD, UK
Valli, Miss S.	Butterworth & Co. (Publishers) Ltd, Borough Green, Kent, UK
Van Steveninck, Prof. R.	School of Agriculture, La Trobe University, Bundoora, Victoria, Australia
Vince-Prue, Dr D.	Glasshouse Crops Research Institute, Worthing Road, Rustington, Littlehampton, West Sussex, UK
Whittingham, Prof. C. P.	Rothamsted Experimental Station, Harpenden, Hertfordshire, UK
Whittington, Prof. W. J.	University of Nottingham, School of Agriculture, Sutton Bonington, Loughborough LE12 5RD, UK
Wild, Dr A.	Department of Soil Science, University of Reading, London Road, Reading RG1 5AQ, UK
Wildon, Dr D. C.	School of Biological Sciences, University of East Anglia, Norwich NR4 7TJ, UK
Wilson, Miss H. M. B.	Imperial Chemical Industries, Agricultural Division, Jealotts Hill Research Station, Bracknell, Berkshire, UK
Wilson, J. A.	Department of Biological Sciences, University of Lancaster, Bailrigg, Lancaster, UK
Withers, A. C.	Glasshouse Crops Research Institute, Worthing Road, Rustington, Littlehampton, West Sussex, UK
Woledge, Dr J.	The Grassland Research Institute, Hurley, Maidenhead, Berkshire, UK
Woolhouse, Prof. H.	Department of Plant Sciences, The University, Leeds LS2 9T, UK
Wyn Jones, Dr G. D.	Department of Biochemistry and Soil Science, University College of North Wales, Bangor, Gwynedd, UK
Yarwood, Miss M.	Jealotts Hill Research Station, Bracknell, Berkshire, UK

INDEX